MW01598956

# Cucurbituril-based Functional Materials

# Smart Materials

*Series editors:*
Hans-Jörg Schneider, *Saarland University, Germany*
Mohsen Shahinpoor, *University of Maine, USA*

*How to obtain future titles on publication:*
A standing order plan is available for this series. A standing order will bring delivery of each new volume immediately on publication.

*For further information please contact:*
Book Sales Department, Royal Society of Chemistry, Thomas Graham House, Science Park, Milton Road, Cambridge, CB4 0WF, UK
Telephone: +44 (0)1223 420066, Fax: +44 (0)1223 420247
Email: booksales@rsc.org
Visit our website at www.rsc.org/books

# Cucurbituril-based Functional Materials

Edited by

**Dönüs Tuncel**
*Bilkent University, Turkey*
*Email: dtuncel@fen.bilkent.edu.tr*

ROYAL SOCIETY
OF **CHEMISTRY**

Smart Materials No. 36

Print ISBN: 978-1-78801-488-5
PDF ISBN: 978-1-78801-595-0
EPUB ISBN: 978-1-78801-875-3
Print ISSN: 2046-0066
Electronic ISSN: 2046-0074

A catalogue record for this book is available from the British Library

The Royal Society of Chemistry is a charity, registered in England and Wales, Number 207890, and a company incorporated in England by Royal Charter (Registered No. RC000524), registered office: Burlington House, Piccadilly, London W1J 0BA, UK, Telephone: +44 (0) 20 7437 8656.

For further information see our web site at www.rsc.org

Printed in the United Kingdom by CPI Group (UK) Ltd, Croydon, CR0 4YY, UK

# Preface

Smart materials constructed through supramolecular assemblies have been receiving considerable attention because of their potential applications, which include self-healing materials, energy storage, photonic devices, sensors and theranostics. Host–guest chemistry of various macrocyclic receptors with organic guests provides a unique way to control the tailor-made nanoarchitectures for the formation of predesigned functional materials. In this context, cucurbituril is a very interesting macrocycle and can participate in the formation of various supramolecular assemblies having unprecedented properties. Rich host–guest chemistry, through distinctive molecular recognition ability and high selectivity of CB homologues toward to various guests with different shapes and sizes, plays a very important role in the synthesis of functional supramolecular materials.

This book *Cucurbituril-based Function Materials* provides a comprehensive overview of this fascinating macrocycle, cucurbituril (CB) homologue- and derivative-based supramolecular materials. Chapters cover the synthesis, properties and applications of CB-based smart materials and nanostructures. With contributions from key researchers, this book will be of interest to students and researchers working in materials science, as well as those working on cucurbituril-based materials in organic, supramolecular and polymer chemistry.

I would like to take this opportunity to thank all the corresponding authors as well as their coauthors for their contribution and fruitful collaboration in the preparation of this book. I would also like to gratefully acknowledge the referees who kindly reviewed the chapters and provided valuable feedback.

I would like to thank also Prof. Dr Hans-Jörg Schneider and Dr Leanne Marle for their kind invitation to edit this book. I would like to gratefully

Smart Materials No. 36
Cucurbituril-based Functional Materials
Edited by Dönüs Tuncel
© The Royal Society of Chemistry 2020
Published by the Royal Society of Chemistry, www.rsc.org

acknowledge Dr Robin Driscoll and Connor Sheppard, as well as the production team at the Royal Society of Chemistry for their generous help during the preparation and publication of this book.

My special thanks also go to my research group and my family for all their support.

Dönüs Tuncel
Bilkent University, Turkey

# Contents

Smart Materials No. 36
Cucurbituril-based Functional Materials
Edited by Dönüs Tuncel
© The Royal Society of Chemistry 2020
Published by the Royal Society of Chemistry, www.rsc.org

CHAPTER 1

# Introduction: Cucurbituril-containing Functional Materials in the Context of Smart Materials

DÖNÜS TUNCEL[a,b]

[a] Department of Chemistry, Bilkent University, Ankara 06800, Turkey;
[b] UNAM – National Nanotechnology Research Center, Institute of Materials Science and Nanotechnology, Bilkent University, Ankara 06800, Turkey
Email: dtuncel@fen.bilkent.edu.tr

## 1.1 Smart Materials in General

Smart materials can sense and react to external stimuli by adapting their existing properties to new ones.[1-3] They have ability to gather together many important and interesting features so that these features may act in accord in order to achieve important tasks. Applied external stimuli can be generalized as chemical, physical and biological. The most frequently used chemical stimuli are pH, ionic strengths, moisture, redox and competitive guests. Stress, temperature, electric and magnetic fields, and light can be given as examples of physical stimuli, while biochemical stimuli include antigens, enzymes, ligands or other biochemical agents.

The responses against external stimuli can be physical (changes in the color, shape, size *etc.*), chemical (changes in the functional groups, reactivity, hydrophilicity *etc.*) or changes in mechanical properties. Because the

Smart Materials No. 36
Cucurbituril-based Functional Materials
Edited by Dönüs Tuncel
© The Royal Society of Chemistry 2020
Published by the Royal Society of Chemistry, www.rsc.org

responses give rise to large macroscopic changes that are reversible and can be repeated many times, these materials can be used in a wide range of applications such as sensors, actuators, artificial muscles, catalysis, smart interfaces, tissue engineering, biosensors, diagnostics, drug delivery, robotics and biomimetics.

## 1.2 Supramolecular Smart Materials

Supramolecular smart materials constructed through noncovalent interactions, such as hydrogen bonding, π–π stacking and hydrophobic interactions, have a dynamic nature and can undergo spontaneous and continuous assembly/disassembly processes under certain triggers.[4–8] Owing to the dynamic and reversible nature of noncovalent interactions, these materials have the ability to adapt to their environment and possess a wide range of interesting features, including degradability, shape-memory and self-healing. The selection of the noncovalent interactions and building blocks used in the construction of the supramolecular materials are crucial as the nature of noncovalent interactions and the structure of the building blocks determine the degree of stimuli-responsiveness and the smartness of the resulting materials. Therefore, the supramolecular approach allows the design of materials by incorporating the desired functionalities and features for targeted applications.

### 1.2.1 Cucurbituril Containing Smart Materials

In the context of the supramolecular approach, host–guest inclusion complexes stabilized by hydrophobic effect are quite attractive inasmuch as the high selectivity between the host and guest molecules provides dynamic but strong interactions.[9–13] Using this approach, small building blocks can be brought together to construct more complicated structures with desirable topological diversity and programmable functions for specific applications. Their assembly and the disassembly processes in general can be controlled by applying appropriate triggers.

The most common macrocycles used as host molecules are cyclodextrins,[14–16] pillarenes,[17] calixarenes[18] and cucurbiturils,[19–23] which can accommodate guest molecules in their cavities on the basis of complementary shape and size. Among them, cucurbit[n]urils (CB[n]) are a relatively new class of macrocycles with versatile recognition properties and an ability to accommodate different organic guest molecules in aqueous solution with exceptionally high binding constants.[19–23]

In recent years, CB[n]-containing functional materials have been receiving increasing attention due to their versatile applications in areas including but not limited to theranostics, photonics, self-healing, sensing and catalysis.[24–28] The abundant host–guest cucurbituril (CB) homologues allow one to prepare a variety of functional and smart materials including molecular switches, reversible, stimuli-responsive hydrogels, porous organic frameworks, nanoparticles, micelles, vesicles, colloidosomes. Because its cavity is larger than other homologues, CB[8] is highly appealing for complexation-induced nanostructure formation. In this regard, the ability of CB[8] to form ternary complexes with suitably sized and functionalized guests has been extensively utilized in the preparation of nanoparticles, stimuli-responsive reversible micelles, vesicles and capsules for the encapsulation, delivery and controlled release of drugs.

In this regard, this book aims to provide a comprehensive overview of cucurbituril-based functional materials. It starts with Chapter 2, on CB chemistry, which provides a detailed account of the synthesis, isolation, formation mechanisms, and structural and physical properties of CB homologues, derivatives and functional CBs. The discussion also contains some elements of the historical background of CBs.

Chapter 3 summarizes the key roles of the carbonyl groups containing portals of CBs on the host–guest inclusion complex formation. Due to the presence of carbonyl groups, the portals of CBs are partially negatively charged, and, accordingly, this chapter presents research work showing that electrostatic interactions may have considerable effects on the stability of complexes formed by CB hosts.

In Chapter 4, the preparation of cucurbituril-based pseudorotaxanes, rotaxanes and polyrotaxanes is discussed with selected examples. This includes notions of self-sorting, which enables the setup of homo- and hetero(pseudo)rotaxanes. The implications of thermodynamic and kinetic control are briefly showcased as well. In the main part, these assemblies are discussed in the context of stimuli-responsive systems, whose supramolecular chemistry and functionality can be controlled by using chemical inputs (pH, ions), redox signals or light. Finally, some applications are highlighted, such as drug delivery and molecular information processing.

Chapter 5 deals with hybrid nanomaterials of CBs with metals (Au, Ag, Pd, Pt), lanthanides and silica. These nanomaterials are either prepared in the presence of CBs (CB-assisted synthesis) or have been first prepared before CBs are used as capping agents to stabilize the resulting nanostructures. The plasmonic and SERS applications of CB-Au and CB-Ag hybrid nanoparticles and electrocatalytic applications of CB-Pt and CB-Pd are also presented.

Chapters 6, 7 and 8 give an overview on the stimuli-responsive supramolecular assemblies constructed mainly through host–guest complexation of CB homologues. Particularly, the ternary complex formation ability of CB[8], with suitably sized electron donor and acceptor guests, has been utilized in those assemblies. Chapter 6 focuses on the preparation and properties of stimuli-responsive reversible hydrogels and their applications in various areas, including drug delivery, wound dressing and healing, tissue engineering, diagnostic devices, wood conservation, adhesives, stretchable and wearable electronics, injection and printing substances. In the follow-up chapter (Chapter 7), the ability of CB[8] to form ternary complexes with suitably sized and functionalized guests has been demonstrated in the preparation of single-chain nanoparticles, stimuli-responsive reversible micelles and pH-responsive prodrug micelles for the encapsulation, delivery and controlled release of drugs and bioactive nanostructures based on peptide amphiphile vesicles, as well as reversible and stimuli-responsive microcapsules prepared to make use of the advantages of microfluidics. Nanostructures prepared from functionalized CB derivatives are also discussed.

Chapter 8 continues the discussion on supramolecular frameworks formed through the ternary complexes of CBs. The authors demonstrate how self-assembly provides a straightforward strategy for the construction of water-soluble porous supramolecular organic frameworks (SOFs) from rationally designed rigid multitopic molecular components and CB[8]. Using this strategy, a variety of two-dimensional honeycomb, square, and rhombic SOFs have been constructed, some of which exhibit interesting absorption and sensing functions.

The discussion on CB-containing nanomaterials continues in Chapter 10, but the nanomaterials covered in this chapter are prepared through the cross-linking of functionalized CBs, not through ternary complex formation. This chapter presents how functionalized CB[6]-based nanocapsules are constructed by a direct, one-pot method without using any preorganized structure, emulsifier or template, and the cavity of CBs in these nanocapsules is further available for molecular recognition. Their applications in various areas include heterogeneous catalysis, drug delivery and *in vivo* imaging.

Chapters 9, 11 and 12 demonstrate the applications of CB homologues and derivatives. Chapter 9 presents the research works on the interactions of CB homologues and derivatives with biomolecules and drugs because many promising discoveries of supramolecular interactions between CBs and biomolecules and small organic drug molecules have emerged with potential implications in the field of pharmaceutical sciences, which has become one of the most significant areas of potential applications for CBs. This chapter covers the noncovalent interactions of peptides, proteins and drug molecules with CB homologues and derivatives, in addition to discussing the ability of CBs to modulate the functions and bioactivities of these species through host–guest chemistry and the potential impact of CBs on protein enrichment, together with other relevant topics.

Chapter 11 covers some of the recent works on the molecular properties of cucurbituril-based supramolecular functional assemblies of a few organic dyes and of polyoxometalates that have technological and biological importance, especially in the domains of aqueous dye laser, light-emitting devices, photofunctional devices, energy storage devices, molecular architectures, supramolecular catalysts, radiotracer separation, antibacterial agents and drug delivery vehicles.

Chapter 12 deals with the supramolecular assemblies of CBs with conjugated polymers and porphyrins. The effect of CBs on the photophysical properties of conjugated polymers is discussed, and the optoelectronic application of CB-containing π-conjugated materials is demonstrated. Moreover, the application of CB-based photosensitizers in photodynamic therapy is summarized.

# Acknowledgements

We gratefully acknowledge the financial support of the Scientific and Technological Research Council of Turkey (TUBITAK) grant number 215Z035.

# References

1. (a) M. Wun-Fogle, *Materials for Smart Systems*, Cambridge University Press, 2014; (b) M. R. Aguilar and J. S. Roman, *Smart Polymers and Their Applications*, Woodhead Publishing, 2014; (c) Q. Li, *Intelligent Stimuli-responsive Material*, Wiley, 2013; (d) *Handbook of Stimuli-responsive Materials*, ed. M. H. Urban, Wiley-VCH, Weinheim, 2011; (e) *Intelligent Materials*, ed. M. Shahinpoor and H.-J. Schneider, Royal Society of Chemistry, Cambridge, UK, 2007.
2. (a) A. Grinthala and J. Aizenberg, *Chem. Soc. Rev.*, 2013, **42**, 7072–7085; (b) L. Zhai, *Chem. Soc. Rev.*, 2013, **42**, 7148–7160; (c) D. Habault, H. Zhang and Y. Zhao, *Chem. Soc. Rev.*, 2013, **42**, 7244–7256; (d) A. DoÄNring, W. Birnbaum and D. Kuckling, *Chem. Soc. Rev.*, 2013, **42**, 7391–7420; (e) J. Zhuang, M. R. Gordon, J. Ventura, L. Li and S. Thayumanavan, *Chem. Soc. Rev.*, 2013, **42**, 7421–7435.
3. (a) R. B. Grubbs and Z. Sun, *Chem. Soc. Rev.*, 2013, **42**, 7436–7445; (b) F. D. Jochumab and P. Theato, *Chem. Soc. Rev.*, 2013, **42**, 7468–7483; (c) M. Molina, M. Asadian-Birjand, J. Balach, J. Bergueiro, E. Miceliac and M. Calderoaln, *Chem. Soc. Rev.*, 2015, **44**, 6161–6186.
4. (a) J.-M. Lehn, *Supramolecular Chemistry: Concepts and Perspectives*, VCH, Weinheim, 1995; (b) J. L. Atwood, J. E. D. Davies, D. D. MacNicol, F. Vogtle and J.-M. Lehn, *Comprehensive Supramolecular Chemistry*, Pergamon, Oxford, 1996; (c) J. Steed and J. L. Atwood, *Supramolecular Chemistry*, Wiley, New York, 2000.
5. (a) J.-M. Lehn, *Angew. Chem., Int. Ed.*, 1990, **29**, 1304; (b) J.-M. Lehn, *Chem. Soc. Rev.*, 2007, **36**, 151.
6. (a) X. Yan, F. Wang, B. Zheng and F. Huang, *Chem. Soc. Rev.*, 2012, **41**, 6042–6065; (b) C. Heinzmann, C. Weder and L. Montero de Espinosa, *Chem. Soc. Rev.*, 2016, **45**, 342–358; (c) O. J. G. M. Goor, S. I. S. Hendrikse, P. Y. W. Dankers and E. W. Meijer, *Chem. Soc. Rev.*, 2017, **46**, 6621–6637.
7. L. Voorhaar and R. Hoogenboom, *Chem. Soc. Rev.*, 2016, **45**, 4013–4031.
8. R. Dong, Y. Pang, Y. Su and X. Zhu, *Biomater. Sci.*, 2015, **3**, 937–954.
9. X. Ma and H. Tian, *Acc. Chem. Res.*, 2014, **47**, 1971.
10. P. Wei, X. Yan and F. Huang, *Chem. Soc. Rev.*, 2015, **44**, 815.
11. J. Boekhoven and S. I. Stupp, *Adv. Mater.*, 2014, **26**, 1642.
12. X. Yan, F. Wang, B. Zheng and F. Huang, *Chem. Soc. Rev.*, 2012, **41**, 6042.
13. G. Yu, K. Jie and F. Huang, *Chem. Rev.*, 2015, **115**, 7240.
14. Q.-D. Hu, G.-P. Tang and P. K. Chu, *Acc. Chem. Res.*, 2014, **47**, 2017.
15. A. Harada, Y. Takashima and M. Nakahata, *Acc. Chem. Res.*, 2014, **47**, 2128.

16. Y. Kang, K. Guo, B. J. Li and S. Zhang, *Chem. Commun.*, 2014, **50**, 11083.
17. J. Murray, K. Kim, T. Ogoshi, W. Yao and B. C. Gibb, *Chem. Soc. Rev.*, 2017, **46**, 2479.
18. D.-S. Guo and Y. Liu, *Acc. Chem. Res.*, 2014, **47**, 1925.
19. (a) R. Behrend, E. Meyer and F. Rusche, *Justus Liebigs Ann. Chem.*, 1905, **339**, 1; (b) W. L. Mock, in *Comprehensive Supramolecular Chemistry*, ed. F. Vogtle, Pergamon Press, Oxford, 1996, vol. 2, p. 477; (c) W. A. Freeman, W. L. Mock and N.-Y. Shih, *J. Am. Chem. Soc.*, 1981, **103**, 7367; (d) Y. M. Jeon, H. Kim, D. Whang and K. Kim, *J. Am. Chem. Soc.*, 1996, **118**, 9790; (e) D. Whang, Y. M. Jeon, J. Heo and K. Kim, *J. Am. Chem. Soc.*, 1996, **118**, 11333; (f) I.-S. Kim, J. Jung, S.-Y. Kim, E. Lee, J.-K. Kang, S. Sakamoto, K. Yamaguchi and K. Kim, *J. Am. Chem. Soc.*, 2000, **122**, 540.
20. D. Shetty, J. K. Khedkar, K. M. Park and K. Kim, *Chem. Soc. Rev.*, 2015, **44**, 8747.
21. S. J. Barrow, S. Kasera, M. J. Rowland, J. del Barrio and O. A. Scherman, *Chem. Rev.*, 2015, **115**, 12320.
22. L. Isaacs, *Acc. Chem. Res.*, 2014, **47**, 2052.
23. K. I. Assaf and W. M. Nau, *Chem. Soc. Rev.*, 2015, **44**, 394.
24. K. Kim, N. Selvapalam, Y. H. Ko, K. M. Park, D. Kim and J. Kim, *Chem. Soc. Rev.*, 2007, **36**, 267.
25. S. Gürbüz, M. Idris and D. Tuncel, *Org. Biomol. Chem.*, 2015, **13**, 330.
26. H. Zou, J. Liu, Y. Li, X. Li and X. Wang, *Small*, 2018, **14**, 1802234.
27. C. Stoffelen and J. Huskens, *Small*, 2016, **12**, 96.
28. A. Koc and D. Tuncel, *Isr. J. Chem.*, 2018, **58**, 334.

CHAPTER 2

# Cucurbituril Homologues and Derivatives: Syntheses and Functionalization

AHMET KOC[a] AND DÖNÜS TUNCEL*[a,b]

[a] Department of Chemistry, Bilkent University, Ankara 06800, Turkey;
[b] UNAM – National Nanotechnology Research Center, Institute of Materials Science and Nanotechnology, Bilkent University, Ankara 06800, Turkey
*Email: dtuncel@fen.bilkent.edu.tr

## 2.1 Milestones in the Timeline of Cucurbiturils

### 2.1.1 1900s

Cucurbiturils have attracted chemists' attention in the last couple of decades, but their first synthesis dates back to the beginning of the previous century. In 1904, Eberhard Meyer, a PhD student in Heidelberg University in Germany, reported his studies on the condensation reactions of urea with glyoxal and glycoluril with formaldehyde in his thesis.[1] Significant results obtained from the acid-catalyzed condensation of glycoluril with formaldehyde were then published by Meyer and his advisor Robert Behrend in 1905.[2] The reaction was carried out in two steps in which glycoluril and formaldehyde first reacted in the presence of dilute HCl, and then the precipitate yielded was refluxed in concentrated $H_2SO_4$ at 110 °C to afford the product. The condensation product was defined as a white, amorphous powder that is highly hygroscopic but that does not lose its texture upon absorbing water. Behrend *et al.* studied the complexation behavior of the

Smart Materials No. 36
Cucurbituril-based Functional Materials
Edited by Dönüs Tuncel
© The Royal Society of Chemistry 2020
Published by the Royal Society of Chemistry, www.rsc.org

product with various salts and organic molecules and found that it dissolves well in water in the presence of cationic species and weakly in dilute acid and base. They proposed that the product could be formed by the condensation of three molecules of glycoluril with four formaldehyde molecules, as in this reaction:

$$3C_4H_6N_4O_2 + 4CH_2O \xrightarrow{\Delta} C_{16}H_{18}N_{12}O_6 + 4H_2O$$

### 2.1.2 1980s

After the preliminary work on the synthesis of cucurbiturils by Behrend *et al.*, the field stayed dormant for 76 years. In 1981, William L. Mock and coworkers revisited the synthetic procedure reported by Behrend *et al.* and easily obtained the same compound in slightly less yield than the stated yield (40–70%).[3] They crystallized the compound from a sulfuric acid solution of calcium sulfate. Their investigations focused on the structural elucidation of the compound. They took advantage of infrared and $^1$H NMR spectroscopies and of X-ray crystallography to be the first chemists to come up with the unique pumpkin-shaped macrocyclic hexameric structure of cucurbit[6]uril, CB[6], which is composed of two hydrophilic portals decorated with carbonyls and a hydrophobic cavity (Figure 2.1). It was realized later on that the chemical equation proposed by Behrend *et al.* could only represent the formation of a piece of glycoluril oligomer that must further oligomerize and then cyclize at some point to afford the macrocyclic hexameric CB[6]. The correct chemical equation for the formation of CB[6] is then:

$$6C_4H_6N_4O_2 + 12CH_2O \xrightarrow{\Delta} C_{36}H_{36}N_{24}O_{12} + 12H_2O$$

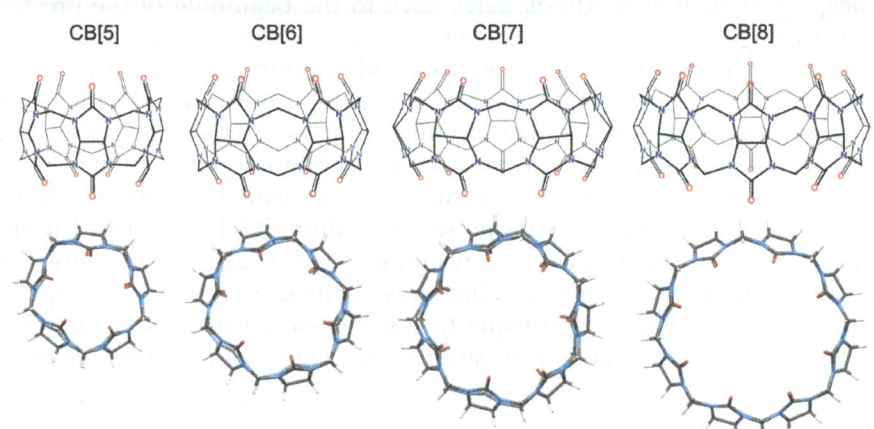

**Figure 2.1** Chemical and crystal structures of the most common CB[*n*] homologues, *n* = 5–8. Color codes: gray: C, white: H, blue: N, red: O. Crystal structure data were obtained from Cambridge Crystallographic Data Centre.

Mock *et al.* gave the fancy name of "cucurbituril" to this compound since the shape of this particular macrocycle resembles that of a pumpkin, which is in the *Cucurbitacea* botanic family. During the 1980s, Mock and coworkers studied the formation of inclusion compounds between CB[6] and alkylammonium ions and determined the dissociation constants of these complexes through [1]H NMR experiments.[4–7] They also performed the first study demonstrating the catalytic behavior of CB[6] in azide–alkyne 1,3-dipolar cycloaddition reactions.[8]

## 2.1.3 1990s

In 1992, a new perspective on cucurbituril chemistry was introduced by Fraser Stoddart and coworkers. They achieved the synthesis of a pentameric cucurbituril derivative, decamethylcucurbit[5]uril, by refluxing dimethylglycoluril and formaldehyde mixture in the presence of concentrated hydrochloric acid.[9] This was the first study representing a fully substituted cucurbituril at the equatorial position. The possibility of synthesizing new cucurbituril homologues and derivatives was greatly promoted by this work. During the 1990s, the construction of CB[6]-based molecular switches[10] and the use of CB[6] as a molecular container for reversible host–guest chemistry[11,12] and as a building block for oligo- and polyrotaxanes[13–18] were demonstrated as well.

## 2.1.4 2000s

The syntheses, isolation and full characterization of pentameric (CB[5]), heptameric (CB[7]) and octameric (CB[8]) homologues were first reported by Kimoon Kim and coworkers in 2000.[19] They noticed that varying the reaction parameters (acid, temperature and duration) could have a significant impact on the yields of different homologues, but eventually all the CB[*n*] forms exist as a mixture in which CB[6] is the major species. This cornerstone study in the cucurbituril timeline has catalyzed the progress in cucurbituril chemistry in the last two decades. In 2001, Anthony Day *et al.* conducted a more detailed study on the controlling parameters in the synthesis of cucurbituril homologues and their formation mechanism.[20] They isolated a new supramolecular system that they called *molecular gyroscane* (CB[5]@CB[10]), formed by the inclusion of CB[5] into CB[10] during cucurbituril synthesis.[21]

CB[*n*] homologues (*n* = 5–8) are insoluble in all organic solvents. In 2001, Kim *et al.* successfully synthesized fully substituted cyclohexanoCB[5] (Ch$_5$CB[5]) and cyclohexanoCB[6] (Ch$_6$CB[6]) through the reaction of cyclohexaneglycoluril with formaldehyde using a mixture of HCl and H$_2$SO$_4$ as catalyst.[22] Cyclohexane-substituted derivatives had good solubility in organic solvents, such as DMSO and methanol, and their water solubility was enhanced significantly compared to that of the parent homologues. A year later, Eiichi Nakamura and coworkers reported for the first time the synthesis of a partially substituted cucurbituril derivative, diphenyl CB[6], by hetero-oligomerization of 1 equivalent glycoluril with 5 equivalents of diphenyl glycoluril.[23]

Although the heterooligomerization method has been used for the synthesis of cucurbituril derivatives, significant difficulties are associated with the formation of a complex mixture of cucurbituril homologues and derivatives, which makes it extremely hard to isolate the desired product. Furthermore, the addition of functional substituents could not be achieved in this way. A direct functionalization method would significantly reduce the complexity in product isolation and give much better yields. With this motivation, in 2003, Kim *et al.* achieved the perhydroxylation of CB[$n$]s ($n = 5$–8) by reacting CB[$n$] with $K_2S_2O_8$ in water at 85 °C for 6 h.[24] Since perhydroxyCB[6] was soluble in DMSO and DMF, its further modifications to various functional groups were achieved easily. In the same year, Day *et al.* introduced a method to achieve higher orders of partial substitution and successfully isolated symmetrical hexamethylCB[6].[25] Lyle Isaacs and co-workers synthesized new glycoluril derivatives bearing electron-withdrawing substituents such as carboxylate, acid, ester, amide and imide.[26] The same research group then synthesized glycoluril oligomers–bearing reactive substituents and used them for synthesizing CB[$n$] analogues ($n = 5$–7) having unusual sizes, shapes and colors.[27]

It is well-known that negatively charged carbonyl portals of cucurbiturils have a strong affinity for metal ions through ion–dipole interactions. When the carbonyl portals of CB[$n$]s are blocked by cations, the inclusion of organic molecules inside the hydrophobic cavity of CB[$n$] is impossible. In 2004, Yuji Miyahara and coworkers synthesized and isolated hemiCB[6] and hemiCB[12] by condensation of ethyleneurea with formaldehyde at pH = 2 (Figure 2.2).[28] Since there is only one carbonyl portal in hemiCB[$n$]s, the binding of organic molecules at the hydrophobic ethylene region is not restricted by the capture of a cation on the carbonyl side. In the same year, Day *et al.* published the synthesis of tetramethylCB[6] and showed its host–guest complexation with 2,2′-bipyridine.[29] Ehud Keinan and coworkers demonstrated the separation and purification of CB[6] derivatives *via* affinity chromatography using aminopentylaminomethylated cross-linked polystyrene beads.[30] They revisited the reaction mixture of Stoddart *et al.*[9] and isolated dodecamethylCB[6] in 0.2% yield using this separation method.

Isaacs *et al.* had previously studied the effect of using diastereomeric glycoluril units in the mechanism of diastereomeric CB[$n$] formation.[31,32] In 2005, Isaacs's and Kim's groups collaboratively identified the diastereomers of CB[6] and CB[7] that are inverted CB[6] (*i*CB[6]) and inverted CB[7] (*i*CB[7]) by 2D NMR studies and X-ray crystallography.[33] These diastereomers are called inverted because one of their glycoluril units faces opposite to the others (Figure 2.2). Again in 2005, Isaacs *et al.* managed to eject CB[5] in the structure of the molecular gyroscane, CB[5]@CB[10], and obtained pure CB[10].[34] This was done by forming binary and ternary complexes of CB[5] and CB[10] with a suitably sized guest and subsequent decomplexation of

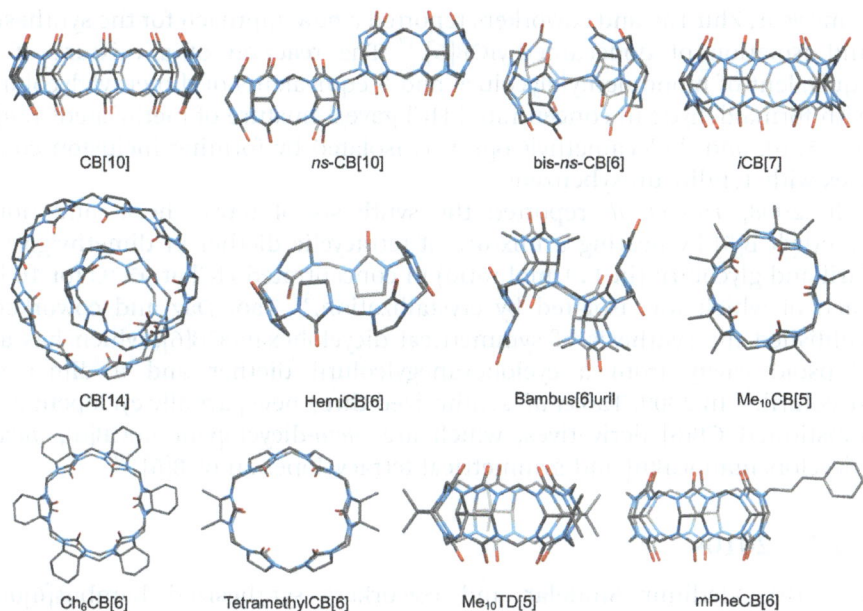

CB[10]      *ns*-CB[10]      bis-*ns*-CB[6]      *i*CB[7]

CB[14]      HemiCB[6]      Bambus[6]uril      Me$_{10}$CB[5]

Ch$_6$CB[6]      TetramethylCB[6]      Me$_{10}$TD[5]      mPheCB[6]

**Figure 2.2** Crystal structures of various CB[*n*] analogues and derivatives. Color codes: gray: C, blue: N, red: O. Hydrogens are omitted for clarity. Crystal structure data were obtained from Cambridge Crystallographic Data Centre.

the guest from the cavity of CB[10] under forcing conditions. They also studied the complexation of free CB[10] with a variety of compounds and demonstrated that it can play a key role in molecular devices and biomimetic systems.

Inspired by the discovery of *i*CB[6] and *i*CB[7] as viable reaction intermediates, Isaacs *et al.* investigated the occurrence of other kinetic products from cucurbituril synthesis. In 2006, they reported the discovery of nor-seco-CB[10] (*ns*-CB[10]), which was isolated in 15% yield when glycoluril and formaldehyde was heated in concentrated HCl at 50 °C.[35] Its structure was elucidated by 2D NMR spectroscopy and X-ray crystallography (Figure 2.2). Binding of various guests, including alkyl and adamantyl amines, ferrocenes, viologens, dyes and amino acids, to *ns*-CB[10] in the form of ternary complexes was shown.

The following year, Isaacs *et al.* tried a different reaction condition in which glycoluril reacts with 1.5 equivalents of paraformaldehyde in concentrated HCl at 80 °C.[36] They isolated a glycoluril trimer and a chiral (±) bis-*ns*-CB[6] (Figure 2.2). They determined its cavity size as between that of CB[6] and CB[7], as well as a binding constant with various suitably sized guests. They also conducted a series of $^1$H NMR experiments to observe how (±) bis-*ns*-CB[6] exhibits diastereoselective recognition toward enantiomeric guests, including amino acids, amino alcohols and a *meso*-diamine. In the same year, Zhu Tao and coworkers reported a new approach for the synthesis and isolation of dodecamethylCB[6].[37] The reaction of a mixture of 1 equivalent of monomethylglycoluril and 5 equivalents of dimethylglycoluril with formaldehyde in concentrated HCl gave a mixture of methylatedCB[*n*]s (*n* = 5, 6), and dodecamethylCB[6] was isolated by forming inclusion complex with 1,4-dihydroxybenzene.

In 2008, Tao *et al.* reported the synthesis of tetra-, hexa- and non-amethylCB[5] by heating a mixture of tetracyclic diether of dimethylglycoluril and glycoluril (in 1 : 1 mol ratio) in concentrated HCl at 95 °C for 10 h, each of which was isolated by crystallization.[38] Tao, Day and coworkers published the synthesis of symmetrical dicyclohexaneCB[6], which has an ellipsoid cavity from a cyclohexanegylcoluril diether and a dimer of glycoluril.[39] In 2009, Tao *et al.* synthesized three new partially cyclopentane-substituted CB[6] derivatives, which are *meta*-dicyclopentanoCB[6], *meta*-tricyclopentanoCB[6] and symmetrical tetracyclopentanoCB[6].[40]

### 2.1.5   2010s

In 2010, Vladimir Sindelar and coworkers synthesized bambus[6]uril through the HCl-catalyzed condensation of 2,4-dimethylglycoluril with formaldehyde.[41] Structural elucidation was performed by X-ray, and [1]H NMR analyses (Figure 2.2). It had good binding with halide anions *via* C–H⋯X[−] hydrogen bonds, and its Cl[−] adduct showed high solubility in chloroform and methanol.

Synthesis of substituted CB[*n*] derivatives bearing a single reactive moiety had been a long-time desire of supramolecular chemists because doing further chemistry with a monofunctionalized CB[*n*] derivative would yield more well-defined and controlled structures. In 2011, De-Qi Yuan and co-workers reported the synthesis of monomethylCB[6] in 14% yield by the acid-catalyzed condensation of a mixture of glycoluril and 3a-methylglycoluril with paraformaldehyde at 95 °C for 24 h.[42] Isaacs *et al.* came up with a novel strategy for the preparation of monofunctionalized CB[6] derivatives. They first conducted the templated synthesis of methylene-bridged glycoluril hexamer and then reacted this hexamer with nitro- or carboxylic acid-substituted phthalaldehydes in concentrated HCl at room temperature to yield monosubstituted CB[6]s that are ready for further functionalizations.[43] They also studied the templating effects of various ions in the macrocyclization step.

Oren Scherman and coworkers achieved for the first time the direct monofunctionalization of a CB[*n*] derivative in 2012.[44] They synthesized monohydroxylated CB[6] *via* oxidation of CB[6] by 1 equivalent of ammonium persulfate in the presence of a bisimidazolium guest and converted it to monopropargyloxylated CB[6], which successfully underwent copper-catalyzed azide–alkyne cycloaddition (CuAAC) with an azide-containing compound. Isaacs *et al.* used their building block approach to synthesize a monofunctionalized CB[6] derivative–bearing propargyloxy moiety, which then reacts with an azido amine compound through CuAAC.[45] The resulting compound self-assembles in water in the shape of a cyclic daisy chain. In another study by Isaacs *et al.*, a building block approach was again employed, but this time glycoluril hexamer reacted with monosubstituted glycoluril bis(cyclic ether), yielding monoalkyl chloride or azide-functionalized CB[7].[46]

In 2013, Kim *et al.* slightly modified their procedure for perhydroxylating CB[*n*] and obtained monohydroxyCB[7] in 27% yield, which was then allyl-oxylated and attached to a thiolated silicon surface through thiol–ene photo-click reaction to afford a supramolecular velcro.[47] In the same year, Isaacs *et al.* synthesized two dimeric CB[6] compounds by acid-catalyzed condensation of hexameric glycoluril with benzene or biphenyl tetra-aldehydes and demonstrated the formation of ladder-like supramolecular assemblies in the presence of oligoviologen in water.[48] Riina Aav and co-workers synthesized enantiomerically pure (*S*)- and (*R*)-cyclohexylhemiCB[6] in good yields and studied their structural and binding properties.[49] Tao *et al.* reported the isolation of 360° twisted CB[14] (Figure 2.2).[50] Sindelar *et al.* synthesized norbornahemiCB[*n*]s (*n* = 4–8), the 6-membered one being the major product, through acid-catalyzed condensation of norbornane-fused ethyleneurea and paraformaldehyde.[51] A year later, the same research group reported the synthesis and isolation of a single methylene-bridge–substituted CB[6] (mPheCB[6]) by heating a mixture of 3-phenylpropionaldehyde, formaldehyde and glycoluril to 90 °C in concentrated HCl for 24 h (Figure 2.2).[52]

In 2015, He Tian and coworkers synthesized a new CB[5] analogue, $Me_{10}TD[5]$, by condensing dimethylpropanediurea and formaldehyde in concentrated HCl at 95 °C for 24 h.[53] The compound was soluble in water and formed a linear supramolecular polymer with 1,4-xylene diamine dihydrochloride. At about the same time, Sindelar *et al.* reported the same compound under pressoCB[5] name (Figure 2.2).[54] A facile method for the synthesis of monohydroxyCB[*n*]s (*n* = 5, 6, 7, 8) *via* photochemical oxidation in very high yields was introduced by David Bardelang, Olivier Ouari and coworkers.[55] However, one year later, the work was corrected as giving much lower but still reasonable yields of monohydroxyCB[*n*]s.[56]

In 2016, Tao *et al.* reported the synthesis, isolation and structural confirmation of CB[13] and CB[15], two new twisted members of the CB[*n*] family.[57] Sindelar *et al.* synthesized a new CB[5] derivative with four dimethylpropanediurea and one glycoluril unit that shows cation-modulated self-assembly in methanol.[58]

In 2017, Hang Cong, Zhu Tao and coworkers prepared mono-, di- and trihydroxylated hexamethylCB[6]s *via* photochemical oxidation and converted them to allyloxylated derivatives for further applications.[59] More recently, synthesis of monopropargyloxylatedCB[7] and its covalent conjugation with different compounds *via* CuAAC reaction in the construction of supramolecular assemblies for various applications have been demonstrated.[60–62]

## 2.2    Synthesis of Unsubstituted CB[*n*] Homologues

The original procedure for the unknowing synthesis of CB[6] consisted of two steps: (1) glycoluril and an excess of formaldehyde were treated with dilute HCl, yielding a precipitate; (2) this precipitate was solubilized in concentrated $H_2SO_4$ at 110 °C, which was then diluted with water and crystallized to give CB[6] (Scheme 2.1).[2]

Kim and coworkers altered the reaction parameters in cucurbituril synthesis and diversified the CB[*n*] family with new penta-, hepta- and octameric homologues.[19] Performing the reaction in 9 M $H_2SO_4$ for 36 h at 75–100 °C gives a mixture of isolable CB[*n*] homologues (*n* = 5–8, 10), the content of which is 60% of CB[6], 20% of CB[7], 10% of CB[5] and 10% of other homologues as confirmed by mass spectrometry (Scheme 2.2). In the same crude reaction mixture, they detected trace amounts of CB[9] and CB[11] by mass spectrometry too, but these homologues have not been isolated. CB[6] precipitates out of the mixture as CB[5] and CB[7] dissolved in an acetone/ water mixture. To the filtrate that contains CB[5] and CB[7], first methanol and then acetone are added to precipitate CB[7] and CB[5], respectively. Isolated homologues are further purified by recrystallizations. Kim *et al.* achieved the isolation and purification of CB[8] from a different CB[*n*] mixture, which was obtained by hydrothermal synthesis in a high-pressure reactor in concentrated HCl at 115 °C for 24 h.[19] Concentration of CB[8] in this reaction mixture is slightly higher as compared to that resulting from the previous synthetic procedure. Crystals of CB[8] are isolated after leaving the reaction mixture overnight at room temperature, which are further purified by washing with acetone/water mixture and crystallizing in 6 M $H_2SO_4$.

Day *et al.* studied the influential parameters in the synthesis of CB[*n*] homologues in more detail such as reactant concentration, temperature and kind of acid.[20] For example, when glycoluril concentration is fixed around 155–190 mg mL$^{-1}$ in acid, the use of 9 M $H_2SO_4$, 9 M HCl, concentrated HCl and 6 M $HBF_4$ does not significantly change the product distribution with CB[6] being the major product, CB[5] and CB[7] almost in equal proportions and CB[8] as the minor species. However, the use of concentrated $H_2SO_4$ yields predominantly CB[6] and a small portion of CB[5]. Additionally, HCl concentrations less than 5 M do not catalyze the cyclization of glycoluril oligomers, and therefore no CB[*n*] can be obtained. They found out that performing CB[*n*] synthesis with glycoluril concentrations as low as 125 µg mL$^{-1}$ produces only the smaller homologues, CB[5] and CB[6]. When glycoluril concentration in the reaction medium is increased to 1.7 g mL$^{-1}$, the yield of higher homologues (*n* = 7–8) also increases. As for the reaction temperature, 100 °C was the optimal temperature for the formation of CB[*n*]s in concentrated HCl, but 50 °C also works if the reaction is continued long enough (>1 month). Keeping the reaction for 2 months in concentrated $H_2SO_4$ at room temperature yields CB[6] in quantitative yield, but using concentrated HCl under the same conditions does not give any CB[*n*]s.[20]

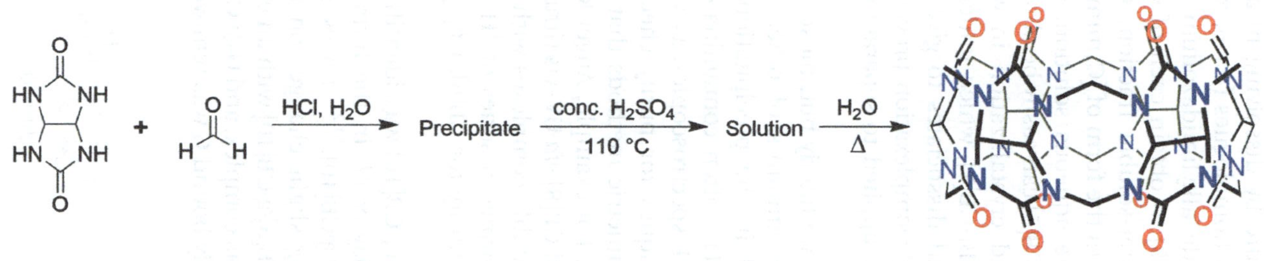

**Scheme 2.1** Behrend's procedure for the synthesis and crystalization of CB[6].

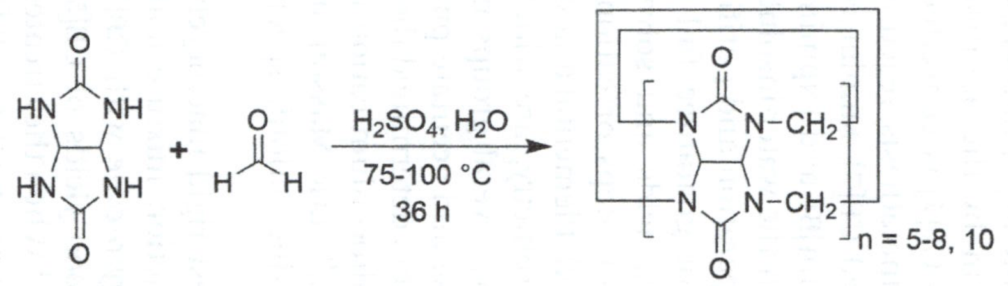

**Scheme 2.2** Kim's procedure for the synthesis of CB[n] (n = 5–8, 10).

A practical method for the separation of CB[7] from the whole reaction mixture is to solubilize CB[7] in hot 20% aqueous glycerol and then precipitate it in methanol.[20,63,64] Odd-numbered CB[$n$] homologues ($n = 5, 7$) can be separated from the crude reaction mixture by dissolving in a sufficient amount of water, leaving even-numbered homologues ($n = 6, 8$) undissolved. Scherman *et al.* employed tailor-made alkylimidazolium ionic liquids that bind only with CB[6], CB[7] or CB[8] to obtain each of CB[$n$] ($n = 5$–8) in pure form (Figure 2.3).[65,66] In water-soluble fraction, CB[7] complexes with [$C_4$mim]Br and then precipitates in the form of [$C_4$mim]PF$_6$-CB[7] complex upon addition of NH$_4$PF$_6$ to the aqueous solution. Once the precipitate of [$C_4$mim]PF$_6$-CB[7] is collected, CB[5] can easily be obtained in pure form by the addition of methanol and crystallization in water. [$C_4$mim]PF$_6$-CB[7] is converted back to [$C_4$mim]Br-CB[7], which undergoes solid-state metathesis reaction with NH$_4$PF$_6$ and dissociates to give pure uncomplexed CB[7]. Similar complexation and decomplexation procedures using [$C_2$mim]Br and [Npmim]Br salt guests are applied to separate CB[6] and CB[8] in the water-insoluble fraction.

These separation and purification techniques are handy but, most of the time, do not guarantee full purification. Trace amounts of other CB[$n$] homologues, acids and solvents that are used in the precipitation and crystallization steps or cationic species like NH4$^+$ often contaminate the desired CB[$n$]. Elemental analysis, $^1$H and $^{13}$C NMR spectroscopies, as well as mass spectrometry, are widely preferred techniques for purity checking. Additionally, several groups have published titrimetric methods that offer more precise and accurate purity determination. For example, Angel Kaifer and coworkers determined the purity of CB[7] and CB[8] *via* UV–vis titrations with a cobaltocenium cation that forms highly stable complexes with both compounds.[67] Eric Masson and coworkers performed a series of $^1$H NMR titrations using cationic species having strong affinity for CB[6], CB[7] and CB[8].

The largest rigid macrocyclic CB[$n$] homologue, CB[10] was identified in a crude reaction mixture and crystallized by Day *et al.* in the form of a *molecular gyroscane* with CB[5] inside it, CB[5]@CB[10].[20,21] Isaacs *et al.* obtained good yields of CB[5]@CB[10] by doing slight changes on Day's procedure.[34] When they treated the solution of CB[5]@CB[10] with an excess of a suitable guest (G), G@CB[5] precipitated out as complex, and G$_2$@CB[10] formed as a ternary inclusion complex. The weakly bound G was removed by

[$C_2$mim]Br         [$C_4$mim]Br         [Npmim]Br

$Ka$ ([$C_2$mim]Br-CB[6]) = $10^4$ M$^{-1}$    $Ka$ ([$C_4$mim]Br-CB[7]) = $10^7$ M$^{-1}$    $Ka$ (([Npmim]Br)$_2$-CB[8]) = $10^{10}$ M$^{-2}$

**Figure 2.3** Structures of alkylimidazolium ionic liquids that complexes with CB[6], CB[7] and CB[8] and their binding constants.

washing with an excess of methanol, then free CB[10] was collected after heating in acetic anhydride and washing with DMSO, methanol and water.

The synthesis and isolation of the largest CB[n] homologues (twisted CB[13], CB[14] and CB[15]) were reported by Tao *et al.*[50,57] In the study where CB[14] was detected and isolated, they employed the same reaction conditions as Kim and Day.[19,68,69] For CB[13] and CB[15], they applied milder conditions using 1 equivalent of glycoluril and 2 equivalents of formaldehyde in 9 M HCl stirred at 90 °C for 6 h. Isolation of twisted CB[n] is a lengthy procedure starting with the extraction of water-soluble components from the reaction mixture and separation of each species by Dowex ion-exchange chromatography for several months. It is interesting that all the twisted homologues are soluble in water and DMSO.

Isaacs *et al.* reported the *i*CB[6] and *i*CB[7], inverted members of the CB[n] family, which exist under reaction conditions described by Kim and Day.[19,21,32] Isolation of the inverted members was achieved by fractional crystallization and then size-selective complexation or by using Superdex 30 gel permeation chromatography. As shown in Figure 2.2, the structure contains a single inverted glycoluril unit that faces opposite to the other units. This structure makes them diastereomers of the corresponding CB[n] homologues. The presence of a single inverted glycoluril decreases the cavity volume of *i*CB[6] and *i*CB[7] and causes an induced permanent dipole moment, which therefore alters their recognition properties. When *i*CB[6] and *i*CB[7] were separately treated in concentrated DCl, they produced CB[5], CB[6] and CB[7] in different yields. Thus, the scientists concluded that *i*CB[6] and *i*CB[7] are isolable intermediates in the formation mechanism of CB[n]s.

## 2.3 Mechanistic Studies: How Does CB[n] Form?

Day *et al.* performed the preliminary investigations on the reaction mechanism of CB[n] formation.[20,69] Scheme 2.3 demonstrates the simplified reaction mechanism proposed by Day *et al.* and validated by subsequent studies of other groups. Paraformaldehyde can sometimes be preferred to supply fresh formaldehyde in the reaction medium and to provide a high concentration of the reagents.[20] Isaacs *et al.* have done the most extensive studies to elucidate the mechanism of CB[n] formation.[31,70-73] For a successful CB[n] synthesis, the glycoluril : formaldehyde ratio is kept at 1 : 2, but they used a 1 : 1 ratio in order to obtain a reaction mixture that contains intermediate species.[70] As a result, they isolated and characterized acyclic glycoluril oligomers ranging from dimer to hexamer and observed that all are in an energetically favored C-shaped structure. Then they reacted each of these acyclic oligomers alone or in binary combinations with 2 equivalents of formaldehyde and noticed huge differences in outcomes. For instance, pentamer reacts with 2 equivalents of formaldehyde to give CB[5] in 84% yield. Hexamer, under the same conditions, gives CB[6] exclusively.

It is apparent from these results that pentamer and hexamer have preorganized structures that enable unimolecular cyclization rather than

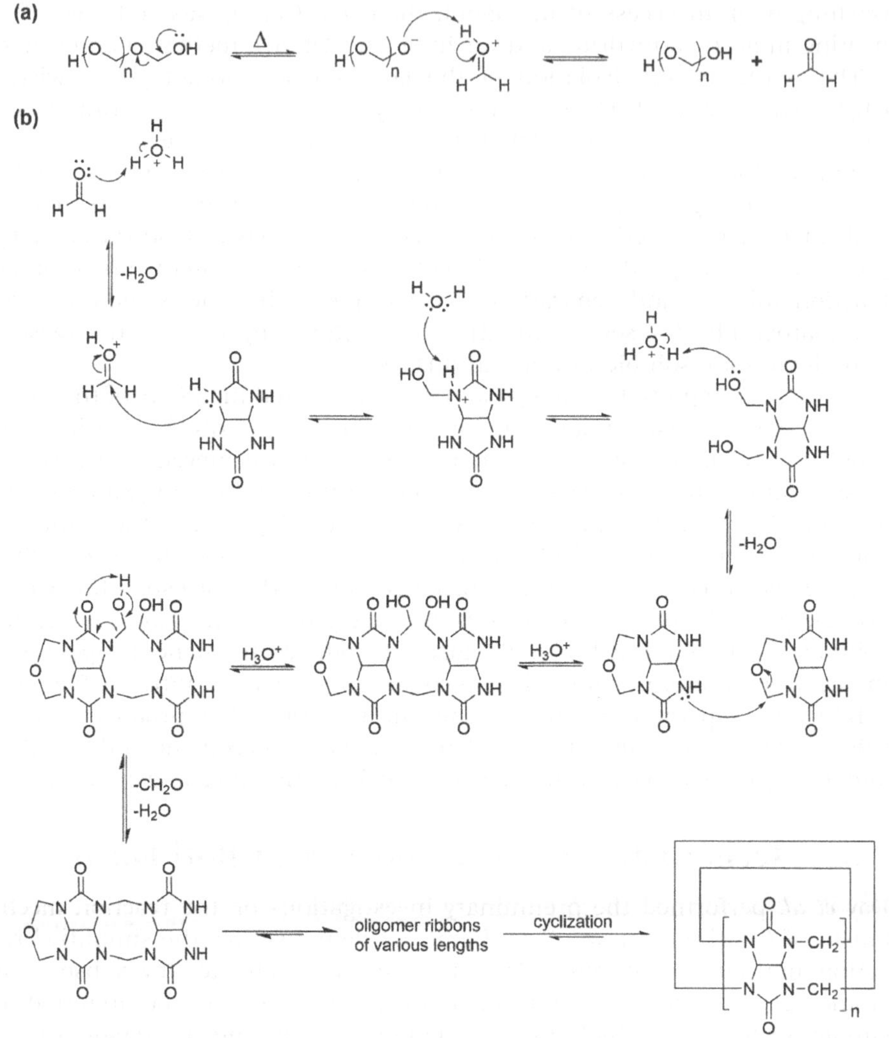

**(a)**

**(b)**

**Scheme 2.3**   (a) Depolymerization of paraformaldehyde into formaldehyde and (b) simplified formation mechanism of CB[*n*].

further oligomerization. When dimer or trimer reacts with 2 equivalents of formaldehyde, CB[6] is obtained in 68% or 75% yields, respectively. When tetramer reacts under the same conditions, it gives 40% of CB[8]. These examples verify that CB[*n*] formation follows a step-growth polymerization mechanism since formation of the even-numbered species are favored in both cases (step-growth polymerization of a single type of oligomer gives an even-numbered species, *e.g.*, $2+2+2=6$, $3+3=6$ or $4+4=8$). However, other CB[*n*] homologues also form besides the major product. This is because the oligomer itself fragments into shorter oligomers during the reaction that can either undergo cyclization or rejoin and oligomerize further.

They also studied the template effects in the formation of CB[$n$] and observed that acyclic pentamer and hexamer bind to the guests, having strong affinity for CB[6] and CB[7].

## 2.4 Structural and Physical Properties of CB[$n$]s

As mentioned in the previous sections, structure of CB[6] was first unraveled by Mock *et al.* in 1981. A strong absorption peak at 1720 cm$^{-1}$ signified the presence of a carbonyl group; thus the glycoluril unit must be retained in the structure of the product. $^1$H NMR spectrum of the product dissolved in 90% deuterated formic acid showed 3 sets of equally intense signals: a singlet at 5.75 ppm was assigned to glycoluril methine hydrogens, and two doublets at 4.43 ppm and 5.97 ppm were assigned to methylene hydrogens that are magnetically nonequivalent due to their endo- and exocyclic positioning. They finally confirmed the structure of the product by X-ray crystallographic analysis as being a cyclic hexamer of glycoluril connected *via* methylene bridges.[3] The unique pumpkin-shaped macrocyclic molecule has two negatively charged carbonyl portals at the top and bottom of the molecule and a hydrophobic inner cavity.

It is a fact that the main members of CB[$n$] family ($n = 5–8$) have quite rigid skeletons, unlike other macrocycles. However, the largest homologues discovered so far, CB[10], CB[13], CB[14] and CB[15], possess structural flexibility. For example, unlike the main members, CB[10] has an elliptic crystal structure, while the rest of the higher CB[$n$] homologues have twisted crystal structures (Figure 2.2). Dimensions of the most common CB[$n$] homologues[19] and inverted CB[$n$]s[33] are given in Table 2.1, in which there are some trends worth mentioning. For instance, for CB[$n$]s ($n = 5–8$, 10), the cavity diameters are ~2 Å larger than the portal diameters, and the difference between the outer and inner diameters is around 8.5–8.8 Å. All given CB[$n$]s, including the inverted ones, have the same height of 9.1 Å. However, cavity diameters for the inverted homologues are smaller than their noninverted counterparts due to the presence of one glycoluril unit whose methine protons are directed into the cavity.

The carbonyl portals on CB[$n$]s are polarized and thus negatively charged, whereas the cavity is strongly nonpolarizable since there are no functional groups or electron pairs in this environment. Therefore, the carbonyl rims can involve ion–dipole interactions or hydrogen bonding, but the nonpolar nature of the cavity allows for only hydrophobic interactions within the cavity.[74] For these reasons, the carbonyl portals bind to cationic species, whereas the hydrophobic cavity can encapsulate hydrophobic compounds. Additionally, the carbonyl portals, being 2 Å narrower than the cavity, create a steric barrier for the guests. The guests, carrying cationic tails and nonpolar cores, usually demonstrate very high binding affinity for CB[$n$]s. Besides the electronic structure of the guest, its size and shape complementarity to CB[$n$] play a key role in complexation strength.[75] For example, CB[5] has such a small portal diameter and cavity volume that it can bind

**Table 2.1**　Structural parameters of main CB[$n$] homologues and inverted CB[$n$]s.

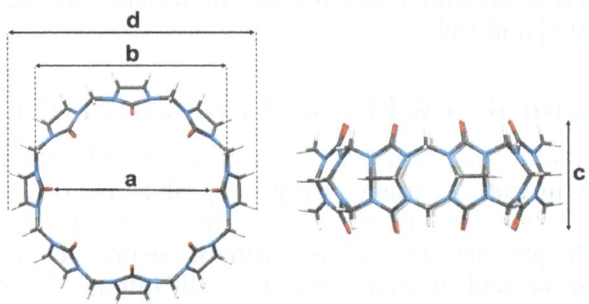

|  | Mr | a[Å] | b[Å] | c[Å] | d[Å] | V[Å] |
|---|---|---|---|---|---|---|
| CB[5] | 830 | 2.4 | 4.4 | 9.1 | 13.1 | 82 |
| CB[6] | 996 | 3.9 | 5.8 | 9.1 | 14.4 | 164 |
| CB[7] | 1163 | 5.4 | 7.3 | 9.1 | 16.0 | 279 |
| CB[8] | 1329 | 6.9 | 8.8 | 9.1 | 17.5 | 479 |
| CB[10] | 1661 | 9.5–10.6 | 11.3–12.4 | 9.1 | 20.0 | 870 |
| iCB[6] | 996 | 4.3–3.9 | 3.8–5.8 | 9.1 | 10.7–14.4 | — |
| iCB[7] | 1163 | 6.7–5.4 | 5.4–7.3 | 9.1 | 11.2–16.0 | — |

only to small cations, such as $Na^+$, $NH_4^+$, $Co^{2+}$, $Fe^{2+}$, or to gases like $O_2$, $N_2$, and $CO_2$.[76–79] CB[8], on the other hand, has enough portal and cavity size to encapsulate quite bulky molecules, including long cationic alkyl chains, fullerene, other macrocyclic guests or adamantane and ferrocene derivatives having positively charged amphiphilic structures.[78–82] One-to-one inclusion complexes of CB[7] with adamantanes,[83] ferrocenes,[84] cobaltocenes[85] and diamantanes[86,87] have binding constants values, $K$, between $10^9$ and $10^{17}$ $M^{-1}$ ($K$ for the biotin-avidin pair, the strongest noncovalent interaction, is $10^{15}$ $M^{-1}$).[74] Accommodation of a guest inside the cavity of CB[$n$]s generally results in noticeable changes in the UV–Vis absorbance and fluorescence and the NMR spectra of the guest. Therefore, the polarity of the CB[$n$] cavity can be assessed by analyzing these spectral shifts. For example, the absorption band of Rhodamine 6G undergoes a red shift upon being encapsulated by CB[7], indicating that the cavity of CB[7] is less polar than water (similar to alcohols).[88]

Water solubility of odd-numbered CB[$n$] homologues, CB[5] and CB[7], is higher (20–30 mM) compared to that of even-numbered ones, CB[6] (0.018 mM) and CB[8] (<0.01 mM).[89,90] In the case of CB[5] and CB[7], molecule–water interactions are stronger than molecule–molecule inter-actions, but for CB[6] and CB[8], the situation is *vice versa*. Therefore, CB[6] and CB[8] form well packed crystals upon drying, while CB[5] and CB[7] are amorphous in dried form.[91–93] However, their solubility increases strik-ingly in concentrated aqueous acids. For example, CB[6] has 61-mM solubility in formic acid/water (1 : 1), and CB[7] has 700-mM solubility in 3 M HCl.[94,97] CB[$n$] homologues have high thermal stability such that TGA results show no decomposition until 420 °C for CB[5], CB[6] and CB[8], while the decomposition of CB[7] starts at a relatively lower temperature of 370 °C.[89]

## 2.5   Synthesis of Substituted CB[*n*] Derivatives

Cucurbit[*n*]urils can be designed and synthesized in several ways to bear various substituents on different regions of the molecule. CB[*n*]s that are fully or partially substituted with alkyl or aryl groups on the equatorial positions or on the methylene bridges have been reported as summarized in Section 2.1. Reactive functional group-bearing CB[*n*]s synthesized either from substituted glycoluril monomers or oligomers or *via* post-functionalization method was also touched upon. The presence of these substituents often results in an increased solubility of the macrocycle in common solvents, especially for CB[6] and CB[8] that are practically insoluble in any solvent. Incorporation of reactive substituents allows for further derivatizations and make it possible to synthesize covalently attached supramolecular assemblies. With better solubility and increased reactivity, the usability of these species in various reactions and applications has been shown. Four main strategies to endow substituents on CB[*n*]s are (1) to use substituted glycoluril, (2) to use substituted glycoluril oligomers, (3) to use aldehydes other than formaldehyde and (4) to directly functionalize already synthesized CB[*n*]s. Pioneering studies regarding the first three methods will be discussed in this section. An expanded discussion of direct functionalization method will be made later in this chapter (Section 2.6).

### 2.5.1   Substituted CB[*n*]s by Reaction of Substituted Glycolurils with Formaldehyde

Let's first get familiar with substituted glycolurils. Glycoluril is the building block of CB[*n*]s and is synthesized from 2 equivalents of urea and glyoxal *via* an acid-catalyzed condensation reaction. If one uses a ketone reagent such as 2,3-butadione or 1,2-cyclohexadione instead of an aldehyde (glyoxal), methine moieties on the resulting glycoluril will possess the corresponding substituents in place of hydrogens (Scheme 2.4).

The very first example of a substituted CB[*n*] was decamethylCB[5], synthesized by Stoddart *et al.* in 1992, although the parent CB[5] had not been yet discovered back then.[9] They reacted dimethylglycoluril and an aqueous solution of formaldehyde in concentrated HCl under reflux for 3 h, which, upon cooling down, gave the water insoluble product in 16% yield. The crystal structure of the compound with $C_2$ symmetry was revealed by X-ray analysis.

Further investigations on the binding properties of decamethylCB[5] was done by several groups.[93–95] Takahiko Inazu and coworkers showed that

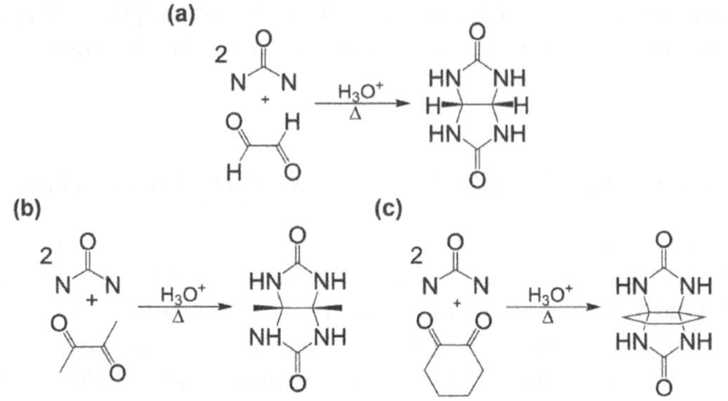

**Scheme 2.4**  Synthesis of (a) glycoluril, (b) dimethylglycoluril and (c) cyclo-hexanoglycoluril.

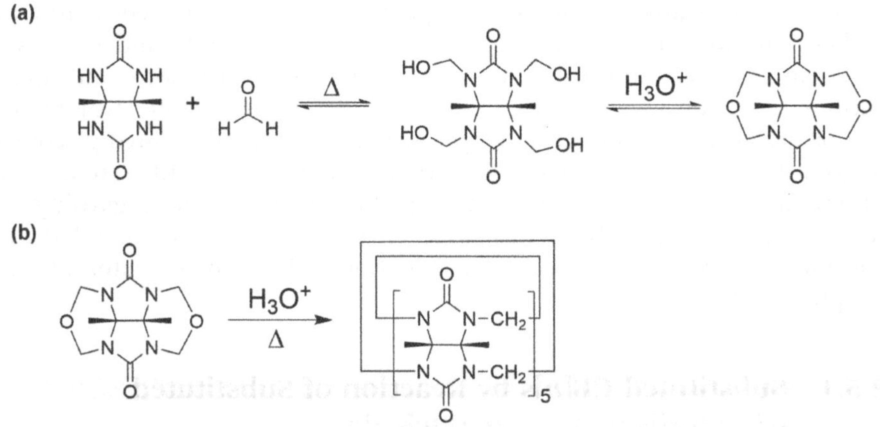

**Scheme 2.5**  Synthesis of (a) dimethylglycoluril tetracyclic diether and (b) deca-methylCB[5] from dimethylglycoluril tetracyclic diether.

decamethylCB[5] can act as a molecular sieve to selectively entrap many gaseous species, including $O_2$, $N_2$, NO, $N_2O$, Ar, CO and $CO_2$, at room temperature.[96] In 2001, Day *et al.* isolated the tetracyclic diether of di-methylglycoluril as an intermediate in the initial stages of a CB[*n*]-forming reaction.[20] Heating this tetracyclic diether for 1.5 h in a strongly acidic medium gave decamethylCB[5] in 85% yield, which is significantly better than that reported by Stoddart (Scheme 2.5).

Day *et al.* modified this procedure to get partially methylated CB[*n*]s. They reacted the tetracyclic diether of dimethylglycoluril with glycoluril in a 1:1 ratio in the presence of 1 equivalent of LiCl in concentrated HCl at 100 °C for 2 h. Their trials showed that the percentage of substituted CB[6] products increased from 56% to 76% if there is LiCl in the reaction medium.

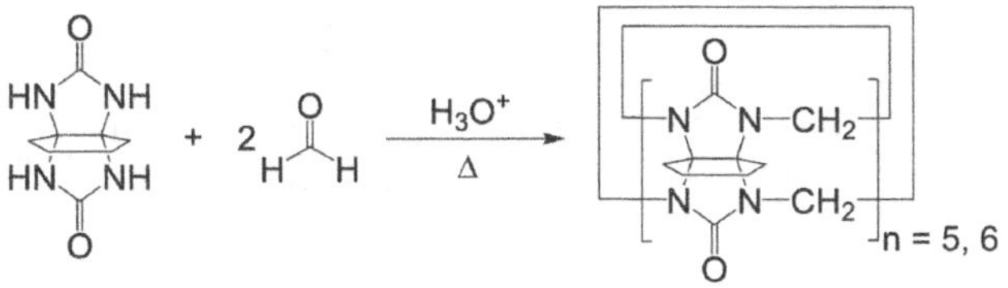

**Scheme 2.6**    Synthesis of cyclohexanoCB[5] and –CB[6] using cyclohexanoglycoluril.

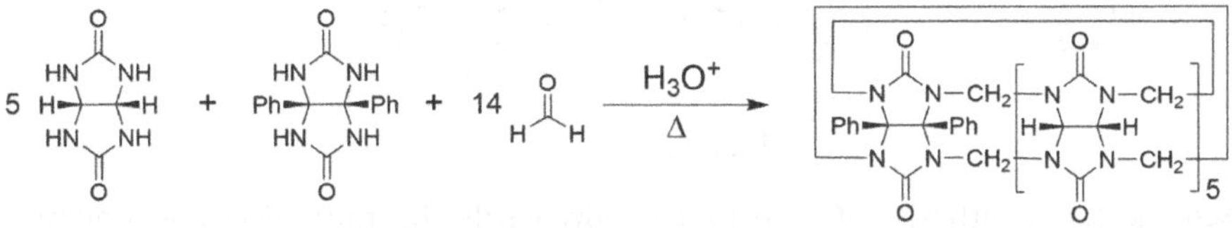

**Scheme 2.7**    Synthesis of diphenylCB[6] by heterooligomerization.

They separated the methylated CB[$n$]s ($n = 5$–7) through cation exchange column chromatography. Out of the mixture of partially methylated CB[6]s, symmetrical hexamethylCB[6] was isolated as the major product after several crystallizations with 1,4-dioxane.[25]

Kim *et al.* refluxed cyclohexanoglycoluril with 2 equivalents of formaldehyde in concentrated HCl and $H_2SO_4$ mixture for 24 h and collected cyclohexanoCB[5] and cyclohexanoCB[6] with 16% and 2% yields, respectively (Scheme 2.6).[22] These substituted derivatives had water solubilities ($\sim 2 \times 10^{-1}$ mol $L^{-1}$) almost as high as that of α-cyclodextrin. They also demonstrated the potential of cyclohexanoCB[5] as an ion-selective electrode for detection of $Pb^{2+}$ and choline.

For the synthesis of partially substituted CB[$n$]s, mixtures of glycolurils and substituted glycolurils can be heterooligomerized. The first example of such compounds was diphenylCB[5], which was synthesized in 30% yield by heating a mixture of glycoluril and diphenyl glycoluril (5:1 molar ratio) with 14 equivalents of formaldehyde in dilute $H_2SO_4$ at 70–95 °C for 36 h (Scheme 2.7).[23] This was also the first example of an aryl-substituted CB[$n$] derivative.

## 2.5.2   Substituted CB[$n$]s from Substituted Glycoluril Oligomers

Day, Isaacs and coworkers synthesized a series of glycoluril monomers bearing electron-withdrawing reactive substituents.[26] They first prepared disodium dihydroxytartrate (DHT) by oxidizing 2,3-dihydroxyfumaric acid with $Br_2$ in glacial acetic acid based on reported procedures.[97,98] Next, DHT was converted to diester-substituted glycoluril (**1**) in two steps.[99] Cyclic diether of diester substituted glycoluril (**2**) was also synthesized. Both **1** and

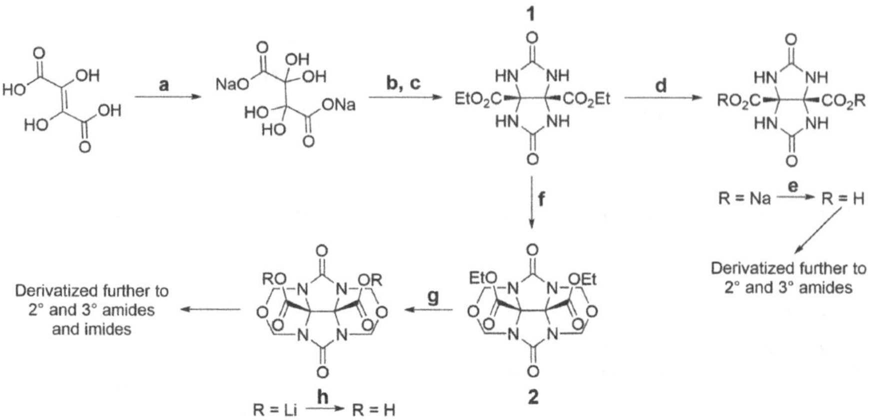

**Scheme 2.8** Synthesis of substituted glycolurils bearing electron-withdrawing groups. *Conditions:* (a) glacial acetic acid, Br$_2$, H$_2$O; (b) EtOH, HCl; (c) benzene, urea, TFA, reflux; (d) NaOH, H$_2$O, 80 °C, 90%; (e) *p*-toluenesulfonic acid, 46%; (f) anhydrous TFA, paraformaldehyde; (g) LiOH, methanol, H$_2$O, 70 °C, 80% and (h) TFA.

2 were then used as precursors to obtain glycolurils and glycoluril cyclic diethers that are substituted in the methine groups with carboxylic acid and its derivatives such as secondary amide, imide and tertiary amide. All the conversions and reaction conditions are given in Scheme 2.8.

Based on the mechanistic investigations they conducted, Isaacs and Day independently hypothesized that when a mixture of glycoluril and cyclic diethers of glycoluril reacts, the building blocks selectively hetero-oligomerize to give CB[*n*]s with certain patterns.[25,31,32] To test this hypothesis, Isaacs *et al.* synthesized a series of cyclic diether glycoluril oligomers, as shown in Scheme 2.9A, and isolated 2, 3C and 4CC in good yields. Derivatives of dimeric 3C with carboxylic acid, amide and imide functional groups were prepared as previously mentioned. Bis(phthalhydrazide) with free ureidyl (NH) groups was used as an alternative to glycoluril in acid-catalyzed condensation reactions with 2, 3C and 4CC. These reactions yielded CB[5], CB[6] and CB[7] analogues, respectively, with unusual sizes, shapes and colors. The reaction of CB[7] with bis(phthalhydrazide) and the X-ray crystal structure of the resulting CB[6] analogue (10) is given in Scheme 2.9B. The structure of 10 is not pumpkin-shaped but rather ellipsoidal whose curvatures around bis(phthalhydrazide) groups are slightly toward the cavity of the macrocycle. Its size differs from that of CB[6] significantly with 5.90×11.15×6.92-Å cavity dimensions. This kind of analogue also fluoresces since phthalhydrazide moieties provide conjugated bonds and their fluorescence properties change upon guest binding.[100–102]

In 2012, Isaacs *et al.* showed the synthesis of a monofunctionalized CB[7] derivative from a glycoluril hexamer and a cyclic diether of a monoalkyl chloride–attached glycoluril.[46] They first synthesized a monosubstituted

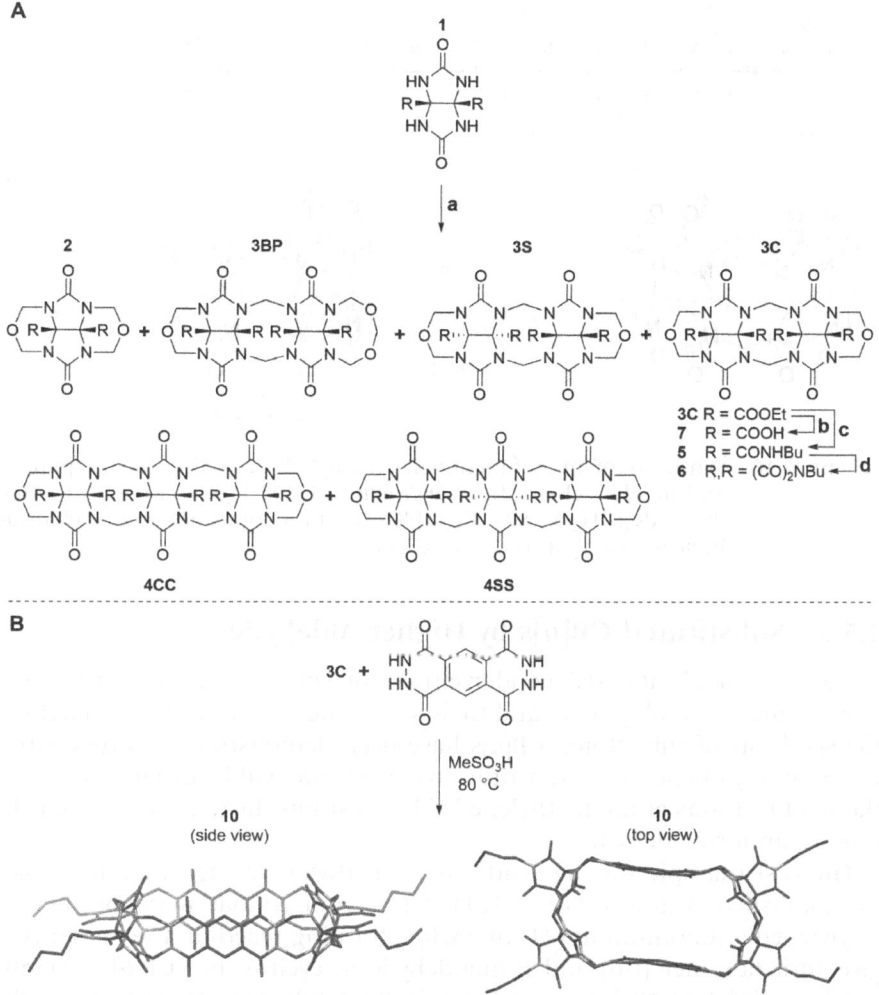

**Scheme 2.9** (A) Controlled oligomerization of **1**. *Conditions:* (a) 1,2-dicholoroethane, PTSA, reflux; (b) LiOH, H₂O, methanol, 89%; (c) *n*-butylamine, 75 °C, 68%; (d) 1,2-dicholoroethane, PTSA, reflux, 39%. (B) synthesis of **10** from **7** and bis(phthalhydrazide). Crystal data of **10** was taken from Cambridge Crystallographic Data Center. Color codes: gray: C, blue: N, red: O. Hydrogens are omitted for clarity.

glycoluril bis(cyclic ether) (G1) and then reacted it with glycoluril hexamer (G6) in 9 M $H_2SO_4$ at 110 °C in the presence of potassium iodide. They isolated the monoalkyl chloride functionalized CB[7] (**1**) in 16% yield after Dowex ion exchange chromatography, followed by conversion from chloride to azide group (**2**) using $NaN_3$ in $H_2O$ at 80 °C with 81% yield. The resulting azide-functionalized CB[7] underwent a CuAAC reaction with propargyl-ammonium chloride using Pericàs's catalyst in $H_2O$ at 50 °C, yielding the compound **3** in 95% yield (Scheme 2.10).

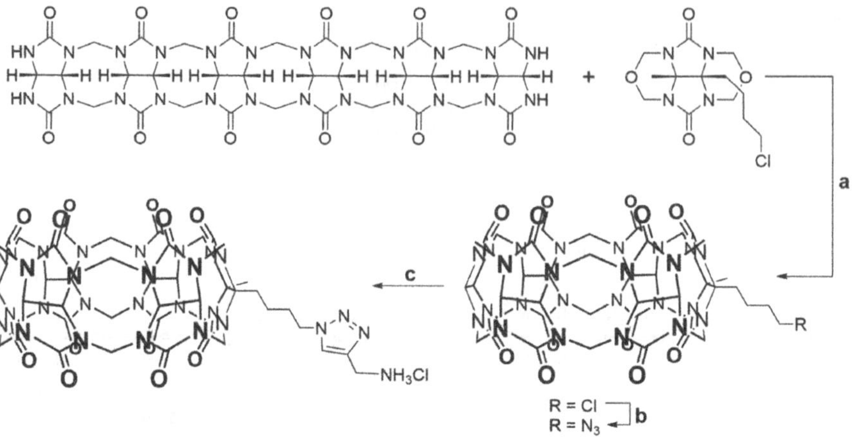

**Scheme 2.10**  Synthesis of monofunctionalized CB[7] by building block approach and its click chemistry. *Conditions:* (a) 9 M $H_2SO_4$, KI, 110 °C, 16%; (b) $NaN_3$, $H_2O$, 80 °C, 81%; (c) propargylammonium chloride, Pericàs's catalyst, $H_2O$, 50 °C, 95%.

### 2.5.3  Substituted CB[*n*]s by Higher Aldehydes

Up to now, acid-catalyzed condensations of substituted glycoluril mono-mers, dimers and oligomers and their substituted cyclic diether versions in the synthesis of substituted CB[*n*]s have been demonstrated. In this part, a different way of endowing substitutions on a CB[*n*] will be discussed, that is, the modifications at the methylene bridge positions by the use of aldehydes other than formaldehyde.

The first example for the synthesis of methylene bridge–modified CB[*n*] was published by Isaacs *et al.* in 2011.[43] They noticed that, in the presence of *p*-xylylenediammonium ion (1) in a CB[*n*]-forming reaction, the tendency of glycoluril hexamer (G6) and formaldehyde to cyclize into CB[6] is signifi-cantly diminished. G6 is thermodynamically stabilized by forming a complex with 1, and precipitation of 1@G6 complex as the reaction occurs also contributes to stability against further transformations. They performed the reaction using less than 2 equivalents of paraformaldehyde (1.67 eq.) per glycoluril unit and 0.1 equivalent of 1 in 12 M HCl at 58 °C for 5 days. 1 decomplexed from G6 after washing with water and subsequently adding 5 M NaOH solution to precipitate free G6. A series of monofunctionalized CB[6] derivatives were then synthesized by reacting G6 with nitro- or carb-oxylic acid-substituted phthalaldehydes and naphthalenedialdehyde, as shown in Scheme 2.11. In another publication, they reported a mono-functionalized CB[6] synthesized *via* condensation of glycoluril hexamer (G6) and 4-hydroxyphthalaldehyde in 9 M $H_2SO_4$ at room temperature for four days, affording 52% product.[45] This CB[6] derivative bearing a reactive –OH moiety was then subjected to nucleophilic substitution with propargyl bromide, resulting in a propargyloxy moiety on the benzene unit.

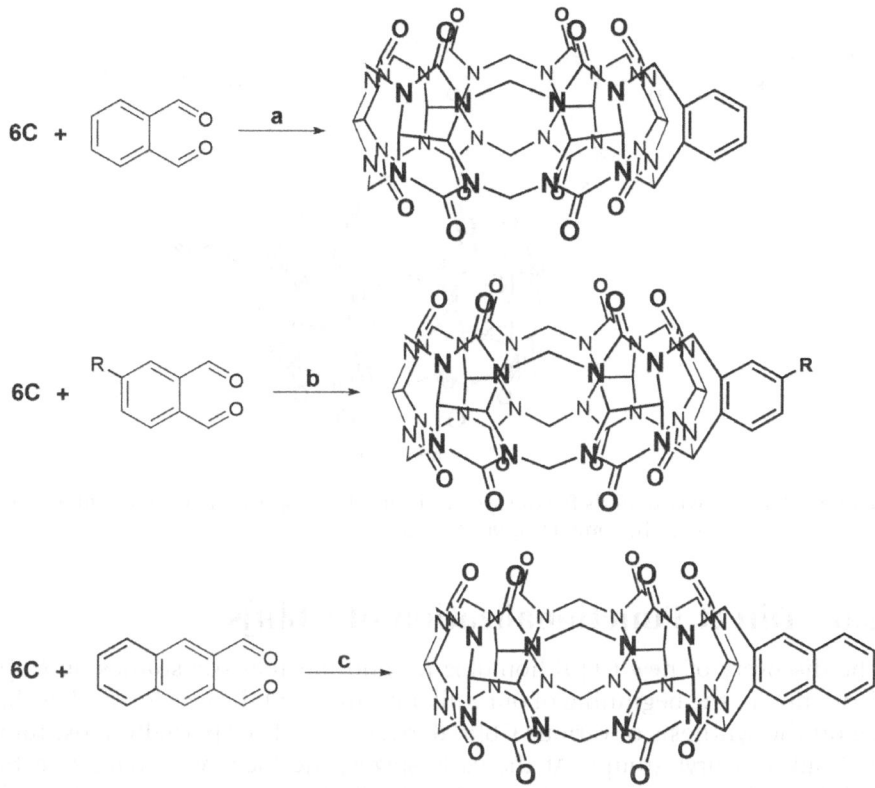

**Scheme 2.11**　Synthesis of monosubstituted CB[6] derivatives from glycoluril hexamer (G6) and phthalaldehydes and naphthalenedialdehyde. *Conditions:* (a) 9 M H$_2$SO$_4$, rt, 36 h, 72%; (b) 9 M H$_2$SO$_4$, rt, 48 h, 56% for COOH, 58% for NO$_2$; (c) conc. HCl, rt, 24 h, 83%.

The propargyloxylated compound was then shown to undergo CuAAC reaction with an azide-functionalized compound.

Introducing a substituent on a single methylene bridge in CB[*n*] was first demonstrated in 2014 by Sindelar *et al.* Equimolar condensation of glycoluril, paraformaldehyde and 3-phenylpropianaldehyde using concentrated HCl as catalyst at 90 °C for 24 h yielded 0.2% of single methylene bridge–modified CB[6] (mPheCB[6]) after column purification.[52] Sindelar *et al.* wanted to devise an alternative method to increase the yield of mPheCB[6] to considerable levels. They were inspired by the Isaacs group's strategy of using glycoluril hexamer (G6) in the synthesis of CB[6] derivatives. They then managed to obtain the pure product in 18% yield after reacting the hexamer (G6) and 3-phenyl-propianaldehyde (in 1 : 8 mol ratio) in concentrated HCl at 90 °C for 3 h. Both methods are illustrated in Scheme 2.12. They also observed that the product self-assembles in solid state to form tetrameric supramolecular systems and that the presence of a substituent in the methylene bridge does not practically alter the binding constants of CB[6] against various alkylammonium salts.

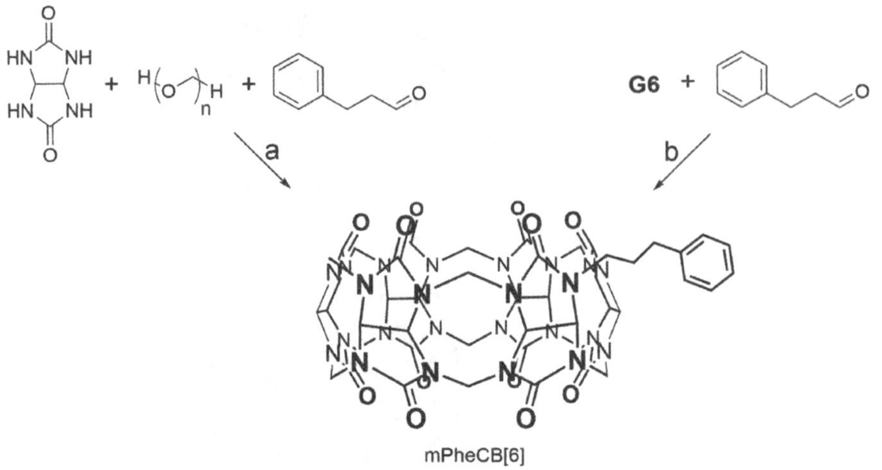

mPheCB[6]

**Scheme 2.12**   Two methods for the synthesis of mPheCB[6]: (a) conc. HCl, 90 °C, 24 h, 0.2%; (b) conc. HCl, 90 °C, 3 h, 18%.

## 2.6   Direct Functionalization of CB[*n*]s

The discovery of new CB[*n*] homologues and mechanistic studies on CB[*n*] formation in the beginning of our millennium paved the way for studies that report the synthesis of various CB[*n*] derivatives fully or partially substituted with alkyl or aryl groups. At the early stages, the focus was mainly on the building block synthesis of substituted CB[*n*]s, as discussed in detail in Section 2.5. Substituted CB[*n*]s generally had improved solubility in organic solvents and water. However, this approach has not been very practical since it requires the isolation of the desired product from a mixture of CB[*n*] homologues and derivatives after lengthy separation processes, often ending up with very low yields of the product. This situation has directed the attention of chemists to a seemingly less troublesome method that is the direct functionalization of CB[*n*]s. The main obstacles for direct functionalization of preformed macrocycles were low solubility of CB[*n*]s in common solvents and their thermal and chemical stabilities, indicating that very harsh conditions are needed in order for them to react with other molecules.

### 2.6.1   Hydroxylation of CB[*n*]s

The introduction of the direct functionalization method that allows for the synthesis of fully functionalized CB[*n*]s ($n = 5$–$8$) from preformed macrocycles was made by Kim *et al.* in 2003.[24]

Each of the CB[*n*]s ($n = 5$–$8$) was separately subjected to oxidation by more than $2n$ ($n = 5$–$8$) equivalents of $K_2S_2O_8$ in $H_2O$ at 85 °C for 6 h (Scheme 2.13). PerhydroxyCB[*n*]s (($OH)_{2n}$CB[*n*]s where $n = 5$–$8$) were obtained after crystallizing the product from the reaction mixture by acetone vapor diffusion and subsequent washing with acetone. The resulting compounds were characterized by NMR spectroscopy, mass spectrometry and elemental analysis. X-ray crystallography analysis of $(OH)_{12}CB[6]$ elucidated the structure of the compound as having 12 hydroxyl units in methine moieties on the equatorial region of the macrocycle. It was also observed that $(OH)_{12}CB[6]$ has almost identical portal and cavity diameters (3.9 Å and 5.5 Å, respectively) with parent CB[6]. Interestingly, the yields of smaller macrocycles, $(OH)_{10}CB[5]$ and $(OH)_{12}CB[6]$, were 45% and 42%, respectively, while the larger macrocycles, $(OH)_{14}CB[7]$ and $(OH)_{16}CB[8]$, were obtained in only 5% and 4%, respectively, which was attributed to the lower chemical stability of the hydroxylated products of the larger homologues. The mechanism of hydroxylation of CB[$n$]s has yet to be fully resolved, but it has been considered to follow a free radical mechanism in which persulfate produces OH radicals that target methine positions on the equatorial region. Although the parent CB[$n$]s are insoluble in organic solvents, perhydroxy derivatives dissolve in DMSO and DMF, which enables one to perform further derivatizations on $(OH)_{2n}CB[n]$. Such examples will be shown in Section 2.6.2.

Direct functionalization of CB[$n$]s to yield perhydroxylated products was a significant progress in CB[$n$] chemistry that has since allowed the use of CB[$n$]s in various applications. However, the availability of many reactive hydroxyl groups on perhydroxyCB[$n$]s often leads to the formation of a mixture of fully or partially derivatized compounds in subsequent reactions.[103,104] For this reason, chemists have put huge efforts into the synthesis of monofunctionalized CB[$n$] derivatives, and plenty of studies have been published on this topic since 2012, which will be discussed hereafter.

In order to obtain well-defined, single-point reactive CB[$n$] derivatives, the reaction conditions for the hydroxylation of CB[$n$]s need to be tamed. Scherman *et al.* performed mass spectrometric analyses during the formation of perhydroxyCB[6] and concluded that the oxidation proceeds in a stepwise fashion. Since CB[6] is practically insoluble in water, the chance of partially hydroxylated CB[6]s with higher water solubilities to undergo further hydroxylation by $K_2S_2O_8$ gets higher. Therefore, they reasoned that before the initiation of the oxidation reaction, CB[6] needs to be solubilized in water as much as possible so as to prevent the formation of multihydroxylated derivatives. On the basis of this argument, in 2012,

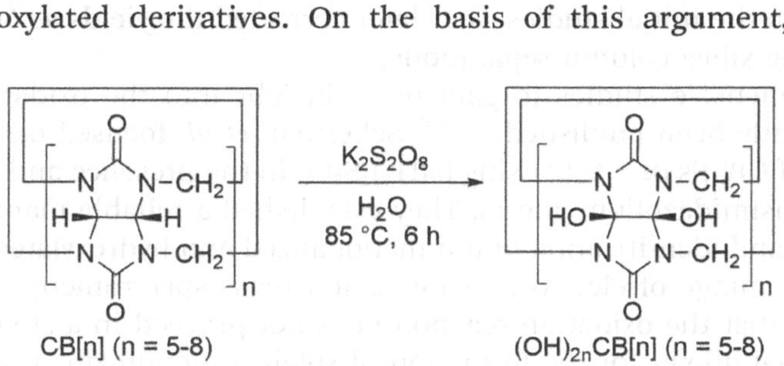

CB[n] (n = 5-8)          (OH)$_{2n}$CB[n] (n = 5-8)

**Scheme 2.13**   Synthesis of perhydroxyCB[$n$]s. *Yields:* $(OH)_{10}CB[5]$: 45%; $(OH)_{12}CB[6]$: 42%; $(OH)_{14}CB[7]$: 5%; $(OH)_{14}CB[8]$: 4%.

Scherman *et al.* became the first scientists reporting the synthesis of monohydroxylated CB[6] ((OH)$_1$CB[6]).[44] In order to solubilize CB[6] in water before the hydroxylation reaction, they first mixed CB[6] and a suitably sized bisimidazolium guest (bis) in 1:1 mol ratio. The bis@CB[6] complex dissolved in water as expected and then reacted with 1 equivalent of (NH$_4$)$_2$S$_2$O$_8$ in H$_2$O at 85 °C for 12 h (Scheme 2.14). The resulting reaction mixture mainly contained bis@CB[6], bis@(OH)$_1$CB[6] and little amount of bis@(OH)$_2$CB[6]. The separation of these species was achieved by passing the concentrated reaction mixture through a reverse-phase microporous resin, MCI GEL CHP20P. Finally, purified bis@(OH)$_1$CB[6] complex was refluxed in DCM-containing NH$_4$PF$_6$ to remove the bisimidazolium guest and obtain pure (OH)$_1$CB[6] in 12% yield. X-ray analysis of the product showed that the portal diameter of (OH)$_1$CB[6] is the same as that of CB[6] (3.9 Å), while its cavity diameter (5.6 Å) is slightly less than that of CB[6] (5.8 Å).

The monohydroxylation of CB[7] was shown in 2013 by Kim *et al.* by reacting CB[7] with an 0.8 equivalent of K$_2$S$_2$O$_8$ in the presence of 6 equivalents of K$_2$SO$_4$ at 85 °C for 12 h under N$_2$ atmosphere (Scheme 2.14).[47] Using a submolar amount of K$_2$S$_2$O$_8$ helped to prevent the formation of multi-hydroxylated derivatives to some extent, but the reaction still yielded a mixture of unfunctionalized and mono-, di- and trihydroxylated CB[7]s. From the reaction mixture, CB[7] and hydroxylated CB[7] were extracted with concentrated HCl and precipitated in methanol. They then isolated each of these compounds by passing the mixture through a sephadex G-15 column, yielding 27% of (OH)$_1$CB[7].

In 2015, Bardelang, Ouari and coworkers introduced a salt-free method that can oxidize all the main CB[$n$] homologues ($n = 5$–8) to yield monohydroxylated CB[$n$]s ($n = 5$–8).[55,56] What they did was basically to oxidize CB[$n$]s photochemically in the presence of 0.5 equivalent of H$_2$O$_2$ under UV radiation for 2–5 h. They proposed that the reaction is initiated by the UV light–triggered formation of OH radicals from H$_2$O$_2$. Studies using [18]O-labeled water revealed that both radical and ionic species are present in the reaction medium, suggesting that quite complex reactions are taking place during the photochemical oxidation. They suggested the reaction mechanism given in Scheme 2.15b. Although yields around 90–100% were reported in the initial studies, they later corrected the yields as 5–37% after performing silica column separations.

Comprehensive studies to gain more insight into the oxidation CB[$n$]s have recently been published.[105,106] Scherman *et al.* focused on the hydroxylation of CB[$n$]s ($n = 6$–8) using (NH$_4$)$_2$S$_2$O$_8$ in the presence and absence of cationic bisimidazolium guests. They established a reliable manner for the detection and identification of unfunctionalized and hydroxylated CB[$n$]s by taking advantage of electrospray-ionization mass spectrometry. They demonstrated that the oxidation reaction does not proceed in a stoichiometric fashion that allows for the formation of solely a monohydroxylated derivative. Therefore, the monohydroxylated products are collected in low yields. However, in the case of CB[7], in the presence of a cationic bisimidazolium

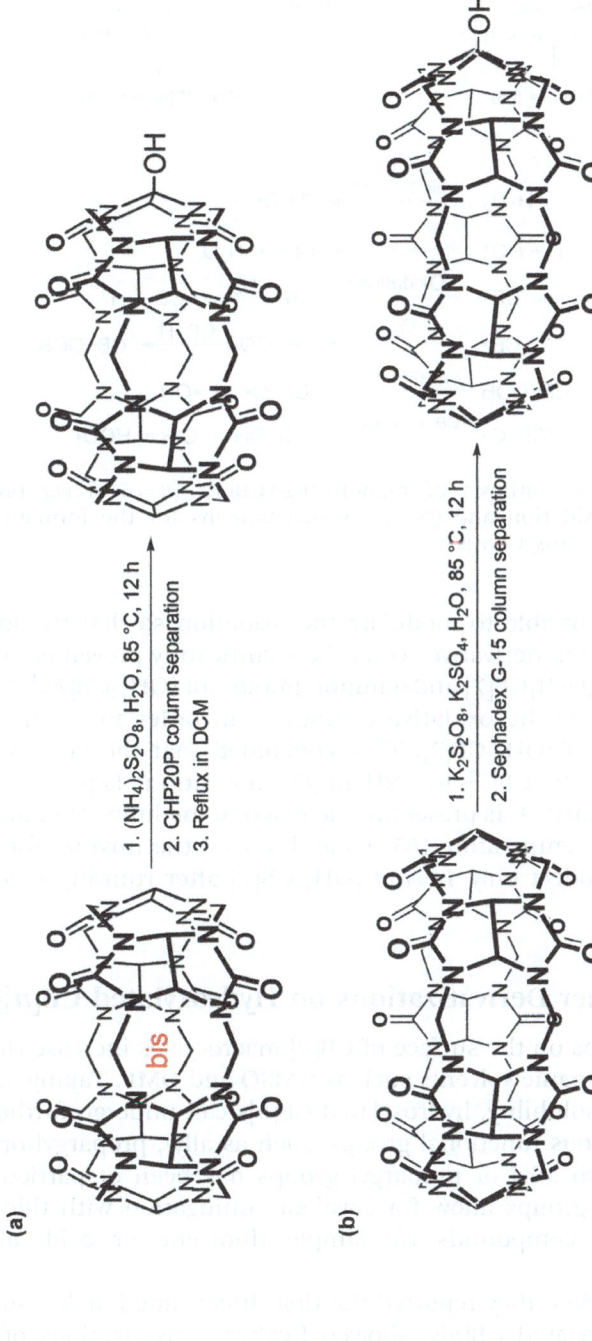

**Scheme 2.14** Synthesis of (a) monohydroxyCB[6] and (b) monohydroxyCB[7].

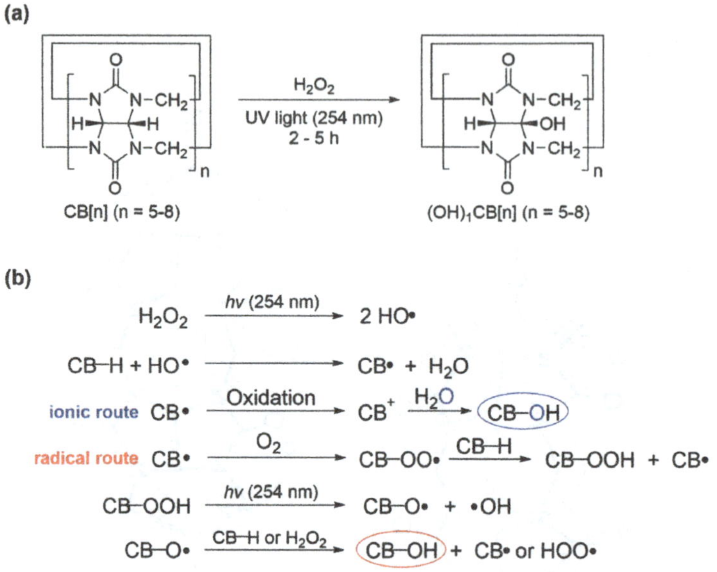

**(a)**

CB[n] (n = 5-8)                                      (OH)₁CB[n] (n = 5-8)

**(b)**

**Scheme 2.15** (a) Synthesis of monohydroxyCB[$n$]s ($n = 5$–8) *via* photochemical oxidation and (b) Likely mechanisms for the formation of mono-hydroxyCB[$n$]s.

guest, it was possible to modulate the oxidation so that the formation of multihydroxylated derivatives could be significantly prevented, yielding un-reacted CB[7], (OH)₁CB[7] and a minor amount of (OH)₂CB[7].[105] Kaifer *et al.* tried to optimize the oxidative conditions in order to get the maximum possible yield of (OH)₁CB[7]. They concluded that the best yields can be obtained when the CB[7]- to- (NH₄)₂S₂O₈ mol ratio is kept at 1:1.5 and 1.5 equivalent of NaHSO₃ is present in the reaction medium. They also applied a lower reaction temperature (65 °C) and stirred the mixture for 12 h. They were able to collect only 14% of (OH)₁CB[7] after running P column with water.[106]

## 2.6.2   Further Derivatizations on Hydroxylated CB[$n$]s

Hydroxyl groups on the surface of CB[$n$] macrocycles increase the solubility in water and organic solvents such as DMSO and DMF. Taking advantage of the improved solubility, hydroxylated CB[$n$]s can undergo further reactions to acquire various functional groups, such as allyl, propargyl or acetyl. De-rivatization into allyl or propargyl groups has been of particular interest because these groups allow for covalent conjugation with thiol- or azide-functionalized compounds *via* simple thiol–ene or azide–alkyne click reactions.

Kim *et al.*, when they reported the first direct functionalization of CB[$n$]s into perhydroxylated CB[$n$]s, showed further derivatizations of them into allyloxylated and acetyloxylated CB[*n*]s too. They first deprotonated the alcohol units on $(OH)_{12}CB[6]$ using bases such as NaH and $Et_3N$ in anhydrous DMSO. Then they added about 24–28 equivalents of allyl bromide or propionic anhydride and stirred the mixtures at room temperature for 6–12 h. The addition of the reaction mixtures into water results in white precipitate, which was collected and washed with ether to obtain pure products. The yields of perallyloxyCB[6] and perpropionyloxy-CB[6] were 67% and 70%, respectively (Scheme 2.16a).[24] They also demonstrated that perallyloxyCB[6] successfully undergoes thiol–ene photo-click reaction with an alkylthiol compound. Helmut Ritter and coworkers synthesized a perpropargyloxyCB[6] and reacted it with azide-functionalized cyclodextrins *via* microwave-assisted click chemistry in the presence of $CuSO_4$ and sodium ascorbate in DMSO.[107] Perfunctionalized CB[*n*] derivatives have been used in various applications ranging from artificial ion channels to detection of plasma membrane proteins.[108–112]

The same kind of derivatizations have also been applied to monohydroxylated CB[*n*]s (Scheme 2.16b). Scherman *et al.* synthesized monopropargyloxyCB[6] by treating bis@(OH)$_1$CB[6] with NaH in dried DMSO followed by the addition of propargyl bromide and stirring the reaction mixture at room temperature for 12 h. Treating the resulting mixture with

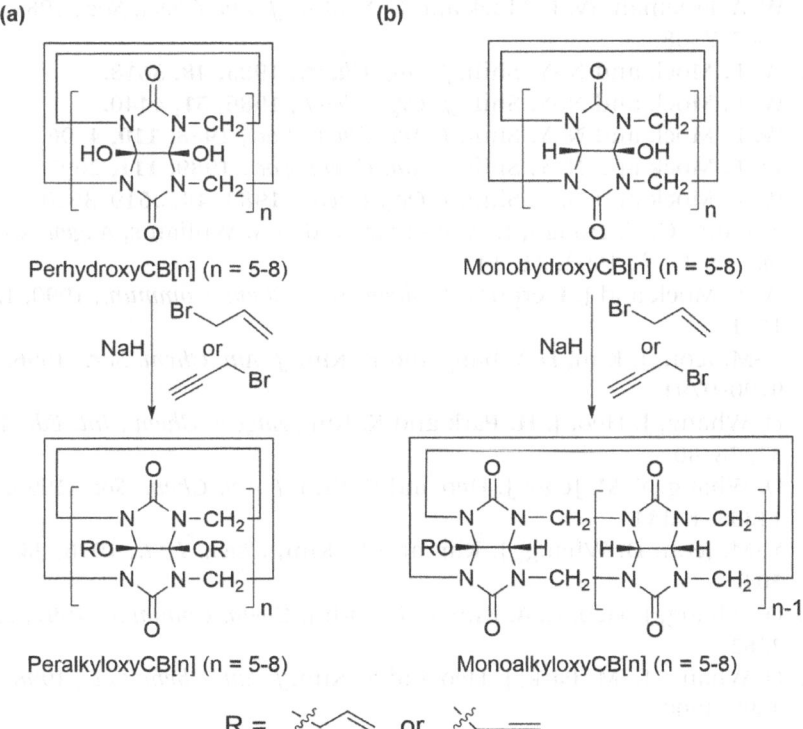

**Scheme 2.16**  Synthesis of (a) peralkyloxyCB[*n*]s and (b) monoalkyloxyCB[*n*]s.

ether and methanol gives a yellow product with 38% yield.[44] With an analogous procedure, Kim *et al.* synthesized monoallyloxyCB[7] in 19% yield and used it in a supramolecular velcro application.[47] They have recently demonstrated that monoallyloxyCB[7] self-assembles in water to form vesicles that can break apart when illuminated by a laser in the presence of gluthathione.[112] More recently, synthesis of monopropargyloxyCB[7] and its covalent conjugation with various compounds have been shown.[60-62] Dönüs Tuncel and coworkers constructed a multifunctional supramolecular assembly having high $^1O_2$ generation efficiency by conjugating monopropargyloxyCB[7] to triglycosylated porphyrin *via* CuAAC reaction.[61] In another study, Andreas Hennig, Werner Nau and coworkers covered the surface of polymeric nanoparticles with CB[7] units *via* the covalent attachment of monopropargyloxyCB[7] through CuAAC reaction and precisely controlled the surface coverage *via* host–guest chemistry.[62]

# References

1. E. Meyer, Ph.D. Thesis, Heidelberg University, 1904.
2. R. Behrend, E. Meyer and F. Rusche, *Liebigs Ann. Chem.*, 1905, **339**, 1–37.
3. W. A. Freeman, W. L. Mock and N. Y. Shih, *J. Am. Chem. Soc.*, 1981, **103**, 7367–7368.
4. W. L. Mock and N.-Y. Shih, *J. Org. Chem.*, 1983, **48**, 3618.
5. W. L. Mock and N.-Y. Shih, *J. Org. Chem.*, 1986, **51**, 4440.
6. W. L. Mock and N.-Y. Shih, *J. Am. Chem. Soc.*, 1988, **110**, 4706.
7. W. L. Mock and N.-Y. Shih, *J. Am. Chem. Soc.*, 1989, **111**, 2697.
8. W. L. Mock and N.-Y. Shih, *J. Org. Chem.*, 1983, **48**, 3619–3620.
9. A. Flinn, G. C. Hough, J. F. Stoddart and D. J. Williams, *Angew. Chem., Int. Ed. Engl.*, 1992, **31**, 1475.
10. W. L. Mock and J. Pierpont, *J. Chem. Soc., Chem. Commun.*, 1990, 1509–1511.
11. Y.-M. Jeon, J. Kim, D. Whang and K. Kim, *J. Am. Chem. Soc.*, 1996, **118**, 9790–9791.
12. D. Whang, J. Heo, J. H. Park and K. Kim, *Angew. Chem., Int. Ed.*, 1998, **37**, 78–80.
13. D. Whang, Y.-M. Jeon, J. Heo and K. Kim, *J. Am. Chem. Soc.*, 1996, **118**, 11333–11334.
14. Y.-M. Jeon, D. Whang, J. Kim and K. Kim, *Chem. Lett.*, 1996, **25**, 503–504.
15. D. Whang, J. Heo, C.-A. Kim and K. Kim, *Chem. Commun.*, 1997, 2361–2362.
16. D. Whang, K.-M. Park, J. Heo and K. Kim, *J. Am. Chem. Soc.*, 1998, **120**, 4899–4900.
17. S.-G. Roh, K.-M. Park, G.-J. Park, S. Sakamoto, K. Yamaguchi and K. Kim, *Angew. Chem., Int. Ed. Engl.*, 1999, **38**, 638–641.

18. D. Tuncel and J. H. G. Steinke, *Chem. Commun.*, 1999, 1509–1510.
19. J. Kim, I.-S. Jung, S.-Y. Kim, E. Lee, J.-K. Kang, S. Sakamoto, K. Yamaguchi and K. Kim, *J. Am. Chem. Soc.*, 2000, **122**, 540–541.
20. A. I. Day, A. P. Arnold, R. J. Blanch and B. Snushall, *J. Org. Chem.*, 2001, **66**, 8094–8100.
21. A. I. Day, R. J. Blanch, A. P. Arnold, S. Lorenzo, G. R. Lewis and I. Dance, *Angew. Chem., Int. Ed.*, 2002, **41**, 275–277.
22. J. Z. Zhao, H.-J. Kim, J. Oh, S.-Y. Kim, J. W. Lee, S. Sakamoto, K. Yamaguchi and K. Kim, *Angew. Chem., Int. Ed.*, 2001, **40**, 4233–4235.
23. H. Isobe, S. Sato and E. Nakamura, *Org. Lett.*, 2002, **4**, 1287–2189.
24. S. Y. Jon, N. Selvapalam, D. H. Oh, J.-K. Kang, S.-Y. Kim, Y. J. Jeon, J. W. Lee and K. Kim, *J. Am. Chem. Soc.*, 2003, **125**, 10186–10187.
25. A. I. Day, A. P. Arnold and R. J. Blanch, *Molecules*, 2003, **8**, 74–84.
26. C. A. Burnett, J. Lagona, A. X. Wu, J. A. Shaw, D. Coady, J. C. Fettinger, A. I. Day and L. Isaacs, *Tetrahedron*, 2003, **59**, 1961–1970.
27. J. Lagona, J. C. Fettinger and L. Isaacs, *Org. Lett.*, 2003, **5**, 3745–3747.
28. Y. Miyahara, K. Goto, M. Oka and T. Inazu, *Angew. Chem., Int. Ed.*, 2004, **43**, 5019–5022.
29. Y. J. Zhao, S. F. Xue, Q. J. Zhu, Z. Tao, J. X. Xhang, Z. B. Wei, L. S. Long, M. L. Hu, H. P. Xiao and A. I. Day, *Chin. Sci. Bull.*, 2004, **49**, 1111–1117.
30. S. Sasmal, M. K. Sinha and E. Keinan, *Org. Lett.*, 2004, **6**, 1225–1228.
31. A. Chakraborty, A. Wu, D. Witt, J. Lagona, J. C. Fettinger and L. Isaacs, *J. Am. Chem. Soc.*, 2002, **124**, 8297–8306.
32. A. Wu, A. Chakraborty, D. Witt, J. Lagona, F. Damkaci, M. A. Ofori, J. K. Chiles, J. C. Fettinger and L. Isaacs, *J. Org. Chem.*, 2002, **67**, 5817–5830.
33. L. Isaacs, S.-K. Park, S. Liu, Y. H. Ko, N. Selvapalam, Y. Kim, H. Kim, P. Y. Zvalij, G.-H. Kim, H.-S. Lee and K. Kim, *J. Am. Chem. Soc.*, 2005, **127**, 18000–18001.
34. S. Liu, P. Y. Zavalij and L. Isaacs, *J. Am. Chem. Soc.*, 2005, **127**, 16798–16799.
35. W.-H. Huang, S. Liu, P. Y. Zavalij and L. Isaacs, *J. Am. Chem. Soc.*, 2006, **128**, 14744–14745.
36. W.-H. Huang, P. Y. Zavalij and L. Isaacs, *Angew. Chem., Int. Ed.*, 2007, **46**, 7425–7427.
37. L.-B. Lu, Y.-Q. Zhang, Q.-J. Zhu, S.-F. Xue and Z. Tao, *Molecules*, 2007, **12**, 716–722.
38. L.-B. Lu, D.-H. Yu, Y.-Q. Zhang, Q.-J. Zhu, S. F. Xue and Z. Tao, *J. Mol. Struct.*, 2008, **885**, 70–75.
39. L. M. Zheng, J. N. Zhu, Y. Q. Zhang, Q. J. Zhu, S. F. Xue, Z. Tao, J. X. Zhang, X. Zhou, Z. B. Wei, L. S. Long and A. I. Day, *Supramol. Chem.*, 2008, **20**, 709–716.

40. L.-H. Wu, X.-L. Ni, F. Wu, Y.-Q. Zhang, Q.-J. Zhu, S.-F. Xue and Z. Tao, *J. Mol. Struct.*, 2009, **920**, 183–188.

41. J. Svec, M. Necas and V. Sindelar, *Angew. Chem., Int. Ed.*, 2010, **49**, 2378–2381.

42. M. M. Ahmed, K. Koga, M. Fukudome, H. Sasaki and D.-Q. Yuan, *Tetrahedron Lett.*, 2011, **52**, 4646–4649.

43. D. Lucas, T. Minami, G. Iannuzzi, L. Cao, J. B. Wittenberg, P. Anzenbacher, Jr. and L. Isaacs, *J. Am. Chem. Soc.*, 2011, **133**, 17966–17976.

44. N. Zhao, G. O. Lloyd and O. A. Scherman, *Chem. Commun.*, 2012, **48**, 3070–3072.

45. L. Cao and L. Isaacs, *Org. Lett.*, 2012, **12**, 3072–3075.

46. B. Vinciguerra, L. Cao, J. R. Cannon, P. Y. Zavalij, C. Fenselau and L. Isaacs, *J. Am. Chem. Soc.*, 2012, **134**, 13133–13140.

47. Y. Ahn, Y. Jang, N. Selvapalam, G. Yun and K. Kim, *Angew. Chem., Int. Ed.*, 2013, **52**, 3140–3144.

48. J. B. Wittenberg, P. Y. Zavalij and L. Isaacs, *Angew. Chem., Int. Ed.*, 2013, **52**, 3690–3694.

49. R. Aav, E. Shmatova, I. Reile, M. Borissova, F. Topic and K. Rissanen, *Org. Lett.*, 2013, **15**, 3786–3789.

50. X.-J. Cheng, L.-L. Liang, K. Chen, N.-N. Ji, X. Xiao, J.-X. Zhang, Y.-Q. Zhang, S.-F. Xue, Q.-J. Zhu, X.-L. Ni and Z. Tao, *Angew. Chem., Int. Ed.*, 2013, **52**, 7252–7255.

51. T. Fiala and V. Sindelar, *Synlett*, 2013, **24**, 2443–2445.

52. L. Gilberg, M. S. A. Khan, M. Enderesova and V. Sindelar, *Org. Lett.*, 2014, **16**, 2446–2449.

53. X. Jiang, X. Yao, X. Huang, Q. Wang and H. Tian, *Chem. Commun.*, 2015, **51**, 2890.

54. L. Ustrnul, P. Kulhanek, T. Lizal and V. Sindelar, *Org. Lett.*, 2015, **17**, 1022–1025.

55. M. M. Ayhan, H. Karoui, M. Hardy, A. Rockenbauer, L. Charles, R. Rosas, K. Udachin, P. Tordo, D. Bardelang and O. Ouari, *J. Am. Chem. Soc.*, 2015, **137**, 10238–10245.

56. M. M. Ayhan, H. Karoui, M. Hardy, A. Rockenbauer, L. Charles, R. Rosas, K. Udachin, P. Tordo, D. Bardelang and O. Ouari, *J. Am. Chem. Soc.*, 2016, **138**, 2060.

57. Q. Li, S.-C. Qiu, J. Zhang, K. Chen, Y. Huang, X. Xiao, Y. Zhang, F. Li, Y.-Q. Zhang, S.-F. Xue, Q.-J. Zhu, Z. Tao, L. F. Lindoy and G. Wei, *Org. Lett.*, 2016, **18**, 4020–4023.

58. L. Ustrnul, M. Babiak, P. Kulhanek and V. Sindelar, *J. Org. Chem.*, 2016, **81**, 6075–6080.

59. X.-X. Wang, K. Chen, F.-F. Shen, Y. Wang, Y.-Q. Zhang, Z. Tao and H. Cong, *Eur. J. Org. Chem.*, 2017, **2017**, 6980–6985.

60. H. Chen, Z. Huang, H. Wu, J.-F. Xu and X. Zhang, *Angew. Chem., Int. Ed.*, 2017, **56**, 16575–16578.

61. A. Koc, R. Khan and D. Tuncel, *Chem. - Eur. J.*, 2018, **24**, 15550–15555.
62. S. Zhang, Z. Dominguez, K. I. Assaf, M. Nilam, T. Thiele, U. Pischel, U. Schedler, W. M. Nau and A. Hennig, *Chem. Sci.*, 2018, **9**, 8575–8581.
63. R. L. Halterman, J. L. Moore and L. M. Mannel, *J. Org. Chem.*, 2008, **73**, 3266–3269.
64. A. Thangavel, A. M. M. Rawashdeh, C. Sotiriou-Leventis and N. Leventis, *Org. Lett.*, 2009, **11**, 1595–1598.
65. D. Jiao, N. Zhao and O. A. Scherman, *Chem. Commun.*, 2010, **46**, 2007–2009.
66. D. Jiao and O. A. Scherman, *Green Chem.*, 2012, **14**, 2445–2449.
67. S. Yi and A. E. Kaifer, *J. Org. Chem.*, 2011, **76**, 10275–10278.
68. A. I. Day, A. P. Arnold and R. J. Blanch, Unisearch Limited, Australia, *PCT Int. Appl.*, WO2000068232A1, 2000.
69. A. I. Day, R. J. Blanch, A. Coe and A. P. Arnold, *J. Inclusion Phenom. Macrocyclic Chem.*, 2002, **43**, 247–250.
70. J. Lagona, J. C. Fettinger and L. Isaacs, *J. Org. Chem.*, 2005, **70**, 10381–10392.
71. W.-H. Huang, P. Y. Zavalij and L. Isaacs, *J. Am. Chem. Soc.*, 2008, **130**, 8446–8454.
72. L. Isaacs, *Chem. Commun.*, 2009, 619–629.
73. L. Isaacs, *Isr. J. Chem.*, 2011, **51**, 578–591.
74. K. I. Assaf and W. M. Nau, *Chem. Soc. Rev.*, 2015, **44**, 394–418.
75. S. J. Barrow, S. Kasera, M. J. Rowland, J. del Barrio and O. A. Scherman, *Chem. Rev.*, 2015, **115**, 12320–12406.
76. H. J. Buschmann, E. Cleve, K. Jansen and E. Schollmeyer, *Anal. Chim. Acta*, 2001, **437**, 157–163.
77. H. J. Buschmann, E. Cleve, K. Jansen, A. Wego and E. Schollmeyer, *J. Inclusion Phenom. Mol. Recognit. Chem.*, 2001, **40**, 117–120.
78. O. A. Gerasko, D. G. Samsonenko and V. P. Fedin, *Russ. Chem. Rev.*, 2002, **71**, 741–760.
79. S. Liu, C. Ruspic, P. Mukhopadhyay, S. Chakrabarti, P. Y. Zavalij and L. Isaacs, *J. Am. Chem. Soc.*, 2005, **127**, 15959–15967.
80. S. Y. Kim, I. S. Jung, E. Lee, J. Kim, S. Sakamoto, K. Yamaguchi and K. Kim, *Angew. Chem., Int. Ed.*, 2001, **40**, 2119–2121.
81. G. Jiang and G. Li, *J. Photochem. Photobiol., B*, 2006, **85**, 223–227.
82. Y. H. Ko, H. Kim, Y. Kim and K. Kim, *Angew. Chem., Int. Ed.*, 2008, **47**, 4106–4109.
83. S. Moghaddam, C. Yang, M. Rekharsky, Y. H. Ko, K. Kim, Y. Inoue and M. K. Gilson, *J. Am. Chem. Soc.*, 2011, **133**, 3570–3581.
84. I. Hwang, K. Baek, M. Jung, Y. Kim, K. M. Park, D.-W. Lee, N. Selvapalam and K. Kim, *J. Am. Chem. Soc.*, 2007, **129**, 4170–4171.
85. D. Sobransingh and A. E. Kaifer, *Langmuir*, 2006, **22**, 10540–10544.
86. L. P. Cao, M. Sekutor, P. Y. Zavalij, K. Mlinaric-Majerski, R. Glaser and L. Isaacs, *Angew. Chem., Int. Ed.*, 2014, **53**, 988–993.

87. M. Sekutor, K. Molcanov, L. P. Cao, L. Isaacs, R. Glaser and K. Mlinaric-Majerski, *Eur. J. Org. Chem.*, 2014, 2533–2542.

88. J. Mohanty and W. M. Nau, *Angew. Chem., Int. Ed.*, 2005, **44**, 3750–3754.

89. J. W. Lee, S. Samal, N. Selvapalam, H.-J. Kim and K. Kim, *Acc. Chem. Res.*, 2003, **36**, 621–630.

90. H.-J. Buschmann, E. Cleve, K. Jansen, A. Wego and E. Schollmeyer, *Mater. Sci. Eng., C*, 2001, **114**, 35–39.

91. W. M. Nau, M. Florea and K. I. Assaf, *Isr. J. Chem.*, 2011, **51**, 559–577.

92. D. Bardelang, K. Udachin, D. M. Leek, J. C. Margeson, G. Chan, C. I. Ratcliffe and J. A. Ripmeester, *Cryst. Growth Des.*, 2011, **1**, 5598–5614.

93. E. Masson, X. Ling, R. Joseph, L. Kyeremeh-Mensah and X. Lu, *RSC Adv.*, 2012, **2**, 1213–1247.

94. J. Lagona, P. Mukhopadhyay, S. Chakrabarti and L. Isaacs, *Angew. Chem., Int. Ed.*, 2005, **44**, 4844–4870.

95. X. X. Zhang, K. E. Krakowiak, G. Xue, J. S. Bradshaw and R. M. Izatt, *Ind. Eng. Chem. Res.*, 2000, **39**, 3516–3520.

96. K. A. Kellersberger, J. D. Anderson, S. M. Ward, K. E. Krakowiak and D. V. Dearden, *J. Am. Chem. Soc.*, 2001, **123**, 11316–11317.

97. K. Jansen, H.-J. Buschmann, A. Wego, D. Döpp, C. Mayer, H.-J. Drexler, H.-J. Holdt and E. Schollmeyer, *J. Inclusion Phenom. Macrocyclic Chem.*, 2001, **39**, 357–363.

98. Y. Miyahara, K. Abe and T. Inazu, *Angew. Chem., Int. Ed.*, 2002, **41**, 3020–3023.

99. H. J. H. Fenton, *J. Chem. Soc.*, 1895, **67**, 48–50.

100. H. J. H. Fenton, *J. Chem. Soc.*, 1905, **87**, 804–818.

101. N. Branda, R. M. Grotzfeld, C. Valdes and J. Rebek, *J. Am. Chem. Soc.*, 1995, **117**, 85–88.

102. B. D. Wagner, P. G. Boland, J. Lagona and L. Isaacs, *J. Phys. Chem. B*, 2005, **109**, 7686–7691.

103. J. Lagona, B. D. Wagner and L. Isaacs, *J. Org. Chem.*, 2006, **71**, 1181.

104. R. H. Gao, L. X. Chen, K. Chen, Z. Tao and X. Xiao, *Coord. Chem. Rev.*, 2017, **348**(1), 24.

105. J. A. McCune, E. Rosta and O. A. Scherman, *Org. Biomol. Chem.*, 2017, **15**, 998–1005.

106. N. Dong, J. He, T. Li, A. Peralta, M. R. Avei, M. Ma and A. E. Kaifer, *J. Org. Chem.*, 2018, **83**, 5467–5473.

107. M. Munteanu, S. Choi and H. Ritter, *Macromolecules*, 2009, **42**, 3887–3891.

108. Y. J. Jeon, H. Kim, S. Jon, N. Selvapalam, D. H. Oh, I. Seo, C.-S. Park, S. R. Jung, D.-S. Koh and K. Kim, *J. Am. Chem. Soc.*, 2004, **126**, 15944–15955.

109. J. Kim, Y. Ahn, K. M. Park, Y. Kim, Y. H. Ko, D. H. Oh and K. Kim, *Angew. Chem., Int. Ed.*, 2007, **46**, 7393–9395.

110. D. Kim, E. Kim, J. Lee, S. Hong, W. Sung, N. Lim, C. G. Park and K. Kim, *J. Am. Chem. Soc.*, 2010, **132**, 9908–9919.

111. D.-W. Lee, K. M. Park, M. Banerjee, S. H. Ha, T. Lee, K. Suh, S. Paul, H. Jung, J. Kim, N. Selvapalam, S. H. Ryu and K. Kim, *Nat. Chem.*, 2011, **3**, 154–159.
112. K. M. Park, K. Baek, Y. H. Ko, A. Shrinidhi, J. Murray, W. H. Jang, K. H. Kim, J.-S. Lee, J. Yoo, S. Kim and K. Kim, *Angew. Chem., Int. Ed.*, 2018, **57**, 3132–3136.

CHAPTER 3

# Key Roles of Cavity Portals in Host–Guest Binding Interactions by Cucurbituril Hosts

ANGEL E. KAIFER

Department of Chemistry, University of Miami, Coral Gables, FL 33124, USA
Email: akaifer@miami.edu

## 3.1 Introduction

The development of supramolecular chemistry has relied heavily on the synthesis of molecular hosts, that is, compounds with the ability to form noncovalent inclusion complexes with suitable guest molecules. Starting with natural compounds, such as the cyclodextrins,[1] supramolecular chemists have prepared and investigated many other host families, such as the calixarenes,[2] cavitands[3] and pillararenes.[4] To a large extent, each of these host families is scalable in the sense that the number of monomeric units composing the macrocyclic ring can be controlled in order to modulate the cavity size of the resulting host. The number ($n$) of monomers is usually indicated in brackets within the name, and thus we refer to calix[$n$]arenes and pillar[$n$]arenes. Cavitands are mostly tetrameric, although some examples of pentameric and hexameric cavitands have been reported.[5]

The chemistry of cucurbiturils was held hostage for decades because of the lack of scalability. At the end of the twentieth century, cucurbit[6]uril (CB[6]) had been known for quite some time, since Mock reported its synthesis and structure in modern times.[6] This host is composed of six glyc#ouril units connected to one another by two rows of methylene bridges (see Figure 3.1); it

Smart Materials No. 36
Cucurbituril-based Functional Materials
Edited by Dönüs Tuncel
© The Royal Society of Chemistry 2020
Published by the Royal Society of Chemistry, www.rsc.org

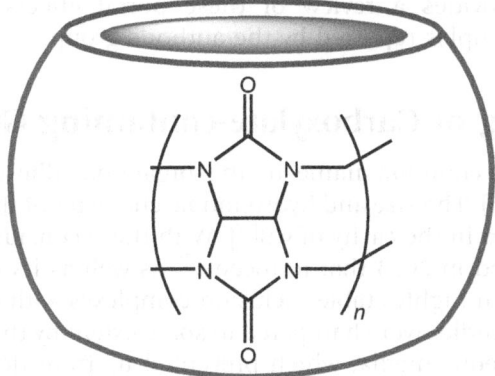

**Figure 3.1**   Structures of cucurbit[$n$]uril hosts.

has a barrel-shaped cavity with two identical openings or portals and an equatorial plane of symmetry, slicing the barrel in two equal halves. However, this host suffers from low solubility in water and most other solvents, which has forced the investigation of its binding properties in very acidic media, such as formic acid solutions.[7] The stagnant chemistry of cucurbiturils was jolted in 2000, when Kim and coworkers first reported the isolation of the heptameric and octameric cucurbiturils.[8] This report was quickly followed by another report authored by Day *et al.*[9] These two publications revolutionized the chemistry of the cucurbit[$n$]urils, leading to an upsurge of publications that is still going strong. Cucurbit[7]uril (CB[7]) and cucurbit[8]uril (CB[8]) have been the two main host compounds responsible for this revolution.[10–13]

The formation of inclusion complexes by CB[7] and CB[8] is generally believed to be driven by hydrophobic effects because the inner cavity of these two hosts is essentially hydrophobic.[14,15] The computation of electrostatic surface potentials shows that the cavity portals accumulate negative charge density, as expected since they are lined by carbonyl oxygens,[16] while positive charge density spreads over the outside surface of the cucurbituril barrel. The carbonyl groups on the portals are instrumental in binding cationic guest species, since ion–dipole interactions may develop between the guest and the host portals. Therefore, some of the best guests for CB[7] are hydrophobic cations, in which the hydrophobic residue fits well inside the cavity, while the positive charges on the guest develop ion–dipole interactions with the carbonyl oxygens on the portals.[17,18] The negative charge density on the portals has also been extensively used to coordinate metal ions that can serve as connectors between cucurbituril units in the solid phase.[19] While cavity inclusion, driven by hydrophobic effects, and interactions between positive charges on the guest and the C=O dipoles lining the portals are certainly the major factors explaining the large stability of many cucurbituril complexes in aqueous solution, the portals may play diverse and somewhat poorly recognized roles. With proper molecular design, it is possible to express unusual properties of these complexes.

This chapter provides a review of these portal effects, with particular emphasis on examples reported by the author's group.

## 3.2   Binding of Carboxylate-containing Guests

Ferrocene is the common name of the organometallic compound cyclopentadienyliron(II). The size and hydrophobic character of this compound are ideal for inclusion in the cavity of CB[7]. With these considerations in mind, my group reported in 2003 that ferrocene,[20] as well as its Co(III) analog, cobaltocenium, form highly stable inclusion complexes with CB[7] in aqueous solution. These studies were hampered to some extent by the low solubility of ferrocene in aqueous media, which prevented us from determining an accurate value for the equilibrium association constant ($K$) with CB[7]. However, we could estimate a lower limit, $K > 4 \times 10^5$ $M^{-1}$. In order to address the ferrocene solubility issue, we moved to simple ferrocenyl derivatives with larger aqueous solubility, such as (ferrocenylmethyl)trimethylammonium, hydroxymethylferrocene and ferrocenecarboxylate (see structures in Figure 3.2).

These three guests were selected to test the effect of the substituent group's charge on the stability of the corresponding CB[7] complexes, and the results, which we published jointly with the groups of Profs. Kim and Inoue, were unexpected.[16] The positively charged (ferrocenylmethyl)-trimethylammonium formed a highly stable complex with CB[7], with $K = 2 \times 10^{12}$ $M^{-1}$ in pure water at 25 °C. Under the same experimental conditions, the neutral hydroxymethylferrocene showed a $K$ value of $3 \times 10^9$ $M^{-1}$, while the negatively charged ferrocenecarboxylate did not bind at all! The high binding affinity between the two first guests and CB[7] is understood as the result of the excellent fit of the hydrophobic ferrocene residue inside the host cavity. We can invoke ion–dipole interactions between the ammonium nitrogen and one of the portals of CB[7] to justify that the positively charged guest forms the most stable complex.[16] In contrast to this, the lack of binding between ferrocenecarboxylate and CB[7] constituted a highly surprising result, given the fact that all three ferrocenyl derivatives were

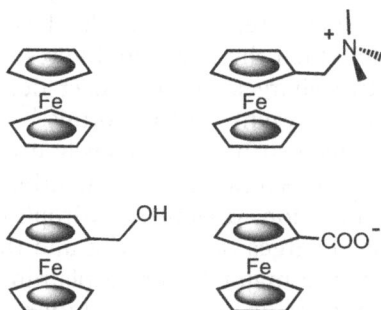

**Figure 3.2**   Structures of ferrocene (clockwise from top left corner): (ferrocenylmethyl)-trimethylammonium, ferrocenecarboxylate, and, hydroxymethylferrocene.

known to bind to another host, β-cyclodextrin, with $K$ values in the range $10^3$–$10^4$ M$^{-1}$.[1] These findings made it clear that the cucurbiturils could effectively discriminate between ferrocene derivatives based on their substituent charges, ranging from no binding interactions to very high binding affinities, in the nanomolar to picomolar regimes. Since the cyclodextrins have no significant charge density on their cavity openings, it was clear that the strong binding differences observed with CB[7] must be related to the negative charge density on the host cavity portals.

In the same work, we published the crystal structure of the ferrocene@CB[7] complex,[16] which shows the coexistence of two complex structures in the same crystal. Both structures are characterized by the lack of alignment between the ferrocene and the host. In other words, the $C_5$ symmetry axis that passes through the centers of the cyclopentadienyl rings and the iron atom in ferrocene lies at an angle with the $C_7$ symmetry axis of the cucurbituril. The two complex structures observed in the crystal differ by the angle between these two axes. However, this crystal structure suggests that, if ferrocenecarboxylate is included inside the cavity of CB[7], the negatively charged carboxylate would be forced into the proximity of several carbonyl oxygens on the portal, resulting in pronounced electrostatic repulsions that would explain the lack of stability of this complex.

Cobaltocenium (see Figure 3.3), as ferrocene, also forms a very stable inclusion complex with CB[7].[20] In 2009, we reported the equilibrium association constant for this complex as $5.7 \times 10^9$ M$^{-1}$ in 50 mM sodium acetate,[21] a medium composition that has become standard for these measurements due to the seminal work of Isaacs and coworkers.[22] The high stability of the cobaltocenium@CB[7] complex was later used to advantage by my group in order to develop a simple titration method to assess the purity of samples of CB[7].[23] The positively charged nature of cobaltocenium seems to play a major role in the high thermodynamic stability of its CB[7] complex, and, in fact, the carboxylate derivative in this case does exhibit a measurable $K$ value of $4.1 \times 10^4$ M$^{-1}$.[21] Comparing these results to those obtained with ferrocene derivatives, it is clearly evident that the positive charge on the cobaltocenium residue favors the binding interaction with CB[7], but the overall thermodynamic stability of the complex with cobaltoceniumcarboxylate is still very modest, again reflecting the deleterious effect of the negatively charged carboxylate.

In a parallel research project at the time, my group prepared dendronized derivatives of ferrocene,[24] cobaltocenium[21] and other redox-active and fluorescent residues.[25] In their water-soluble forms, the surface of these dendrons contained multiple carboxylic acid units, and therefore the

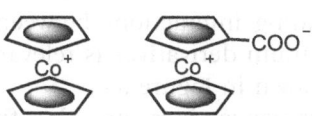

**Figure 3.3**   Structures of cobaltocenium (left) and cobaltoceniumcarboxylate (right).

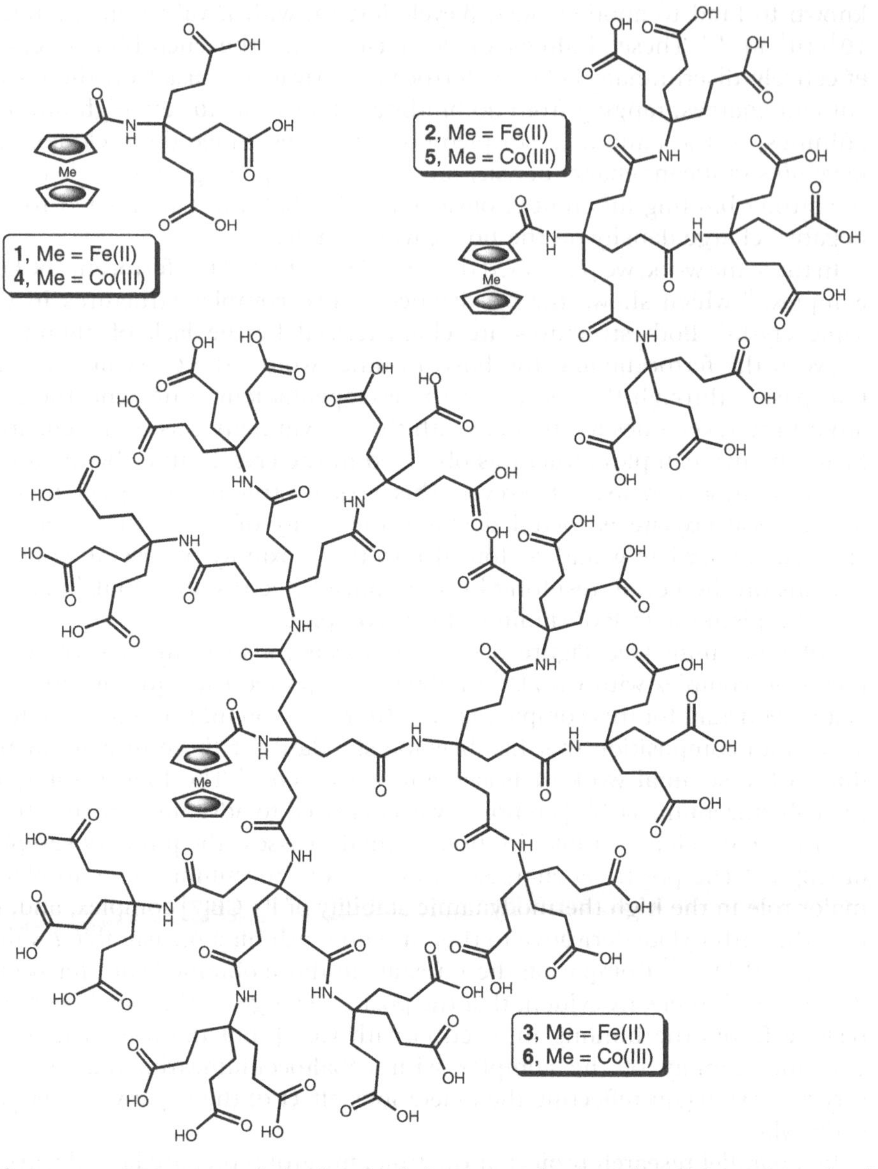

**Figure 3.4** Structures of dendronized ferrocene (1–3) and cobaltocenium (4–6) derivatives. The cobaltocenium derivatives bear a positive charge on the organometallic residue.

investigation of the binding interactions between CB[7] and dendronized ferrocene and cobaltocenium derivatives is relevant here. The structures of these compounds are shown in Figure 3.4.

In analogy with ferrocenecarboxylate, the first-generation ferrocenyl dendrimer **1** is not bound by CB[7] at pH 7, where its three carboxylic acid

**Table 3.1** Equilibrium association constants[a] ($M^{-1}$) at 25 °C in aqueous media between dendronized metallocene derivatives and the CB[7] host.

| Guest | At pH 2 | At pH 7 |
|-------|---------|---------|
| 1 | $3.9 \times 10^4$ | n.b.[b] |
| 2 | $4.2 \times 10^6$ | $3.8 \times 10^5$ |
| 3 | Insol.[c] | $7.7 \times 10^5$ |
| 4 | $3.2 \times 10^5$ | $1.0 \times 10^4$ |
| 5 | $1.1 \times 10^7$ | $3.4 \times 10^6$ |
| 6 | $4.0 \times 10^5$ | $4.0 \times 10^5$ |

[a]Error margin: ±12%.
[b]No binding detected.
[c]Insoluble under these conditions.

groups are negatively charged. In contrast, the same compound is bound at pH 2 because protonation removes the negative charges on the carboxylic acid groups. The intensity of these electrostatic effects decreases as the dendron increases in size.[24] Therefore, the second-generation ferrocenyl dendrimer **2** is bound at both solution pH values, although the binding affinity is higher at pH 2 (Table 3.1). Unfortunately, the complex between **3** and CB[7] is insoluble in pH 2 solution, which prevents us from completing this comparison for the larger ferrocenyl dendrimer. The dendronized cobaltocenium derivatives followed similar binding affinity trends.[21] More stable CB[7] complexes are formed at pH 2 than at pH 7 solutions, presumably reflecting the electrostatic repulsions between the terminal carboxylates on the dendrimer (at pH 7) and the cavity openings on the host. Electrostatics are not enough in this case to prevent the formation of a stable complex between CB[7] and the first-generation compound **4**, even at pH 7 where the three terminal carboxylates on the dendron are fully ionized. A comparison between guests **1** and **4** at neutral pH indicates that the presence of a positive charge on the metallocene binding site stabilizes the CB[7] complex. This is also consistent with the fact that the dendronized cobaltocenium complexes are more stable in all cases than the corresponding ferrocene complexes (Table 3.1). Overall, the binding data set between CB[7] and dendronized metallocenes confirms that electrostatic effects play a significant role in the overall stability of the complexes, and these effects decrease as the size of the dendron increases, moving the carboxylate groups farther away from the host binding site.

The literature contains other examples supporting the negative effects of carboxylate groups on guest binding to the CB[7] host. For instance, Kim and coworkers investigated the binding of amino acids by CB[7] in aqueous solutions at various pH values.[26] With the aromatic amino acids (phenylalanine, tryptophan and tyrosine), they observed a decrease in the CB[7] complex stabilities, going from pH 2 to pH 4, while no further decrease was detected as the pH of the solution increased to 7. These results are consistent with the repulsive interactions between the carboxylate group on the amino acids and the carbonyl portals of CB[7], which would prevent the optimization of the interactions between the host and the guests. These

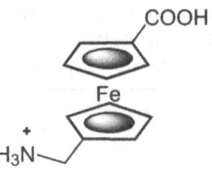

**Figure 3.5** Structure of Sindelar's ferrocene amino acid derivative (fully protonated form).

findings are also consistent with the work published by Nau and coworkers showing the increased binding affinity of amino acids for CB[7] after guest decarboxylation.[27] Also, the extensive work of Urbach and coworkers on the binding of peptides and proteins by CB[7] and CB[8] leads to the same general conclusions, as the aromatic residues adjacent to the amine terminus are preferentially bound by both hosts.[28,29] Finally, Wyman and Macartney have also shown the negative effect of a carboxylate,[30] as compared to its neutral, protonated form, on the binding between cholines and CB[7].

More recently, in collaboration with Sindelar's group, we have shown that a ferrocene amino acid derivative (see Figure 3.5) undergoes pronounced changes in its binding affinity to CB[7] depending on its state of protonation.[31] In its fully protonated state, at low pH, this guest forms a very stable inclusion complex with CB[7] ($K = 5.8 \times 10^6$ M$^{-1}$). However, deprotonation of the carboxylic acid group forces the exclusion of the ferrocenyl residue from the CB[7] cavity, giving rise to an external complex in which the protonated ammonium group still interacts with one of the host portals ($K = 6.8 \times 10^2$ M$^{-1}$). At higher solution pH values, driving the deprotonation of the ammonium group to amine, this external complex further weakens and undergoes complete dissociation.[31]

## 3.3 Pseudorotaxane Switching *via* Proton Transfer on Terminal Carboxylates

The realization that suitably located carboxylates could exert a significant negative effect on guest binding to CB[7] led us to design guest molecules in which these effects could be used to advantage, for instance, to control the primary binding location of the host along an axle-like guest. We reasoned that, with an axle-like guest having two terminal –COOH groups, deprotonation may push the CB[7] to a central binding site away from the terminal –COO$^-$ groups, leading to a controllable or switchable pseudorotaxane system. Based on this general principle, we prepared 4,4′-bipyridinium (viologen) derivatives in which the terminal substituents are aliphatic chains with –COOH end groups.[32] Figure 3.6 provides an illustrative example (viologen guest $7^{2+}$).

From our previous work on the binding of viologen derivatives by CB[7], we had learned that the viologen nucleus is the preferred binding site only when the two *N*-substituents are relatively small, such as methyl or ethyl, or

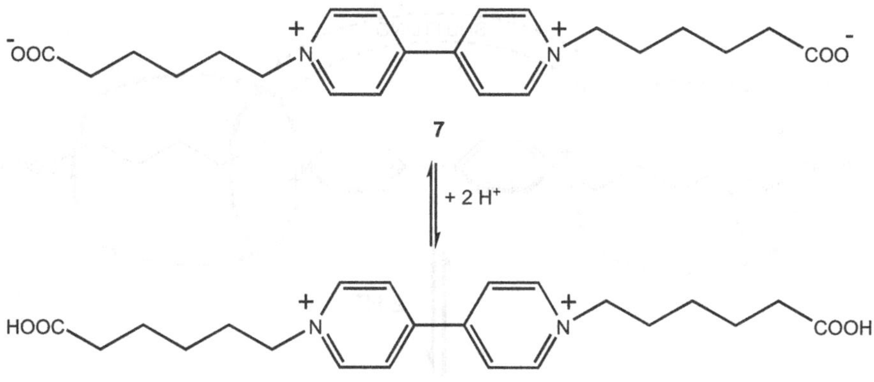

**Figure 3.6**   Structure of viologen $7^{2+}$ and its deprotonated, zwitterionic form **7**.

polar.[33] If the aliphatic *N*-substituents are relatively long (butyl or longer) and lack any polar groups, they become the preferred binding site. Therefore, the binding interactions between guest $7^{2+}$ and CB[7] can be predicted by these simple rules. At acidic pH values, both terminal –COOH groups remain protonated, and the host binds to one of the aliphatic side arms.[32] At the millimolar concentrations used in our NMR experiments, we determined that the intramolecular exchange process in which the CB[7] host shuttles between the two aliphatic side arms of guest $7^{2+}$ is faster than the intermolecular exchange in which CB[7] jumps from one guest molecule to another. NMR and UV–Vis spectroscopic experiments confirmed that at higher solution pH values, upon deprotonation of the terminal –COOH groups, the CB[7] host shifts along the guest axle to occupy the central viologen nucleus.[32] This host movement is attributed to the electrostatic repulsions between the terminal –COO$^-$ groups and the host's oxygen-laced portals. In other words, at low pH, the host–guest system can be understood as a degenerate molecular shuttle, in which the CB[7] jumps back and forth between the two aliphatic side arms. However, at higher pH, the host–guest system does not behave as a molecular shuttle, as the CB[7] remains centered along the axle, encircling the middle viologen nucleus. These changes are fully reproducible, and the pseudorotaxane can be taken through several pH cycles without apparent "fatigue", although we did not investigate this issue over a large number of cycles. Overall, the pseudorotaxane formed by the binding interactions between CB[7] and viologen $7^{2+}$ can be switched between two different states, characterized by different UV–Vis spectra, *via* pH control or proton transfer reactions (see Figure 3.7).

The successful design of this switchable pseudorotaxane led us to question some of the basic interactions involved in its operation. Specifically, we were interested in the thermodynamic and kinetic affects that the terminal anionic carboxylates could exert on the assembly and dissociation of the complex. In joint work with Profs. Serena Silvi (University of Bologna) and

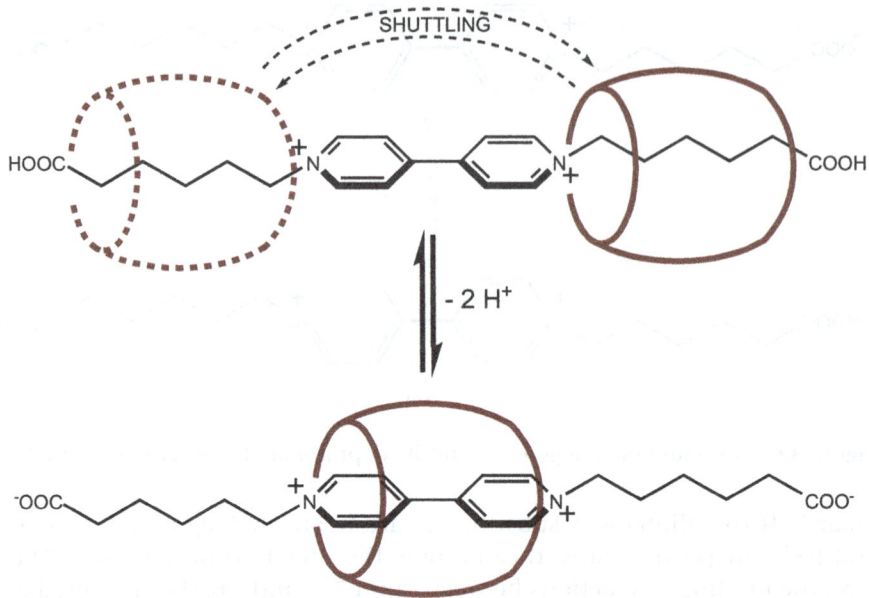

**Figure 3.7** Proton transfer switching of pseudorotaxane formed by CB[7] and viologen derivative $7^{2+}$.

Vladimir Sindelar (Masaryk University), we focused on the thermodynamics and kinetics of the binding interactions between CB[7] and the zwitterionic viologen derivative 7. We found that the carboxylate termini have a relatively small effect on the thermodynamic stability of the complex,[34] as the equilibrium association constant was found to be $1.5 \times 10^5$ $M^{-1}$, a value certainly smaller than the majority of the $K$ values measured between CB[7] and simple viologen derivatives, which are typically at or above $1 \times 10^6$ $M^{-1}$. However, the association and dissociation processes are kinetically very slow, with measured rate constant values of 0.6 $M^{-1}s^{-1}$ and $3.9 \times 10^{-6}$ $s^{-1}$, respectively.[34] These values reflect the fact that complex association, as well as dissociation, requires the passage of one of the terminal carboxylates through the cavity of CB[7].

Our interest in these phenomena continues, and we are currently investigating the kinetics and thermodynamics of CB[7] complexation of viologen and related guests with carboxylate and sulfonate terminal groups.

## 3.4 Cooperative Self-assembly Involving the CB[7] Portals

The examples in the previous sections rely on repulsive electrostatic interactions between anionic groups, such as carboxylates, and the rim of negative charge density associated with the carbonyl oxygens lining the cavity portals of CB[7]. The effectiveness of these repulsive interactions led us to consider the possibility of moving in exactly the opposite direction, that is, using attractive electrostatic interactions to enhance self-assembly, particularly in cases where entropy disfavors assembly. While there are a number of other potential opportunities, we relied on the previous experience of our group with cyclobis(paraquat-*p*-phenylene), a tetracationic cyclophane (BB$^{4+}$ see structure in Figure 3.8), which has been extensively used by Stoddart's group and is colloquially referred to as the "blue box" by researchers in the supramolecular field.[35] The four positive charges on this macrocycle result from the presence of two π electron acceptor bipyridinium (paraquat) groups. While the cavity defined by the bipyridinium groups is ideal for capturing π electron donor groups, the protons on the paraquat are excellent hydrogen-bonding donors. Our hypothesis was that the blue box is ideal to develop side by side attractive interactions with the rim of carbonyl oxygens on each of the CB7 portals, not only because of the four positive charges on this cyclophane but also because of the ability of the aromatic protons on the bipyridinium rings to engage in hydrogen bonding interactions with the carbonyl oxygens on the portals of CB[7].

In order to assist the side-to-side self-assembly of BB$^{4+}$ and CB[7], we designed a simple guest (GH$_2$$^{2+}$, see Figure 3.9 for structure) containing a central *p*-dioxybenzene moiety and two aliphatic side arms terminated in amine groups. The design of the guest anticipates that the central dioxybenzene site, as a good π donor, will be encircled by BB$^{4+}$, generating an inclusion complex with well-defined charge transfer interactions between the paraquat groups on the tetracationic cyclophane host and the included dioxybenzene residue. On the other hand, we also expected that the side arms on the guest may serve as binding sites for CB[7] hosts. Ideally, the guest will bring together a molecule of BB$^{4+}$ and two molecules of CB[7], in a quaternary self-assembly process, which is expected to be disfavored on entropic grounds. The favorable lateral interactions that we anticipated

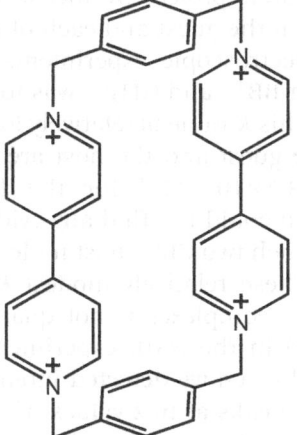

**Figure 3.8** Structure of the "blue box" receptor (BB$^{4+}$).

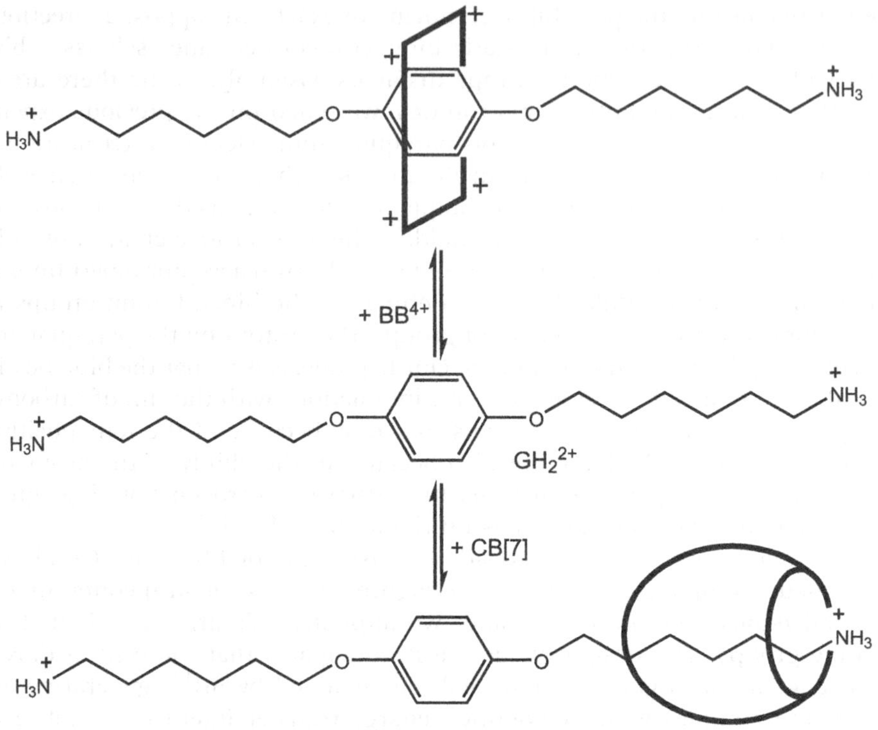

**Figure 3.9** Association equilibria between guest $GH_2^{2+}$ and the two receptor hosts, $BB^{4+}$ and CB[7].

between the hosts may compensate and prevail over the entropic consider-ations. Two key questions at this stage were: Will the quaternary complex form at all? If so, will there be any cooperativity in the self-assembly process?

We initiated this research project by measuring the two 1:1 (binary) binding constants between the guest and each of the hosts, $BB^{4+}$ and CB[7] in separate $^1$H NMR spectroscopic experiments.[36] The association equi-librium constant between $BB^{4+}$ and $GH_2^{2+}$ was found to be $6.4 \times 10^2$ M$^{-1}$ in 50-mM sodium acetate. This $K$ value is relatively low and probably depressed by the fact that both the guest and the host are cationic species. We also measured a value of $3.2 \times 10^3$ M$^{-1}$ for the formation of the binary $GH_2^{2+} \cdot$CB[7] complex and could not find any evidence for the formation of the ternary complex in which two CB[7] host molecules would interact with a single guest molecule. These relatively modest $K$ values indicate that the formation of these binary complexes is not quantitative at the millimolar concentration levels used in the NMR experiments. Given these facts, we were pleasantly surprised when we detected intense electrospray ionization (ESI) mass spectrometric peaks at m/z values of 526.0378 (calcd. 526.0351) and 631.0420 (calcd. 631.0408) corresponding to the quaternary complex with 6 and 5 positive charges, respectively, from an aqueous solution containing all three components.[36] This was our first experimental result strongly suggesting that the formation of the quaternary complex is very efficient and proceeds in cooperative fashion. We also carried out a number of additional $^1$H NMR spectroscopic experiments, and the results indicate that the quaternary complex $GH_2^{2+} \cdot BB^{4+} \cdot (CB[7])_2$ forms quantitatively at the millimolar concentrations prevalent in these experiments. Identical results were obtained from NMR diffusion coefficient measurements.

We tried very hard to crystallize the quaternary complex but could not isolate single crystals suitable for X-ray diffraction experiments. In order to gather additional evidence for the cooperative formation of the quaternary complex, we collaborated with the group of a Miami colleague, Prof. Rajeev Prabhakar, to carry out a detailed computational investigation of this system, using DFT methods (M06-2X functional and 6-31G(d) basis set). The results of this computational work demonstrated that the binding of $BB^{4+}$ to the $GH_2^{2+}$ guest enhances the binding affinity for the CB[7] host molecules, as compared to that shown by the guest on its own. Similarly, binding of CB[7] to $GH_2^{2+}$ facilitates the binding of $BB^{4+}$, suggesting the presence of strong cooperative forces leading to the effective formation of the quaternary complex.[36]

These computational studies also provide details on the origin of the favorable lateral interactions between the host molecules in the quaternary complex. A careful analysis of the energy-minimized structure for the quaternary complex shows the presence of 10 hydrogen bonds between each of the two "paraquat" sides of $BB^{4+}$ and the adjacent CB[7] molecules (see Figure 3.10). The total of 20 hydrogen bonds formed through these side-by-side attractive interactions among the central $BB^{4+}$ and the two terminal CB[7] host molecules are very important to explain the overall thermodynamic stability of the quaternary complex and how entropic hindrance is overcome in its formation.

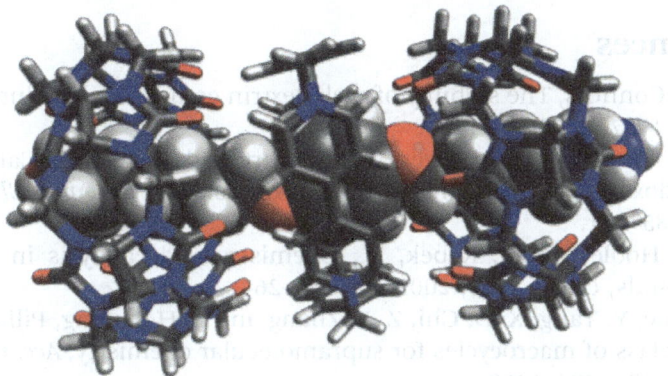

**Figure 3.10**  Energy-minimized structure of the quaternary complex $GH_2^{2+} \cdot BB^{4+} \cdot (CB[7])_2$.

## 3.5  Conclusions

The work summarized in this chapter highlights the fact that electrostatic interactions between certain guests and the oxygen-laced portals of CB[7] may play important roles regarding the stabilization or destabilization of its complexes. The negative charge density accumulated on the cavity portal rims often assists in the stabilization of complexes formed by positively charged guests. However, the presence of negative charges on the guest leads to complex destabilization, ranging from complete—that is, the complex fails to assemble—to partial destabilization, depending on the relative proximity of the guest anionic groups to the host portals. These electrostatic repulsive interactions may be used to advantage in order to control or switch the average location of CB[7] in pseudorotaxanes formed with axle-like guests terminated in carboxylic acid groups. Electrostatic interactions may also be used very effectively to develop favorable lateral interactions with other hosts, leading to systems exhibiting cooperative self-assembly. Overall, this work shows that, while the cucurbituril hosts are very versatile aqueous receptors and their binding properties are primarily dependent on hydrophobic effects, fine-tuning of the electrostatic interactions involving the portal rims provides additional tools to control their complexes in highly effective ways. The extension of these electrostatic interactions to higher cucurbit[$n$]urils ($n \geq 8$) may lead to a number of additional interesting findings.

## Acknowledgements

The author is grateful to the U.S. National Science Foundation for the long-standing support of this research and to the many undergraduate and graduate students, postdoctoral associates and collaborators who did most of the work described here and whose names are listed in the corresponding references.

## References

1. K. A. Connors, The stability of cyclodextrin complexes in solution, *Chem. Rev.*, 1997, **97**, 1325–1357.
2. C. D. Gutsche, J. S. Rogers, D. Stewart and K. A. See, Calixarenes—Paradoxes and paradigms in molecular baskets, *Pure Appl. Chem.*, 1990, **62**, 485–491.
3. R. J. Hooley and J. Rebek, Jr., Chemistry and catalysis in functional cavitands, *Chem. Biol.*, 2009, **16**, 255–264.
4. M. Xue, Y. Yang, X. D. Chi, Z. B. Zhang and F. H. Huang, Pillararenes: A new class of macrocycles for supramolecular chemistry, *Acc. Chem. Res.*, 2012, **45**, 1294–1308.
5. C. Naumann, E. Roman, C. Peinador, T. Ren, B. O. Patrick, A. E. Kaifer and C. J. Sherman, Expanding cavitand chemistry: the preparation

and characterization of [n]cavitands with n ≥ 4, *Chem. - Eur. J.*, 2001, 7, 1637–1645.

6. W. A. Freeman, W. L. Mock and N. Y. Shih, Cucurbituril, *J. Am. Chem. Soc.*, 1981, **103**, 7367–7368.

7. W. L. Mock and N.-Y. Shih, Organic ligand-receptor interactions between cucurbituril and alkylammonium ions, *J. Am. Chem. Soc.*, 1988, **110**, 4706–4710.

8. J. Kim, I. S. Jung, S. Y. Kim, E. Lee, J. K. Kang, S. Sakamoto, K. Yamaguchi and K. Kim, New cucurbituril homologues: Syntheses, isolation, characterization, and X-ray crystal structures of cucurbit[*n*]uril (n = 5, 7, and 8), *J. Am. Chem. Soc.*, 2000, **122**, 540–541.

9. A. Day, A. P. Arnold, R. J. Blanch and B. Snushall, Controlling factors in the synthesis of cucurbituril and its homologues, *J. Org. Chem.*, 2001, **66**, 8094–8100.

10. J. Lagona, P. Mukhopadhyay, S. Chakrabarti and L. Isaacs, The cucurbit[*n*]uril family, *Angew. Chem., Int. Ed. Engl.*, 2005, **44**, 4844–4870.

11. E. Masson, X. Ling, R. Joseph, L. Kyeremeh-Mensah and X. Lu, Cucurbituril chemistry: a tale of supramolecular success, *RSC Adv.*, 2012, **2**, 1213–1247.

12. S. J. Barrow, S. Kasera, M. J. Rowland, J. del Barrio and O. A. Scherman, Cucurbituril-based molecular recognition, *Chem. Rev.*, 2015, **115**, 12320–12406.

13. K. M. Park, J. Murray and K. Kim, Ultrastable artificial binding pairs as a supramolecular latching system: A next generation chemical tool for proteomics, *Acc. Chem. Res.*, 2017, **50**, 644–646.

14. F. Biedermann, V. D. Uzunova, O. A. Scherman and W. M. Nau, A. De Simone, Release of high-energy water as an essential driving force for the high-affinity binding of cucurbit[*n*]urils, *J. Am. Chem. Soc.*, 2012, **134**, 15318–15323.

15. F. Biedermann, M. Vendruscolo, O. A. Scherman, A. De Simone and W. M. Nau, Cucurbit[8]uril and blue-box: High-energy water release overwhelms electrostatic interactions, *J. Am. Chem. Soc.*, 2013, **135**, 14879–14888.

16. W. S. Jeon, K. Moon, S. H. Park, H. Chun, Y. H. Ko, J. Y. Lee, E. S. Lee, S. Samal, N. Selvapalam, M. V. Rekharsky, V. Sindelar, D. Sobransingh, Y. Inoue, A. E. Kaifer and K. Kim, Complexation of ferrocene derivatives by the cucurbit[7]uril host: A comparative study of the cucurbituril and cyclodextrin host families, *J. Am. Chem. Soc.*, 2005, **127**, 12984–12989.

17. L. P. Cao, M. Sekutor, P. Y. Zavalij, K. Mlinaric-Majerski, R. Glaser and L. Isaacs, Cucurbit[7]uril-guest pair with an attomolar dissociation constant, *Angew. Chem., Int. Ed.*, 2014, **53**, 988–993.

18. M. V. Rekharsky, T. Mori, C. Yang, Y. H. Ko, N. Selvapalam, H. Kim, D. Sobransingh, A. E. Kaifer, S. Liu, L. Isaacs, W. Chen, S. Moghaddam, M. K. Gilson, K. Kim and Y. Inoue, A synthetic host-guest system achieves avidin-biotin affinity by overcoming enthalpy-entropy compensation, *Proc. Nat. Acad. Sci. U. S. A.*, 2007, **104**, 20737–20742.

19. J. Lu, J. X. Lin, M. N. Cao and R. Cao, Cucurbituril: A promising organic building block for the design of coordination compounds and beyond, *Coord. Chem. Rev.*, 2013, **257**, 1334–1356.

20. W. Ong and A. E. Kaifer, Unusual electrochemical properties of the inclusion complexes of ferrocenium and cobaltocenium with cucurbit[7]uril, *Organometallics*, 2003, **22**, 4181–4183.

21. D. Sobransingh and A. E. Kaifer, New dendrimers containing a single cobaltocenium unit covalently attached to the apical position of Newkome dendrons: Electrochemistry and guest binding interactions with cucurbit[7]uril, *Langmuir*, 2006, **22**, 10540–10544.

22. S. Liu, C. Ruspic, P. Mukhopadhyay, S. Chakrabarti, P. Y. Zavalij and L. Isaacs, The cucurbit[n]uril family: Prime components for self-sorting systems, *J. Am. Chem. Soc.*, 2005, **127**, 15959–15967.

23. S. Yi and A. E. Kaifer, Determination of the purity of cucurbit[n]uril (n = 7, 8) host samples, *J. Org. Chem.*, 2011, **76**, 10275–10278.

24. D. Sobransingh and A. E. Kaifer, Binding interactions between the host cucurbit[7]uril and dendrimer guests containing a single ferrocenyl residue, *Chem. Commu.*, 2005, 5071–5073.

25. W. Ong, M. Gomez-Kaifer and A. E. Kaifer, Dendrimers as guests in molecular recognition phenomena, *Chem. Commun.*, 2004, 1677–1683.

26. J. W. Lee, H. H. L. Lee, Y. H. Ko, K. Kim and H. I. Kim, Deciphering the specific high-affinity binding of cucurbit[7]uril to amino acids in water, *J. Phys. Chem. B*, 2015, **119**, 4628–4636.

27. D. M. Bailey, A. Hennig, V. D. Uzunova and W. M. Nau, Supramolecular tandem enzyme assays for multiparameter sensor arrays and enantiomeric excess determination of amino acids, *Chem.-Eur. J.*, 2008, **14**, 6069–6077.

28. L. A. Logsdon, C. L. Schardon, V. Ramalingam, S. K. Kwee and A. R. Urbach, Nanomolar binding of peptides containing noncanonical amino acids by a synthetic receptor, *J. Am. Chem. Soc.*, 2011, **133**, 17087–17092.

29. J. M. Chinai, A. B. Taylor, L. M. Ryno, N. D. Hargreaves, C. A. Morris, P. J. Hart and A. R. Urbach, Molecular recognition of insulin by a synthetic receptor, *J. Am. Chem. Soc.*, 2011, **133**, 8810–8813.

30. I. W. Wyman and D. H. Macartney, Cucurbit[7]uril host-guest complexes of cholines and phosphonium cholines in aqueous solution, *Org. Biomol. Chem.*, 2010, **8**, 253–260.

31. L. Mikulu, R. Michalicova, V. Iglesias, M. A. Yawer, A. E. Kaifer, P. Lubal and V. Sindelar, pH Control on the Sequential Uptake and Release of Organic Cations by Cucurbit[7]uril, *Chem.-Eur. J.*, 2017, **23**, 2350–2355.

32. V. Sindelar, S. Silvi and A. E. Kaifer, Switching a molecular shuttle on and off: Simple, pH-controlled pseudorotaxanes based on cucurbit[7]uril, *Chem. Commun.*, 2006, 2185–2187.

33. K. Moon and A. E. Kaifer, Modes of binding interaction between viologen guests and the cucurbit[7]uril host, *Org. Lett.*, 2004, **6**, 185–188.

34. A. E. Kaifer, W. Li, S. Silvi and V. Sindelar, Pronounced pH effects on the kinetics of cucurbit[7]uril-based pseudorotaxane formation and dissociation, *Chem. Comm.*, 2012, **48**, 6693–6695.
35. P. L. Anelli, P. R. Ashton, R. Ballardini, V. Balzani, M. Delgado, M. T. Gandolfi, T. T. Goodnow, A. E. Kaifer, D. Philp, M. Pietraszkiewicz, L. Prodi, M. V. Reddington, A. M. Z. Slawin, N. Spencer, J. F. Stoddart, C. Vicent and D. J. Williams, Molecular meccano.1: [2]Rotaxanes and a [2]catenane made to order, *J. Am. Chem. Soc.*, 1992, **114**, 193–218.
36. M. H. Tootoonchi, G. Sharma, J. Calles, R. Prabhakar and A. E. Kaifer, Cooperative self-assembly of a quaternary complex formed by two cucurbit[7]uril hosts, cyclobis(paraquat-p-phenylene), and a "designer" guest, *Angew. Chem., Int. Ed.*, 2016, **55**, 11507–11511.

CHAPTER 4

# Rotaxanes and Polyrotaxanes

## N. BASÍLIO[*a] AND U. PISCHEL[*b]

[a] Universidade NOVA de Lisboa, Laboratório Associado para a Química Verde (LAQV), Rede de Química e Tecnologia (REQUIMTE), Departamento de Química, Faculdade de Ciências e Tecnologia, 2829-516 Caparica, Portugal; [b] University of Huelva, CIQSO–Centre for Research in Sustainable Chemistry and Department of Chemistry, Campus de El Carmen, E-21071 Huelva, Spain
*Email: nuno.basilio@fct.unl.pt; uwe.pischel@diq.uhu.es

## 4.1 Introduction

The dynamic nature of the noncovalent interactions established between macrocycles and axle molecules in rotaxanes and pseudorotaxanes offers the possibility of exerting precise control over their association/dissociation and/or co-conformational rearrangements with external stimuli. This potential has been widely explored to develop supramolecular switches with implications in the development of molecular machines and devices for applications in controlled drug release, information storage, artificial molecular muscles, stimuli-responsive catalysts, *etc.*[1]

The successful design of stimuli-responsive rotaxanes and pseudorotaxanes requires detailed knowledge of the structural, thermodynamic and kinetic properties of these supramolecular ensembles. The most common strategies toward (pseudo)rotaxane-based switches are (1) pseudorotaxanes whose thermodynamic stability can be modified through the application of external stimuli, allowing the triggered dissociation or association of the assembly, and (2) bistable (pseudo)rotaxanes comprising axle molecules

Smart Materials No. 36
Cucurbituril-based Functional Materials
Edited by Dönüs Tuncel
© The Royal Society of Chemistry 2020
Published by the Royal Society of Chemistry, www.rsc.org

with different binding sites of switchable affinity, allowing the controlled translocation of the macrocycle between binding sites.

Cucurbiturils (CBs) are attractive macrocycles to be employed in the development of (pseudo)rotaxane-based supramolecular switches, owing to their outstanding recognition properties.[2] As these macrocyclic receptors usually display higher affinity for positively charged guests (*i.e.*, for a family of structurally similar guests, the positively charged analogues are bound more tightly by CBs), stimuli, such as pH and redox, that are applied to change the overall charge of binding sites of CB (pseudo)rotaxanes are frequently explored. Nevertheless, other stimuli such as ions or light, directed to perturb not only charge but also conformation, have been investigated as well.

In this chapter, we will cover selected aspects of the synthesis of (pseudo)rotaxanes containing CB macrocycles and then put a strong focus on the design and implementation of stimuli-responsive supramolecular assemblies. The chosen examples are presented to illustrate the diverse implicated functional aspects, such as the choice of stimulus type and application (*e.g.*, release, switching, supramolecular logic, *etc.*). By no means does the limited space here allow us to cover the topic in the most comprehensive manner. However, the interested reader is referred to a series of review articles that complement the information contained in this chapter.[3-12]

## 4.2  Chemical Design of CB-based [*N*]Rotaxanes and [*N*]Pseudorotaxanes

Formulated in a puristic way, any 1 : 1 host–guest complex, including those of CBs, could be understood as a [2]pseudorotaxane. This is especially true when the CB macrocycle is located over the central part of a long-chain guest, corresponding to the classic wheel-axle picture. Examples for this situation include complexes of polyamines with CB[6] (cucurbit[6]uril). Such constructs can be immobilized on Au surfaces[13] or used to construct supramolecular valves for the stimuli-responsive delivery of cargo entrapped in the pores of silica nanoparticles (see Section 4.3).[14] Based on the differential binding properties of CBs,[9] axles that function as guests can be designed to include various binding stations for one or various CB homologues, thereby providing a means for the supramolecular organization of the macrocyclic rings in a preprogrammed fashion. In the following paragraphs, a few examples for this strategy are discussed.

The Tuncel Group developed the heteroditopic guest **1**, containing binding sites for CB[6] and CB[8] (cucurbit[8]uril) (see Figure 4.1).[15] The terminal di(ammoniumalkyl)-1,2,3-triazoles, derived from click chemistry, bind CB[6], while the central dodecyl spacer provides a complexation motif for CB[8]. By mixing 2 equivalents of CB[6] with the preformed CB[8]-**1** complex, a hetero[4]pseudorotaxane is formed, which however, converts to the

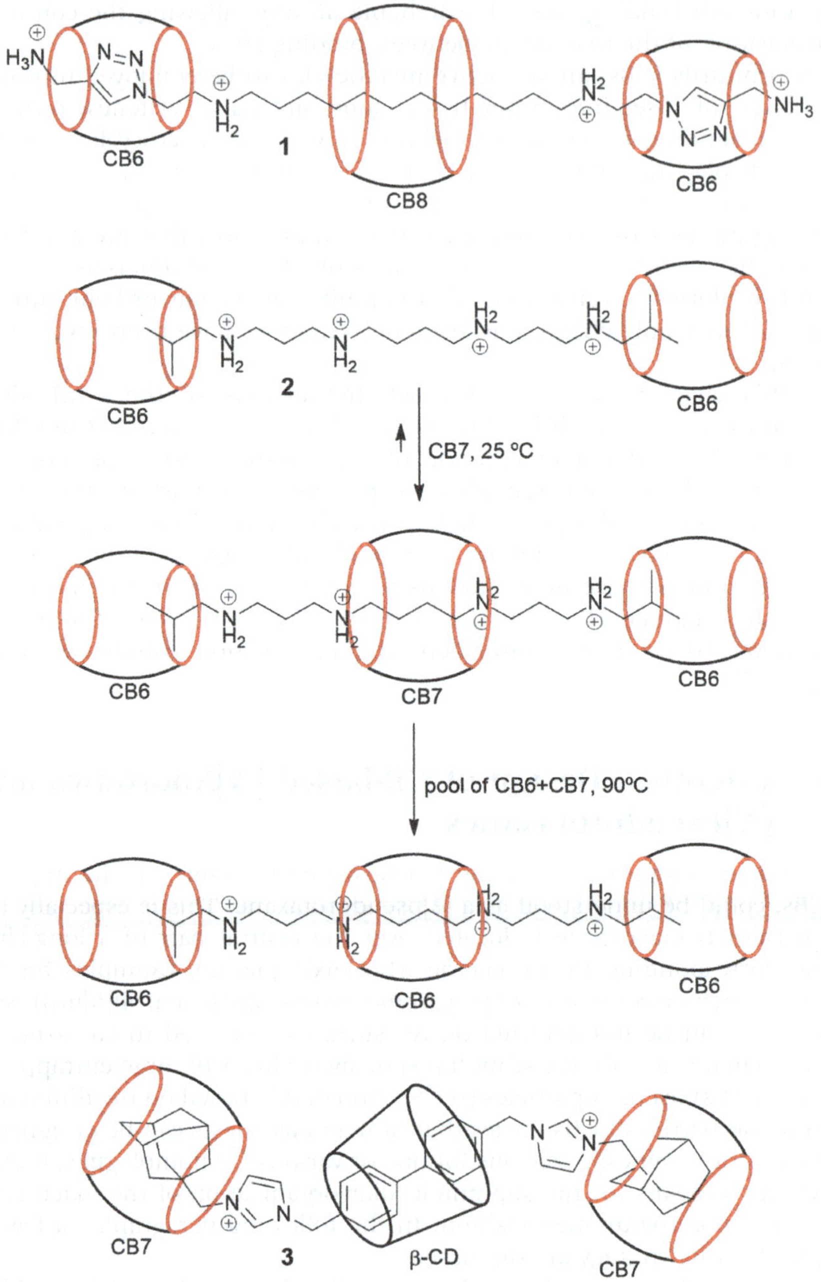

**Figure 4.1** Structures of selected pseudorotaxanes with the ditopic guests **1–3**.

[3]pseudorotaxane (CB[6])$_2$-**1** over time. A similar result is obtained when all components (**1**, CB[6], and CB[8]) are mixed simultaneously. Interestingly, adding excess CB[8] to the [3]pseudorotaxane (CB[6])$_2$-**1** does not yield the

hetero[4]pseudorotaxane. Hence, CB[6] is acting practically as a blocking unit, and the outcome of the self-sorting process depends on the sequence in which the different macrocycles are added. The same group reported also homo[5]pseudorotaxanes and [5]rotaxanes by organizing the di(ammoniumalkyl)-1,2,3-triazole binding sites around a porphyrin template and using again CB[6] as macrocycle.[16]

A related example comes from the Masson Group. They employed a spermine-derived axle (**2**) with two isobutyl endgroups (Figure 4.1).[17] On offering CB[6] and CB[7] (cucurbit[7]uril) at room temperature, the kinetically controlled formation of a hetero[4]pseudorotaxane is observed, having a CB[7] macrocycle organized over the central 1,4-diammoniumbutane section and two CB[6] hosts capping the isobutylamine moieties. However, on heating to 90 °C in the presence of a pool of CB[6] and CB[7], the hetero[4]pseudorotaxane converts into the thermodynamically favored homo[4]pseudorotaxane with all three binding sites occupied by CB[6]. This is an excellent example of how to explore kinetic *versus* thermodynamic self-sorting for the tailored assembly of CB-derived pseudorotaxanes.

The Vicha Group reported an interesting variation (**3**) of these strategies, employing multitopic guests that contain adamantyl endgroups (see Figure 4.1).[18] The latter are well-known for forming highly stable host–guest complexes with CB[7]. The central 4,4'-biphenylene bisimidazolium unit is used to bind β-cyclodextrin (β-CD) as wheel component. The resulting assembly can be understood as hetero[4]rotaxane, as the bulky terminal CB[7] complexes are practically nondissociable stoppers. Likewise, adding covalently organized bulky stopper groups yields [2]rotaxanes, as shown by the Urbach Group for a methylviologen-derived axle and CB[8] as macrocyclic wheel.[19]

In all these examples, the axle unit is composed of a single covalently organized strand. However, taking advantage of the capacity of the larger CB[8] to form homo- or heteroternary complexes by including two guests, researchers have come up with designs that use the CB simultaneously as a wheel and template to assemble the axle in supramolecular fashion. Early examples for this strategy were contributed by the Schalley Group and the Liu Group (see Figure 4.2). The Schalley Group designed the guest axles **4** and **5** that contain electron-accepting methylviologen units and either 2-alkoxy- or 2,6-dialkoxy-naphthalenes as electron-donating units.[20] These units can organize through the formation of charge-transfer (CT) complexes, using CB[8] as template. For example and in line with the previously discussed systems, the two binding sites of axle **4** can be used to form a hetero[3]-pseudorotaxane with one CB[8] (having a methylviologen-2,6-dialkoxy-lnaphthalene CT pair as guest) and one CB[7] (forming a complex with the other methylviologen unit) *via* integrative self-sorting. Upon combining the two axles (**4** and **5**) with CB[8] as the only macrocycle, a hetero[5]-pseudorotaxane architecture is observed (see Figure 4.2). This assembly combines three CB[8] macrocycles and the two axles (**4** and **5**) in a necklace-type arrangement. The axles organize face to face through their electronically complementary units, promoted by the CB.

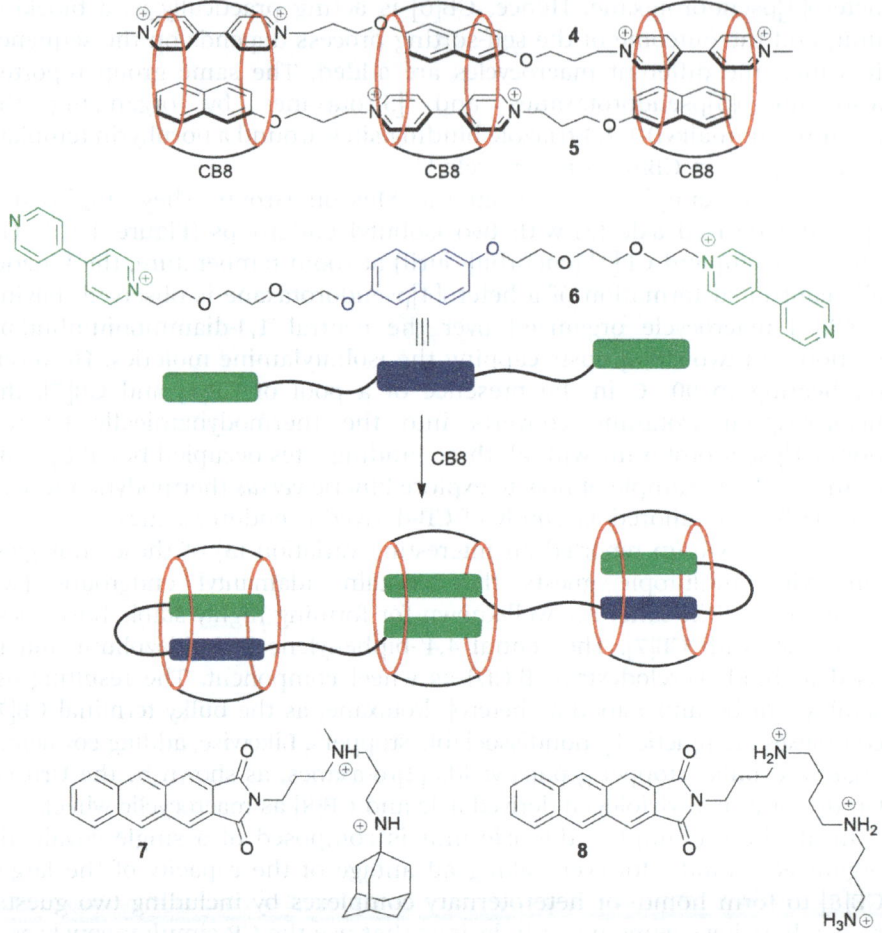

**Figure 4.2**  Pseudorotaxanes based on the formation of ternary complexes of the guests 4–8 with CB[8].

The Liu Group designed a similar system (**6**) that integrates a central 2,6-dialkoxynaphthalene and two protonable pyridyl-pyridinium units at the terminal positions (see Figure 4.2.)[21] On addition of CB[8] to **6**, they obtained a homo[5]pseudorotaxane where the axle is truly self-assembled by forming a 1:2 homoternary complex between two nonprotonated pyridyl-pyridinium units of two molecules of **6** and one CB[8] (Figure 4.2). The other two CB[8] macrocycles are organized by complexation of a similar CT pair, as previously discussed for the example from the Schalley Group. The [5]pseudo-rotaxane can be dissociated and converted into a homo[3]pseudorotaxane by the addition of an acid, thereby building on the pH-dependent complexation of the pyridyl-pyridinium units (see more examples for pH-dependent switching of CB-based pseudorotaxanes in Section 4.3 of this chapter).

In a recent work, the Pischel Group has also shown that the formation of 1 : 2 homoternary CB[8] complexes with an anthracene guest can be exploited to self-assemble the axle of homo- and hetero [5]pseudorotaxanes.[22] For this purpose, the heteroditopic anthracene guests **7** and **8** with additional aminoadamantane or spermidine binding motifs were designed (see Figure 4.2). The addition of solely CB[8] to **7** yields a homo[5]pseudorotaxane with three macrocycles. On the other hand, in the presence of both macrocycle homologues CB[7] and CB[8], the corresponding hetero[5]pseudorotaxane is obtained. The central part of this assembly is formed by the mentioned 1 : 2 homoternary anthracene-CB[8] complex, while the CB[7] forms strong inclusion complexes with the adamantane moieties. The formation of this pseudorotaxane is independent of the addition sequence of the macrocycles. Likewise, the heteroditopic ligand **8** was used for the preparation of the corresponding hetero[5]pseudorotaxane with terminal CB[6] complexes instead of CB[7].

The principles discussed so far were successfully integrated by the Schalley Group for the realization of a cascade that involves the stepwise stimuli-induced transformation between different CB-based pseudorotaxane architectures.[23] They designed a molecular axle **9** that contains a total of six stations, based on methylviologens and 2-alkoxynaphthalenes, prone to undergo CT complex formation (as previously discussed) in different binding situations (see Figure 4.3).

The cascade consists of five steps and is initiated by the addition of 4 equivalents of CB[7], which forms 1 : 1 complexes with the four methylviologen units of **9**. This leads to the homo[5]pseudorotaxane (CB[7])$_4$-**9**. In the second step, 2 equivalents of CB[8] are added, which initiates the competitive displacement of 2 CB[7] and the CT-mediated formation of the hetero[5]pseudorotaxane (CB[8])$_2$-(CB[7])$_2$-**9**. The third step builds on the reductive formation of methylviologen radical cations, which are well-known for forming homodimers inside the CB[8] cavity. Thus, the selective one-electron reduction of the four methylviologens and the concomitant displacement of CB[7] on addition of the strongly binding guest 1,6-diammoniumhexane yields re-folding of the axle, promoting the formation of a homo[3]pseudorotaxane. This process can be reverted by reoxidation and addition of CB[6] (fourth step). Finally, in the fifth step, the two central CB[7] macrocycles are replaced by CB[8], through a multidisplacement procedure using an aminoadamantane guest. This results in the homo[5]-pseudorotaxane (CB[8])$_4$-**9**.

In many of these examples, CB[8] was used as template to preorganize the molecular components of the axle in a noncovalent fashion. However, the use of CBs as templates is also known in the context of the catalyzed formation of covalent bonds. The pioneering example of a CB-promoted reaction is that of a 1,3-dipolar cycloaddition between an azide and a terminal alkine, published by the Mock Group in the 1980s.[24] They found that this click-type reaction is accelerated by CB[6] with a factor of 55000 as compared to the bimolecular reaction in absence of the macrocycle. Later on, other

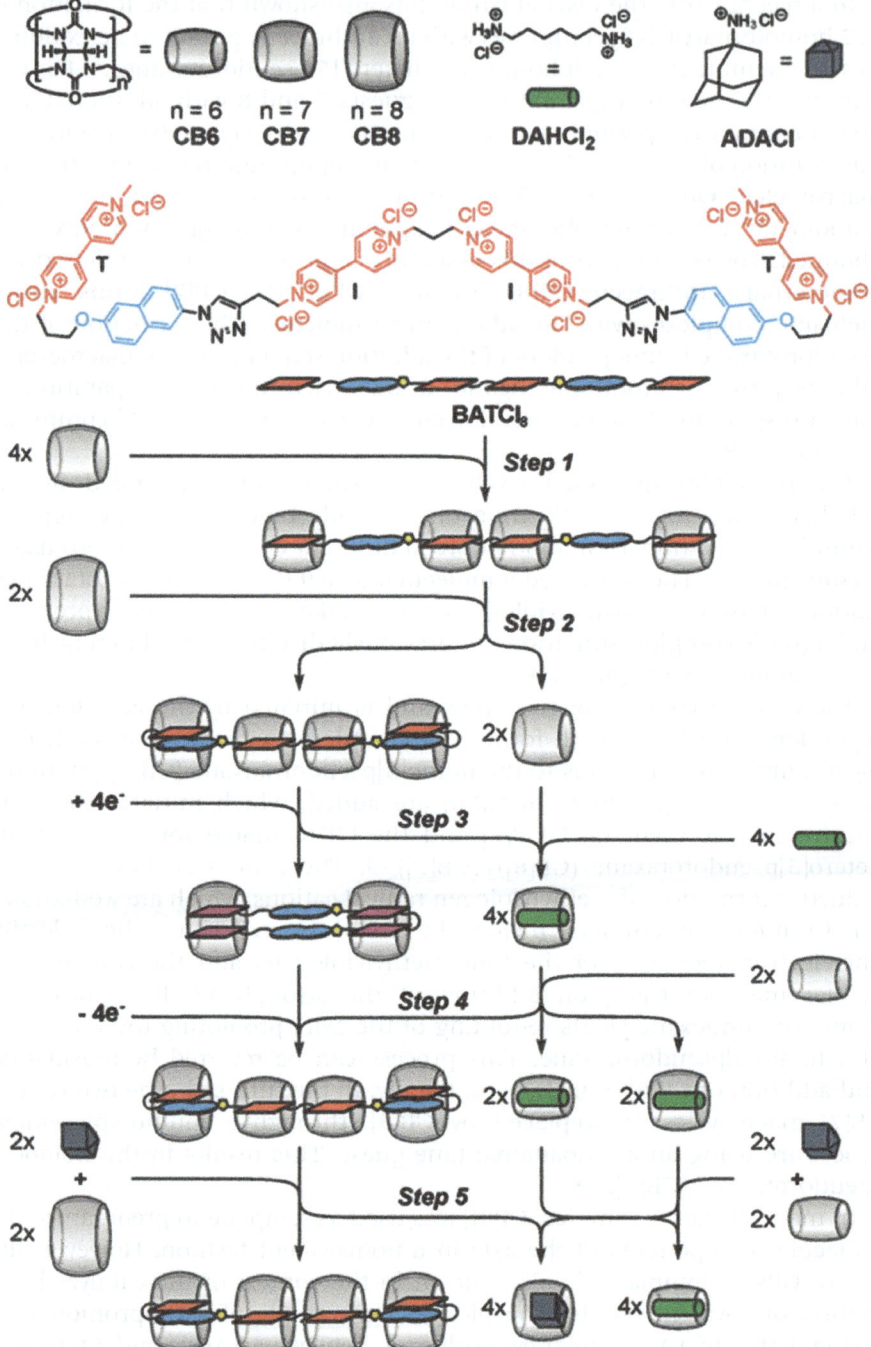

**Figure 4.3** Switching cascade, implemented with multitopic guest **9** (BATCl$_6$ in the original scheme). See description in text.
Reproduced from ref. 23 with permission from the Royal Chemical Society.

examples for CB-promoted catalysis were published, such as the Diels–Alder reaction[25] or photoinduced [2 + 2] cycloadditions.[26] The construction of pseudorotaxanes based on CB-based click reactions has been extensively investigated by the Tuncel Group.[27] The Stoddart Group also explored the CB[6]-catalyzed reactions together with the ability of CBs to undergo cooperative hydrogen bond interactions with cyclodextrins to synthesize [4]heterorotaxanes, polypseudorotaxanes and emissive solid-state materials.[28,29] Recently, Francis and coworkers have developed the CB[6]-promoted azide–alkyne reaction into a very useful and versatile tool in bioconjugation through the formation of protein-rotaxanes.[30]

A specific case of the organization of polypseudorotaxanes is the supramolecular polymerization of small molecular components. Several parameters need to be taken into account when aiming at the formation of supramolecular polymers. These include considerations of binding constants and concentration, the formation of linear chains *versus* cyclic structures, and the binding mode, including complementary binding motifs (see, for example, the previously mentioned CT pairs for the construction of discrete [*n*]pseudorotaxanes).[9] Commonly, the formation of 1 : 1 : 1 hetero-ternary CB[8] complexes with heteroditopic ligands is used for the formation of polypseudorotaxanes.[31] The Kim Group was the first to explore the complexation of heteroditopic ligands with electron donor and acceptor motifs by CB[8]. Depending on the design of the linker holding the donor and the acceptor together, a variety of supramolecular architectures was obtained, often involving cyclic structures, such as those observed for ligand **10** (see Figure 4.4).[32] However, using a gold surface, linear anchored poly-pseudorotaxanes can be produced from **10** and CB[8]. Cyclization can be avoided by incorporating two donor–acceptor pairs in the same ligand, such as the anthracene-methylviologen combination in **11** (Figure 4.4), introduced by Zhang and coworkers.[33] Along these lines of thought, the

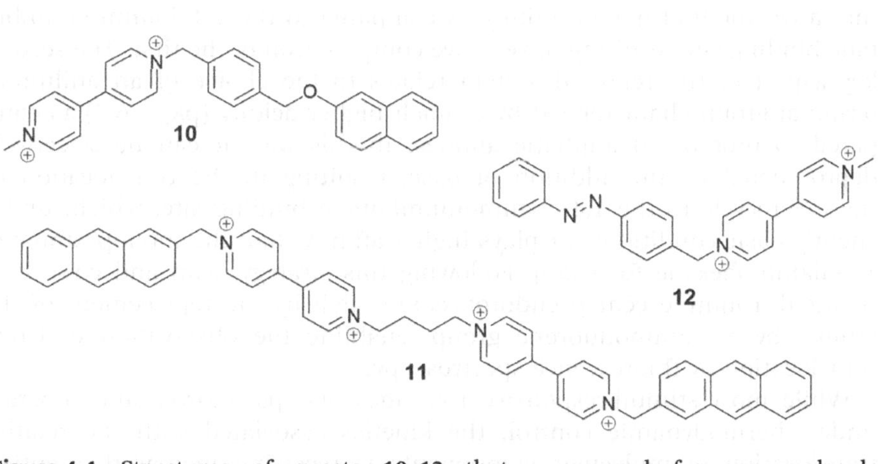

**Figure 4.4** Structures of guests **10–12** that were used for supramolecular polymerization.

Scherman Group published a light-switchable azobenzene-derived variation of this approach (ligand **12**; see structure in Figure 4.4), which enabled them to exert control of depolymerization by *E–Z* photoisomerization of the azo chromophore.[34]

A final example for strategies that are herein meant to illustrate the synthetic flexibility for designing CB-derived polymers, *i.e.*, poly-pseudorotaxanes, builds on the orthogonal combination of host–guest complexation and metal–ligand interaction. This approach was pioneered by the Kim Group and has been used widely since then.[3] Noteworthily, the axle is held together by the metal–ligand interactions. However, the role of the CB-derived host–guest complexes is crucial in that they confer additional stability by increasing linker rigidity and engaging in interactions with the metal centers, mediated, for example, by long-range Coulombic and hydrogen-bonding interactions.[35]

## 4.3  CB-derived Rotaxanes and Pseudorotaxanes as Stimuli-responsive Switches

### 4.3.1  Stimulation by pH Inputs

The variation of pH, building on the acid–base chemistry of guests, is one of the most widely used chemical stimuli to switch CB (pseudo)rotaxanes. In fact, one of the earliest examples of a switchable CB pseudorotaxane was reported by Mock and coworkers, who used pH changes to induce the translocation of CB[6] between two recognition sites (see Figure 4.5A).[36] The triamine axle **13** was designed according to key structural factors that are crucial for the successful operation of the pH-responsive pseudorotaxane. It consists of a 1,6-diammoniumhexane binding site that separates the two ammonium groups at an optimal distance to maximize attractive ion–dipole interactions with the two carbonyl portals of the macrocycle. This translates into a *ca.* 100-fold higher affinity as compared to the 1,4-diammoniumbutane binding site resulting in selective complexation of the first. The second key aspect of the reported system relates to the choice of an anilinium terminal group characterized by a much higher acidity ($pK_a = 6.7$) as compared to protonated aliphatic amines. In this way, it can be selectively deprotonated by the addition of base, resulting in the translocation of the macrocycle to the 1,4-diammoniumbutane binding site, which, under slightly basic conditions, displays higher affinity than the monoprotonated 1,6-diaminohexane for CB[6]. Following this strategy, Kim and coworkers reported a luminescent pseudorotaxane switch by the replacement of the aniline by a 2-aminofluorene group, enabling the observation of CB[6] translocation by fluorescence spectroscopy.[37]

While most stimuli-responsive rotaxanes and pseudorotaxanes operate under thermodynamic control, the kinetics associated with the relative translocation of molecular components is receiving increased attention owing to its critical importance for controlled directional motion.[1]

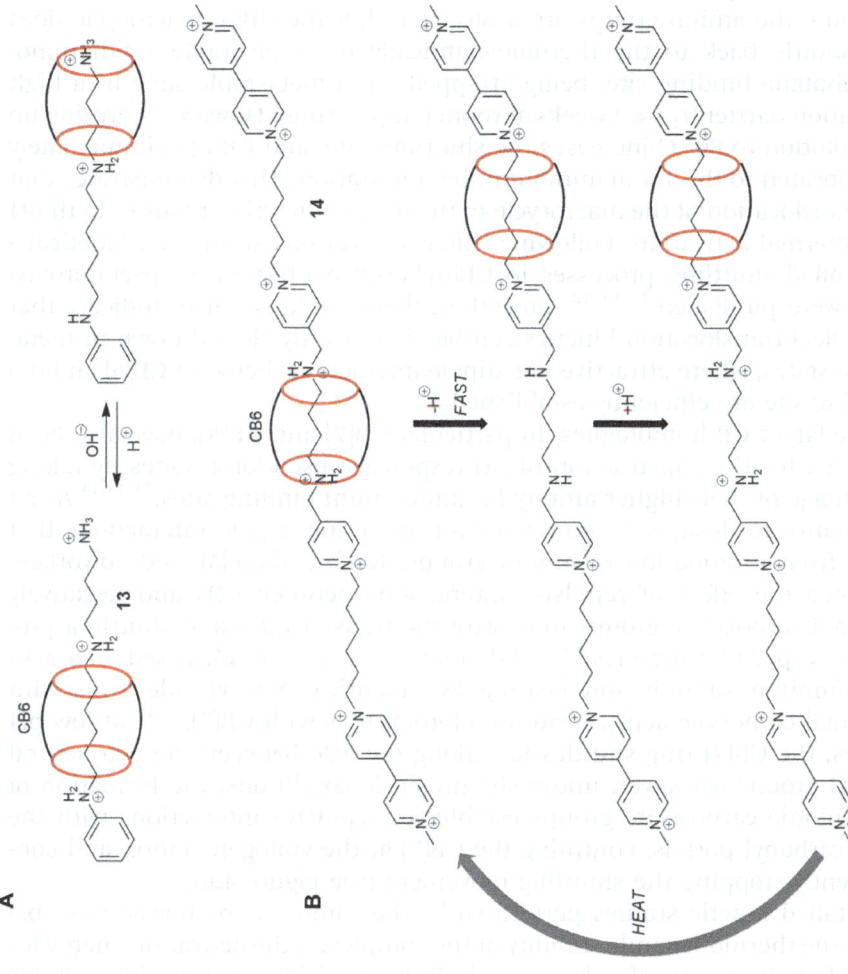

**Figure 4.5** (A) pH-responsive bistable [2]pseudorotaxane obtained from a triamine axle and a CB[6] macrocycle. (B) A kinetically controlled pH-responsive [2]pseudorotaxane switch.

The bistable CB[6]-based [2]pseudorotaxane **14**, depicted in Figure 4.5B, constitutes an illustrative example of a pH-controlled system showing well differentiated kinetics for ring translocation.[38] At neutral pH, the CB[6] macrocycle is predominantly located at the 1,4-diammoniumbutane binding site, owing to its higher affinity as compared to 1,6-di(pyridinium)hexane. Deprotonation of the ammonium groups upon addition of diisopropyl-ethylamine (DIEA) base results in fast translocation of the CB[6] macrocycle to the 1,6-di(pyridinium)hexane binding site. Upon reacidification of the solution, the amino groups are protonated, but the CB[6] macrocycle does not shuttle back to the thermodynamically more favorable 1,4-diammo-niumbutane binding site, being "trapped" in a metastable state by a high activation barrier ($t_{1/2} \approx 2$ weeks at room temperature). However, warming up the solution to 80 °C increases the shuttling rate, and CB[6] is immediately translocated to the 1,4-diammoniumbutane station. This demonstrates that the translocation of the macrocycle in the reverse direction requires both pH and thermal activation. Following this work, various studies on kinetically controlled shuttling processes in CB[6]-based pH-responsive pseudorotax-anes were published.[17,39–42] Altogether, these works seem to indicate that the wheel translocation kinetics can be significantly slowed down in meta-stable states, where attractive ion–dipole interactions between CB[6] and the binding site are efficiently established.

The larger CB homologues, in particular CB[7] and CB[8], have also been explored for the construction of pH-responsive pseudorotaxanes by taking advantage of their higher affinity for ammonium binding sites.[21,43,44] As an alternative to designs based on the attractive ion–dipole interactions that arise from protonation of amino groups, Kaifer, Sindelar and coworkers explored the effect of repulsive interactions between CBs and negatively charged carboxylate groups to control the translocation and shuttling pro-cesses in pseudorotaxanes.[45–47] Axle molecules such as **15**, based on a 4,4'-bipyridinium subunit and bearing two identical *N*-alkyl side arms with terminal carboxylic acids, form pseudorotaxanes with CB[7].[45,46] At low pH values, the CB[7] ring shuttles fast along the axle between the two neutral COOH groups. However, under slightly basic conditions, the formation of the anionic carboxylate groups establishes repulsive interactions with the host carbonyl portals, confining the CB[7] at the viologen station and con-sequently stopping the shuttling movement (see Figure 4.6).

Detailed kinetic studies performed by the same group showed that, be-sides the thermodynamic stability of the complexes, the activation energy for their formation can also be strongly influenced by repulsive interactions established between anionic groups and the carbonyl portals of the CB wheel. The pseudorotaxane is kinetically stable (and therefore can be de-fined as a rotaxane) when the carboxylic groups are ionized, showing an unthreading half-life of *ca.* 49 h, six orders of magnitude slower than observed under acidic conditions (*ca.* 1.5 s).[48] The electrostatic repulsion strategy demonstrated by Kaifer was subsequently explored by other re-searchers to devise pH-responsive CB[7] pseudorotaxanes with fluorescent or

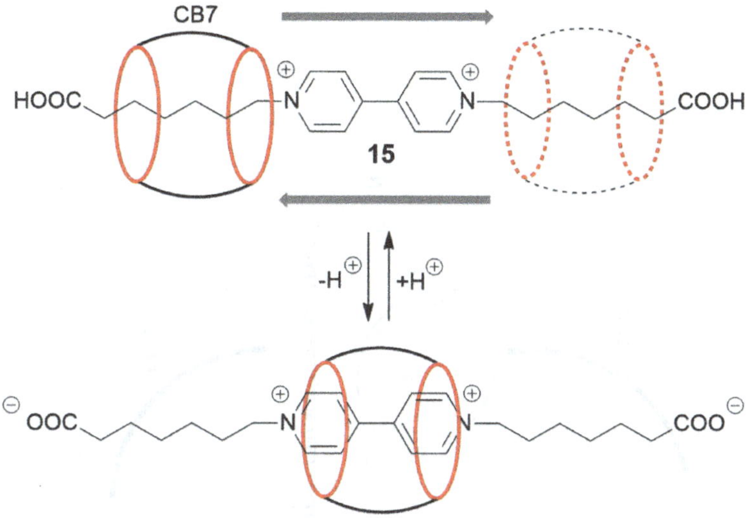

**Figure 4.6** pH-controlled shuttling movement in CB[7]-based pseudorotaxanes.

room-temperature phosphorescent outputs, as reported by the Akkaya and Tian Groups.[49,50]

While the vast majority of pH-responsive units that were chosen for the construction of rotaxanes and pseudorotaxanes display simple proton-transfer reactions, the use of switchable units that undergo complex pH-dependent reversible reactions has potential to increase the functionality of such assemblies. Flavylium molecular switches are characterized by a complex network of reversible reactions with great potential to be employed as switchable binding sites in CB-based (pseudo)rotaxanes. The methylviologen-flavylium dyad **16** was used, together with CB[7] (Figure 4.7), to construct a pH-responsive multistate pseudorotaxane showing inter-conversion rates that vary from hours to seconds, depending on the pH.[51]

The tricationic flavylium form is stable under acidic conditions, and for this species the CB[7] wheel is predominantly located between the benzo-pyrylium and methylviologen units. Upon a pH jump to basic conditions, the ionized *trans*-chalcone species is formed in a few minutes, and the CB[7] wheel is translocated to the methylviologen binding site. Reacidification of the system to a pH of *ca.* 6 leads to the immediate protonation of the phenol group. Under these conditions, the CB[7] remains localized at the dicationic binding site. However, this species is metastable and slowly converts into the flavylium cation in *ca.* 21 h in a process that is accompanied by the trans-location of the CB[7] wheel from the periphery to the middle of the axle. As the chemical process is acid-catalyzed, the rate of the reaction and con-sequently the co-conformational interconversion can be kinetically tuned from hours at near neutral pH to seconds under very acidic conditions. The system also shows some characteristics that are common to sequential

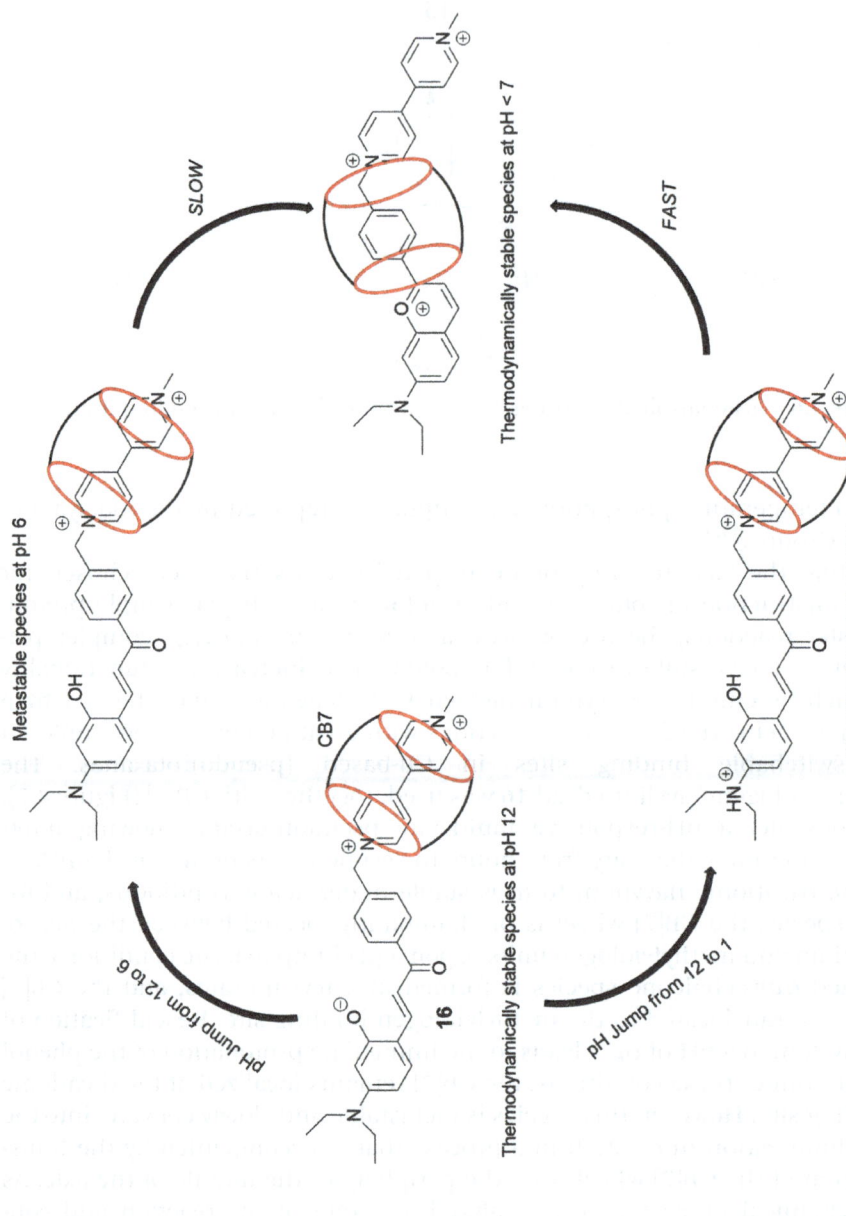

**Figure 4.7**  pH-responsive CB[7] pseudorotaxanes using a flavylium-based multistate axle **16**.

molecular logic gates (systems that are switched depending on the order of input application; see also the more detailed discussion in Section 4.3.6). The output varies depending on the sequence of pH jumps: from pH $=12$ to pH $= 6$ the *trans*-chalcone species is immediately formed, and its absorption spectrum can be read as the output. On the other hand, if the pH is decreased from 12 to 1 and then increased to 6, a different output (the flavylium cation) is observed.

## 4.3.2 Stimulation by Redox Inputs

Electrochemical stimuli have been widely explored to devise switchable CB-derived host–guest complexes, pseudorotaxanes, and rotaxanes. The redox-active dicationic methylviologen (**17**, Figure 4.8) is a prototypical guest for CB[7] and CB[8]. Initially investigated by Kim and coworkers, it was found that this molecule can form stable 1 : 1 complexes with both host molecules.[52,53]

The stability of the host–guest complex formed between CB[7] and **17** shows a modest decrease upon a 1-electron reduction of the guest and a drastic diminution of the binding constant for the fully reduced neutral species, thus enabling control of the association process by means of the application of a redox stimulus. In contrast to CB[7], the 1 : 1 inclusion complex between **17** and CB[8] can be electrochemically switched to a 1 : 2 complex with the radical cation upon the 1-electron reduction.[54] Noteworthy, contrary to CB[7] the stability of the CB[8] complex with the radical cation of **17** is higher than with the fully oxidized methylviologen. Another intriguing feature of these systems is the fact that the 1 : 1 CB[8]-**17** assembly can accommodate a third electron-rich aromatic guest in its cavity to form a 1 : 1 : 1 heteroternary complex.[53] This ability, together with electrochemically triggered formation of the 1 : 2 homoternary complexes with the radical cation, has been widely explored in the development of switchable supramolecular systems.[8]

The cooperative dimerization of the methylviologen radical cation inside the cavity of CB[8] was explored to devise a pseudorotaxane (**18**, Figure 4.8) that can be reversibly interconverted by a redox (or photoredox) stimulus into a molecular loop.[55] The electrochemically triggered interconversion between 1 : 1 : 1 heteroternary and 1 : 2 homoternary complexes was also adopted in other CB[8]-based pseudorotaxanes to switch reversibly between "end-to-interior" (**19**, Figure 4.8) and "end-to-end" loop structures in the same assembly.[56,57] Despite its high stability, the formation of a 1 : 2 homoternary complex between CB[8] and the methylviologen radical cation can be precluded by the judicious design of methylviologen derivatives. For example, a methylviologen-naphthalene dyad (**20**, Figure 4.8), bearing a bulky terminal cationic group, was shown to form a folded intramolecular 1 : 1 CT complex with both $\pi$-acceptor and $\pi$-donor units included in the host cavity. However, upon a 1-electron reduction of the dicationic unit, the folded 1 : 1 complex does not dissociate or unfold, as could be expected, to form

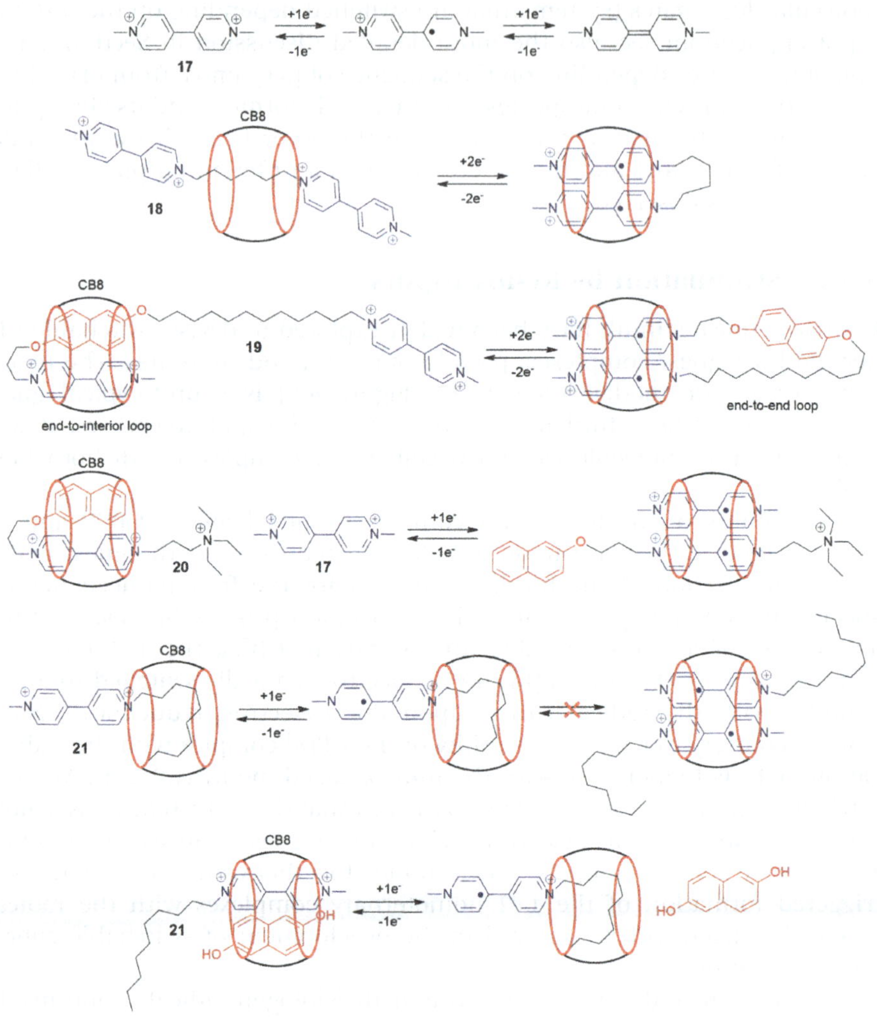

**Figure 4.8**  Redox-responsive CB[8]-methylviologen pseudorotaxanes.

a 1:2 homoternary complex. Interestingly, the dissociation and consequent unfolding of the intramolecular charge-transfer complex can be achieved by a 1-electron reduction in the presence of 1 equivalent of **17**, leading to the formation of a radical-cation heterodimer inside CB[8] and the dislocation of the π-electron–rich 2-alkoxynaphthalene unit from the cavity. The system behaves as a molecular loop lock that can be only unlocked by the redox stimulus in the presence of **17** as the molecular key (*i.e.*, as an AND molecular logic gate).[58] In addition to acceptor–donor dyads, viologen guest molecules bearing long alkyl chains (*e.g.*, methyldecylviologen **21**) can also preclude the formation of the radical cation dimer inside the cavity of

CB[8].[59] Upon encapsulation, the alkyl chain folds inside the cavity to adopt a U-shaped conformation, leaving the viologen units outside the cavity. Upon the 1-electron reduction of the viologen unit, the formation of a 1 : 2 homoternary complex is not observed, thus suggesting that the extraordinarily stable folded conformation is preferred over the CB[8]-assisted dimerization of the radical cation. This observation leads to the development of a molecular "pop-up toy" that reversibly shows and hides the alkyl chain upon application of a redox stimulus. On the one hand, in the presence of the 2,6-dihydroxynaphthalene donor guest, the 1 : 1 : 1 heteroternary complex is readily formed through the simultaneous encapsulation of the donor guest and the viologen acceptor unit inside the cavity. This results in the unfolding of the alkyl chain outside the host cavity. On the other hand, upon the 1-electron reduction of the viologen, however, the donor guest is released, and the 1 : 1 complex is regenerated with the alkyl chain folded back inside the cavity. As an interesting side observation, this conformational switching is accompanied by the catch and release of the donor molecule.

Kaifer and coworkers explored the redox-switchable properties of CB[8]-**17** to promote the dimerization of dendrimers by a redox stimulus.[60,61] The self-assembly process can be reversibly switched between a 1 : 1 stoichiometry and 1 : 2 homoternary complexes when only dendronized viologens and CB[8] species are present in solution or between 1 : 1 : 1 heteroternary and 1 : 2 homoternary complexes in mixtures that also contain π-donor dendrimers. This provides an elegant and effective way to reversibly control the type and size of the self-assembled dendrimers.

Electrochemically active ferrocene derivatives bind to CB[7] with very high affinity. Despite the overall positive charge, upon oxidation to ferrocenium, the stability of the complexes decreases as a result of the lower hydrophobicity of the guest. This differential binding affinity was explored in CB[7] pseudorotaxanes to control the shuttling movements with properly designed ferrocene-containing molecular axles.[46,62] For example, the bis-ferrocenyl–terminated guest **22** (Figure 4.9) was investigated in the presence of 1 equivalent of CB[7]. The redox-induced translocation of the CB[7] wheel from the ferrocenium peripheric binding site to the central docking sites is observed only with high-affinity structural motifs for CB[7] such as *p*-xylylenediammonium and 1,6-diammoniumhexane.

The redox-switchable properties of ferrocene-based pseudorotaxanes were also investigated using cyclohexanocucurbit[6]uril (CyhexCB[6]) wheels, which, contrary to CB[7], show higher affinity for the oxidized ferrocenium form.[63] The contrasting selectivity of CB[7] and CB[6] for the two redox states of the ferrocene-based axle was explored to devise a switchable system, where the CB[6]- and CB[7]-based pseudorotaxanes are reversible interchanged (designated by the authors as a molecular selector).

Similarly to methylviologen, the tetrathiafulvalene (TTF) radical cation also forms stable 1 : 2 host–guest complexes with CB[8]. However, tetrathiafulvalene does not form complexes with CB[8], and the redox switching is limited to the observation of free guest and 1 : 2 host–guest complexes.

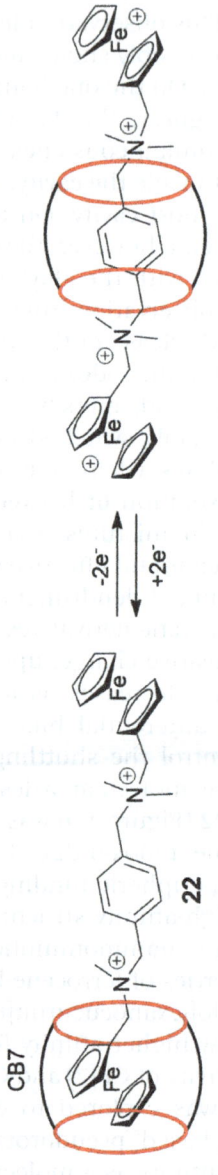

**Figure 4.9** Redox-responsive CB[7] shuttling in ferrocenium-based pseudorotaxanes.

Noteworthy, the formation of the radical cation dimer inside the cavity occurs spontaneously in the presence of $O_2$, allowing isolation and full characterization of the complexes.[64] Interestingly, tetrathiafulvalene, methylviologen and CB[8] can be combined to form a $1:1:1$ heteroternary complex that subsequently can be switched into $1:2$ complexes with methylviologen or TTF radical cations by reduction or oxidation, respectively. In this way a three-state supramolecular switch was demonstrated.[65]

### 4.3.3 Stimulation by Chemical Signals

In contrast to pH and electrochemical stimuli, the use of small molecules, ions or receptors constitutes a less conventional strategy to stimulate CB-based rotaxanes and pseudorotaxanes. Cyclodextrins (CD), for instance, have been used as a switching element to induce either stoichiometric interconversion and/or movement of the CB wheel. The combination of CD and CB macrocycles for heterowheel rotaxanes is particularly attractive owing to frequent cooperative interactions between the two macrocycles that arise, probably, from hydrogen bonding between the hydroxyl groups of the CD and carbonyl portals of CB.[66–69]

The formation of heterowheel pseudorotaxanes between CB[7], α-CD and a viologen-based axle molecule was investigated in detail by the Liu Group.[67] CB[7] was initially complexed with asymmetric *N*-methyl-*N*′-octyl-4,4′-bipyridinium and symmetric *N*,*N*′-dioctyl-4,4′-bipyridinium axles. Under $1:1$ stoichiometric conditions, the CB[7] wheel is mainly located on the octyl side chain. Upon addition of α-CD, the CB[7] molecule is pushed to the viologen binding site owing to the preferential inclusion of the octyl chains by the CD, resulting in the formation of [3]pseudorotaxanes and [4]pseudorotaxanes with the asymmetric and symmetric axles, respectively. In a subsequent work, the formation of heterowheel [3]pseudorotaxane was demonstrated to take place only when two different components are simultaneously introduced in the system.[68] A bipyridinium axle (**23**), containing an adamantane binding site, forms a stable $1:1$ complex with CB[8] (Figure 4.10). The high affinity of the adamantane recognition motifs for CB[8] precludes the displacement of the wheel to the viologen binding site upon addition of a

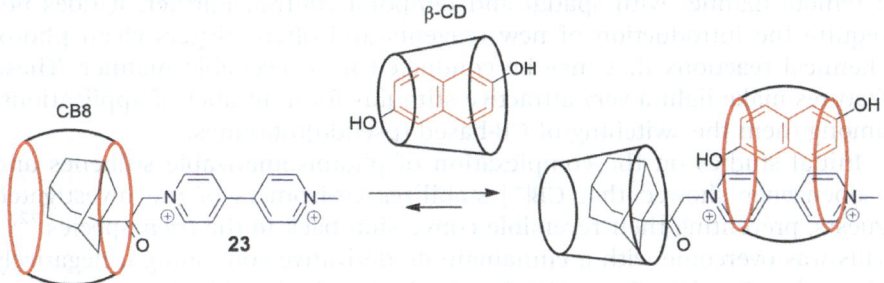

**Figure 4.10** CB[8] shuttling driven by two simultaneous chemical inputs.

competing CD (that also presents moderate affinity for the adamantane site) or of an electron-rich dihydroxynaphthalene donor that is known to increase the stability of CB[8]-viologen adducts through the formation of $1:1:1$ heteroternary complexes. However, when both components are simultaneously added to a solution of the $1:1$ CB[8]-**23** complex, the four molecules self-sort to form a heterowheel pseudorotaxane as shown in Figure 4.10. The high fidelity of this recognition process is thought to result from a combination of CT, ion–dipole, hydrogen-bonding interactions and hydrophobic effects holding together the four-component assembly.

Probably due to the moderate affinity of CBs for metal cations, the utilization of these species as chemical signals to actuate CB-based (pseudo)rotaxanes has been scarcely reported. Masson and coworkers reported a detailed study on the influence of cations on the dethreading rate of molecular axles with benzo-15-crown-5 stoppers from the CB[7] cavity.[70] Half-lives of up to 65 h were observed in the absence of cations at 50 °C, while in the presence of metal cations, rate enhancements up to 500 times were noted. The relative "lubricating" effect of different cations was found to be dependent on several key factors such as the radius, valence, coordination number and hardness. Divalent metal cations, such as $Ba^{2+}$ and $Ca^{2+}$, showed the highest catalytic effect.

More recently, Bardelang and coworkers demonstrated that a pseudorotaxane formed between CB[7] and a viologen-phenylene-imidazole axle (**24**, see Figure 4.11) that can be selectively actuated by $Ag^+$ cations.[71] While in the presence of other salts, the CB[7] wheel remained bound to the methylviologen station; the addition of $Ag^+$ promoted the translocation to the phenylene station, presumably due to simultaneous binding of the cation to one of the CB carbonyl portals and to the imidazole group. The system can be switched back to the original viologen binding site by the addition of NaCl, causing precipitation of AgCl. Additionally, the presence of $Ag^+$ enables the formation of $2:1$ host–guest complexes, allowing the catch and release of a second CB[7] molecule.

## 4.3.4 Photoresponsive CB Assemblies

Light as stimulus offers advantages such as facile application and removal in a remote manner with spatial and temporal control. Further, it does not require the introduction of new reagents and often triggers clean photochemical reactions that may be conducted in a reversible manner. These features make light a very attractive stimulus for a number of applications, among them the switching of CB-based (pseudo)rotaxanes.

Initial studies on the complexation of photoisomerizable stilbenes and azobenzenes showed that CB[7] stabilizes *cis*-isomers of the investigated guests, preventing their reversible conversion back to the *trans*-species.[72,73] This was overcome with a cinnamamide derivative containing a negatively charged carboxylate group attached to the terminal amide that destabilizes the complex with the *cis*-isomer. This allowed photodissociation on

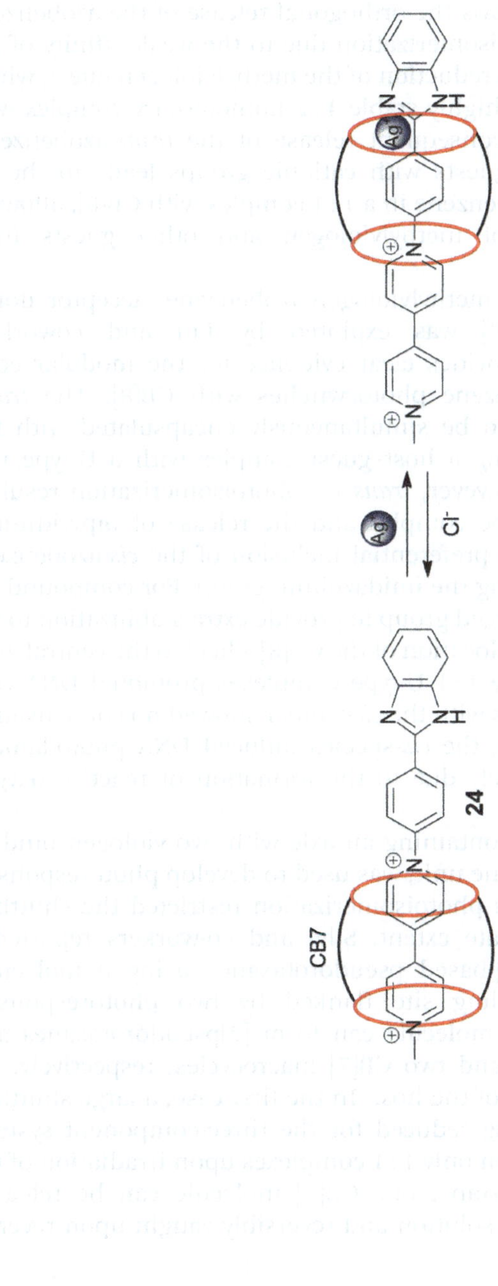

**Figure 4.11** CB[7] shuttling actuated by a silver cation as stimulus.

irradiating the CB[7]-*trans*-cinnamamide complex with 300-nm light and was reverted by irradiation of the liberated *cis*-cinnamamide at 254 nm.[74]

While photoresponsive azobenzene 1 : 1 host–guest complexes with CB[7] were not reported, Scherman and coworkers demonstrated the orthogonal redox- and light-switching of 1 : 1 : 1 heteroternary complexes formed between CB[8], methylviologen **17** and azobenzene derivatives (see Figure 4.12).[75] This system allows the orthogonal release of the azobenzene guest molecule through photoisomerization due to the weak affinity of the *cis*-azobenzene or by 1-electron reduction of the methylviologen guest, which leads to the formation of the highly stable 1 : 2 homoternary complex with the two radical cations and consequent release of the *trans*-azobenzene. Modification of azobenzene guests with cationic groups leads to the increased stability of the *cis*-azobenzene in a 1 : 1 complex with CB[8], allowing the phototriggered release of methylviologen and other guests from heteroternary complexes.[34,76]

The complexation of the methylviologen-azobenzene acceptor–donor dyads **26** and **27** by CB[8] was explored by Liu and coworkers (Figure 4.12).[77] This study provides clear evidence for the modular complexation behavior of azobenzene photoswitches with CB[8]. The *trans*-azobenzene unit was found to be simultaneously encapsulated with the methylviologen moiety, forming a host–guest complex with a U-type molecular loop configuration. However, *trans–cis* photoisomerization resulted in the disruption of the U-type complex and the release of bipyridinium moiety from the cavity due to preferential inclusion of the *cis*-azobenzene unit of compound **26** containing the imidazolium group. For compound **27**, the absence of a positively charged group to provide extra stabilization to the *cis*-species resulted in the translocation of the CB[8] wheel to the central alkyl chain of the axle. Notably, the 1 : 1 U-type complexes promoted DNA condensation, while the complexes with the *cis*-isomer showed no condensation capability. On the other hand, the *cis*-species induced DNA photodamage upon UV light exposition, likely due to the formation of reactive oxygen species.

A CB[7]-based [3]rotaxane, containing an axle with two viologen binding sites separated by an azobenzene unit, was used to develop photoresponsive molecular shuttles.[78] *Cis–trans* photoisomerization restricted the shuttling motion but only to a moderate extent. Silvi and co-workers reported a subsequent example of CB[7]-based pseudorotaxanes using a molecular axle with one viologen binding site flanked by two photoresponsive azobenzene units.[79] The axle molecule can form [2]pseudorotaxanes and [3]pseudorotaxanes with one and two CB[7] macrocycles, respectively, depending on the concentration of the host. In the first case, a large shuttling amplitude was observed, being reduced for the three-component system. Because the *cis*-species can form only 1 : 1 complexes upon irradiation of the *trans*-configured [3]pseudorotaxane, one CB[7] molecule can be released from the assembly to the bulk solution and reversibly caught upon reversal of the photoisomerization.

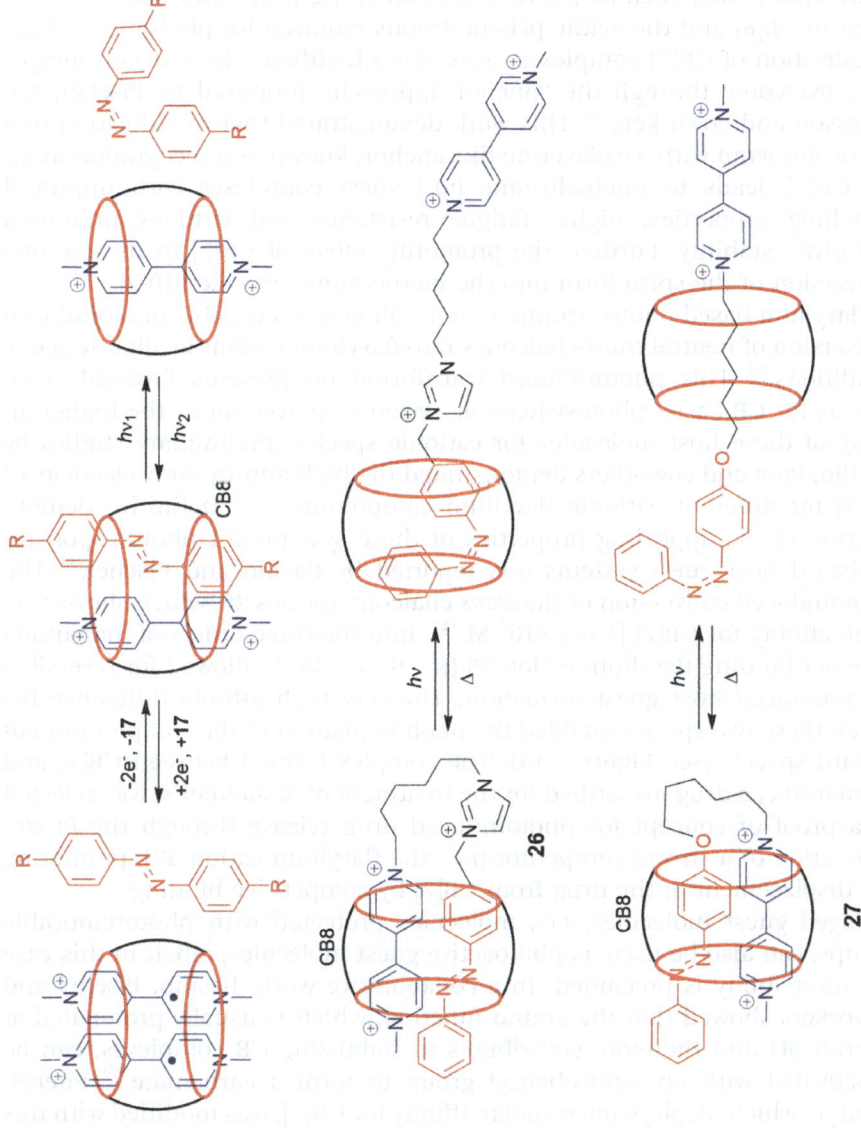

**Figure 4.12** Photoswitchable CB[8]-azobenzene host-guest assemblies.

The complexation of spiropyrans, another important class of photo-chromic molecules, with CBs was also investigated.[80–82] Initial studies carried out by Biczók and coworkers revealed interesting properties of a nitrospiropyran upon inclusion in CBs, such as improved hydrolytic stability and selective binding of the protonated merocyanine form.[80,81] However, some limitations, such as poorer photoswitching properties upon confine-ment in CB[8] and the acidic pH conditions required for photoinduced de-stabilization of CB[7] complexes, were also identified. These disadvantages were overcome through the "anchor approach" proposed by Pischel, An-dréasson and coworkers.[82] This work demonstrated that modification of a nitrospiropyran with a cadaverine-like anchor, known as a recognition motif for CB[7], leads to photochromic host–guest complexes with improved switching properties, higher fatigue resistance and virtually unlimited hydrolytic stability. Further, the promoting effect of CB[7] in the thermal conversion of the spiro form into the merocyanine was identified.

Flavylium-based photochromic systems allow the reversible photoinduced conversion of neutral *trans*-chalcones into flavylium cations at slightly acidic conditions.[83] This photoinduced transformation presents favorable con-ditions for CB-based photoswitchable systems on account of the higher af-finity of these host molecules for cationic species. Preliminary studies by Basílio, Pina and coworkers demonstrated the high affinity and selectivity of CB[7] for different cationic flavylium compounds.[84–87] Definitive demon-stration of the appealing properties of these systems for photoresponsive CB-based host–guest systems was reported by Basílio and Pischel.[88] The photoinduced conversion of the *trans*-chalcone species **28**, which shows very weak affinity for CB[7] ($K = 3 \times 10^2$ M$^{-1}$), into the three-orders-of-magnitude stronger binding flavylium cation **29** ($K = 9 \times 10^5$ M$^{-1}$) allowed for reversible photoinduced host–guest formation. The very high affinity difference be-tween these two species enabled the photoregulation of the complexation of a third species (see Figure 4.13). The complex formed between CB[7] and memantine, a drug prescribed for the treatment of Alzheimer's, was selected as a proof of concept for photoinduced drug release through the *in situ* generation of a strong competitor (*i.e.*, the flavylium cation **29**), promoting the displacement of the drug from CB[7] by competitive binding.

Caged guest molecules, *i.e.*, molecules protected with photoremovable groups, can also be used as photoactive guest molecules, albeit in this case the reversibility is precluded. In a collaborative work, Basílio, Pischel and coworkers showed that the amino function, which is usually protonated at neutral pH and therefore contributes to stabilizing CB complexes, can be deactivated with an *o*-nitrobenzyl group to form a carbamate.[89] Phenyl-alanine, which displays micromolar affinity for CB[7], was modified with this photolabile group, thus precluding the formation of the complexes even at mM concentration. Upon irradiation, the protecting group is removed, and the strongly binding phenylalanine guest is generated. This strategy has the potential to be applied to virtually any guest that is carrying amino groups and therefore may constitute a general approach to developing

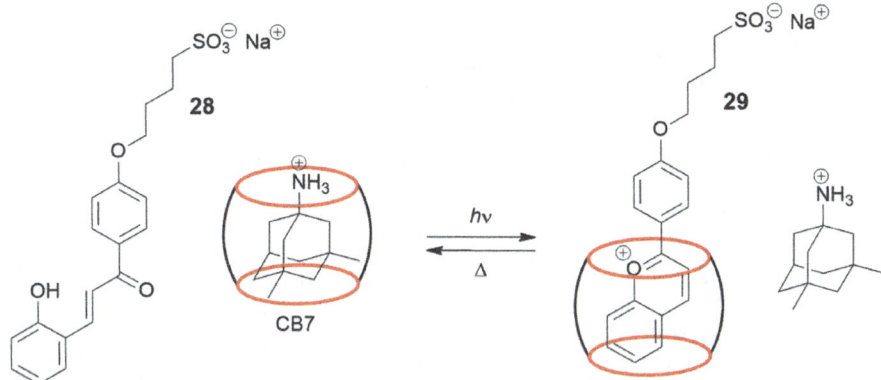

**Figure 4.13** Photoresponsive host–guest complex based on a *trans*-chalcone (**28**)/ flavylium (**29**) photoswitch and its application to the release of memantine.

photoresponsive CB-based supramolecular systems when reversibility is not a requisite.

In addition to directly actuating guest molecules by inducing structural transformations that affect the stability of the complexes, light can be indirectly used to "communicate" with the host–guest complex through a third molecule present in solution (or attached to the host or guest). The most widely used indirect approaches are photoinduced electron transfer (PET) and photoinduced pH jumps. Pischel and coworkers demonstrated that malachite green photobase **30** can be employed to reversibly dissociate a CB[7] inclusion complex through deprotonation of the Hoechst 33258 guest dye (Figure 4.14).[90] The generality of this approach is limited to pH-responsive guest molecules with $pK_a$ values that allow deprotonation in slightly basic conditions (up to *ca.* pH 9).

In a subsequent work, the same group employed a photoacid and a pH-responsive competitor that enables the release of biogenic amines from the CB[7] independently of their $pK_a$.[91] The cascade is initiated with light absorption by *o*-nitrobenzaldehyde **31**, which is transformed into the respective carboxylic acid. Dissociation of this weak acid decreases the pH of the solution, generating (*via* protonation) a strong competitor with a high affinity for CB[7] that induces the release of an encapsulated amine (Figure 4.14). As in the previous example, the pH-responsive competitor was carefully selected to display self-reporting fluorescence emission properties, providing a spectroscopic handle to follow the pH-triggered encapsulation and cargo release process.

One of the first light-responsive CB-based rotaxanes was reported by Kim and coworkers who used [Ru(bpy)$_3$]$^{2+}$ as a sensitizer to reduce a methyl-viologen binding site through photoinduced electron transfer (PET).[55] Besides [Ru(bpy)$_3$]$^{2+}$, other sensitizers can be employed. Peng and coworkers reported a particularly interesting example using donor phenothiazines

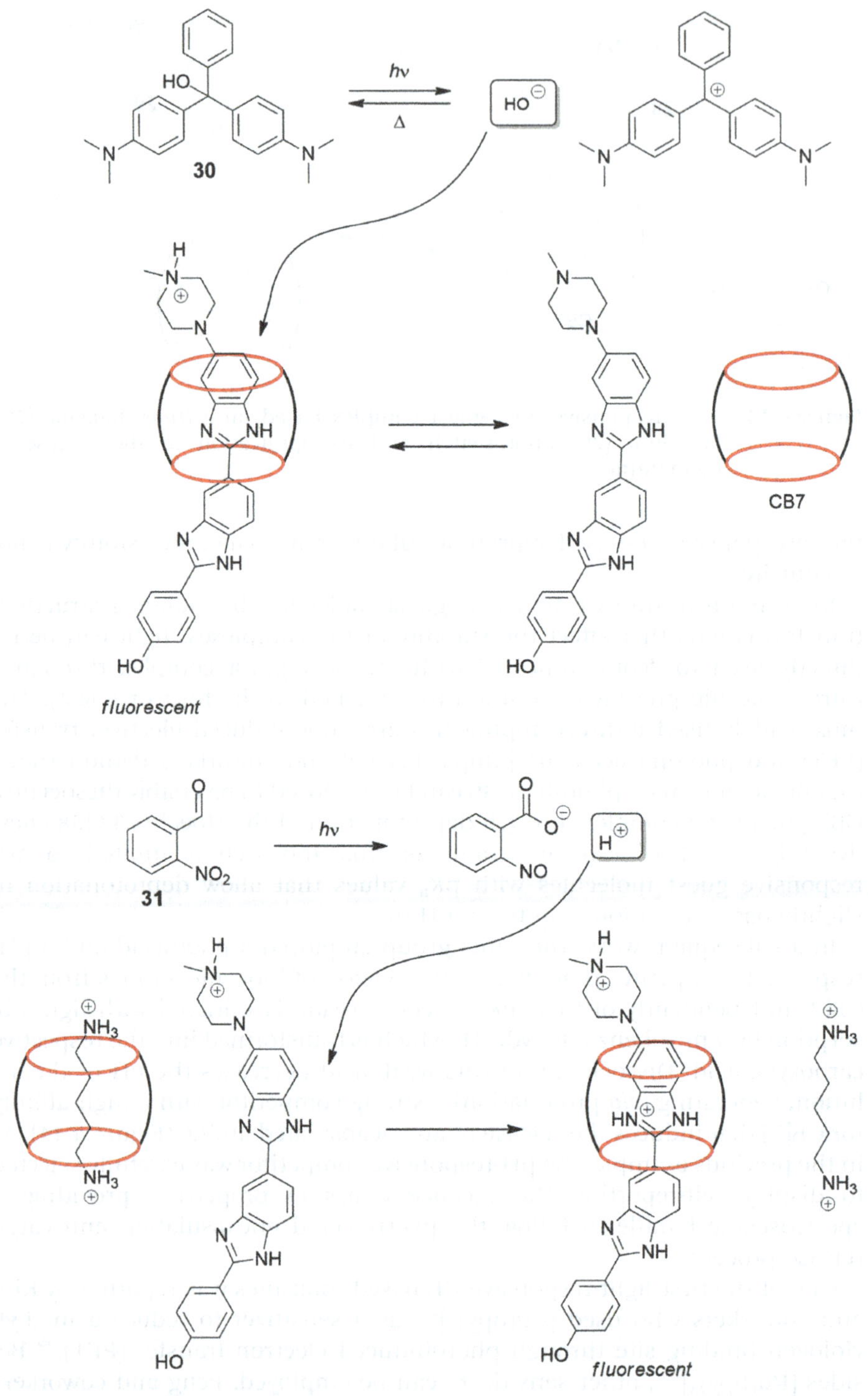

**Figure 4.14** Using photobases or photoacids to switch pH-responsive CB[7] host–guest systems with light.

(PTZ), methylviologen and CB[8] to form a 1:1:1 heteroternary complex with near-infrared absorption. This assembly undergoes PET to reduce the methylviologen, forming the corresponding radical-cation 1:2 homoternary complex under expulsion of the PTZ from the CB[8] cavity.[92]

$[Ru(bpy)_3]^{2+}$ can be used as functional (sensitizer) bulky stoppers for the preparation of CB-based rotaxanes. Sun and coworkers reported a CB[8]-based rotaxane made from a symmetric axle containing a central viologen dication, two butyl spacers and two $[Ru(bpy)_3]^{2+}$ stoppers.[92] This rotaxane was used to demonstrate that the wheel shuttling movement between two peripheric butyl stations can be photochemically halted at the central viologen station by PET from $[Ru(bpy)_3]^2$ stoppers to the dicationic acceptor and to free methylviologen in solution, which is required for dimer formation and stabilization of this central binding site.[93]

An ultrafast photoresponsive molecular shuttle constructed from CB[7] and a styryl(pyridinium) dye was proposed by Chernikova and coworkers.[94] Based on theoretical calculations and steady-state as well as time-resolved optical spectroscopies, the authors proposed that, upon light absorption, the CB[7] wheel reversibly shuttles between two binding sites on a picosecond time scale. In the ground state, CB[7] is mainly localized at the pyridinium binding site. Excitation of the dye through light absorption promotes the formation of an intramolecular CT state with reduced positive charge density on the pyridinium binding site and increased positive character on the donor aryl subunit. This light-induced electronic redistribution was proposed to trigger the shuttling movement of the CB[7] wheel between the two stations. The sub-picosecond displacement of the guest inside the CB[7] wheel is up to six orders of magnitude faster than what was previously reported for other rotaxanes. However, possible reasons for this large difference were not proposed.[95,96]

Basílio and coworkers designed and synthesized a simple *trans*-chalcone axle (**32**), containing a peripheric triethylalkylammonium binding site for CB[7] (Figure 4.15).[97] At pH values around 7–8, the macrocycle binds to this site, forming a [2]pseudorotaxane. Upon light irradiation, the *trans–cis* chalcone photoisomerization produces a metastable state composed by the *cis*-chalcone in fast equilibrium with the hemiketal species formed through a ring-closing tautomerization reaction. Noteworthy, these transformations do not affect the wheel binding site. However, upon irradiation at slightly acidic pH (*ca.* pH 5), the *trans*-chalcone is quantitatively converted into the respective flavylium cation **33** (*via* dehydration of the hemiketal). This transformation is accompanied by the translocation of the CB[7] wheel from the triethylalkylammonium binding site to the 2-phenyl group of the flavylium cation. When kept in the dark, the flavylium cation returns to the *trans*-chalcone species, demonstrating the reversibility of the system. The multistate/multifunctional properties of flavylium compounds displays several appealing features not found in other CB-based light-responsive (pseudo)rotaxanes such as pH-gated photoinduced shuttling; *i.e.*, the photoinduced molecular motion is effective only at slightly acid yet biologically relevant pH. This implies improved fluorescence properties that can be

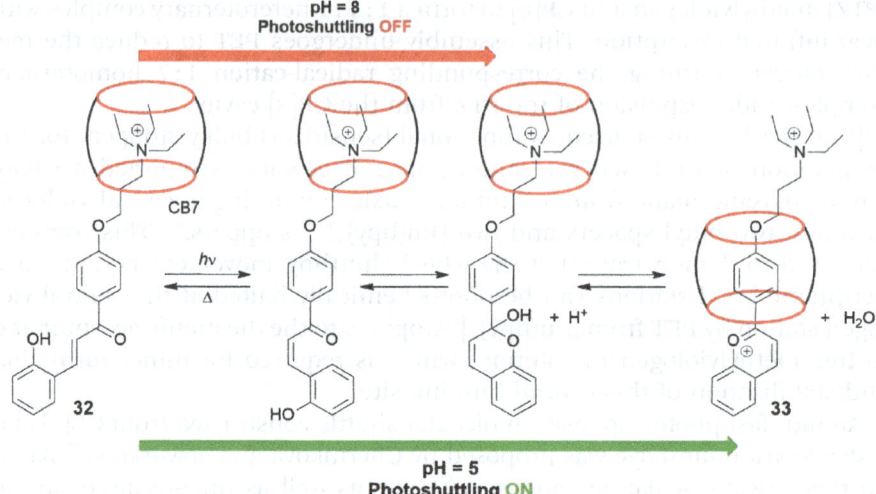

pH = 8
Photoshuttling OFF

CB7

hv
Δ

OH

OH + H⁺

+ H₂O

32

HO

33

pH = 5
Photoshuttling ON

**Figure 4.15**  pH-gated photoresponsive shuttle based on a CB[7]-flavylium [2]pseudorotaxane.

used to report on the shuttling process and the possibility of locking the pseudorotaxane in the flavylium form at low pH (*ca.* pH 2) and unlock it by increasing the pH.

### 4.3.5  Release Applications

The stimuli-responsive properties of CB-based (pseudo)rotaxanes have been explored in different contexts to develop functional smart nanomaterials for the controlled encapsulation and release of cargo molecules. One of the first examples consisted in a self-assembled monolayer (SAM) on a gold surface prepared from a CB[6]-based pseudorotaxane with an axle, containing a 1,4-diammoniumbutane binding site and a 1,2-dithiolane anchoring group.[13] The surface-attached pseudorotaxane was demonstrated to release and re-thread the CB[6] macrocycle upon basification and neutralization, respectively. This reversible process was explored as a stimuli-responsive gate to control the access of $[Fe(CN)_6]^{3-}$ electroactive species to the surface. The gating mechanism arises from the release of bulk CB[6] molecules that creates enough space for the $[Fe(CN)_6]^{3-}$ ions to reach the electrode surface, displaying redox peaks by cyclic voltammetry that cannot be observed before dethreading of the macrocycle.

Probably the most widely explored application of CB-based rotaxanes and pseudorotaxanes for controlled release concerns the use of surface-mounted nanovalves on mesoporous silica nanoparticles (MSN). The strategy exploits the use of axle molecules that are attached to the surface of the MSN. The binding sites must be close enough to the pore so that the cargo encapsulated therein cannot escape due to the steric hindrance imposed by the attached macrocycle. Upon stimulation, the macrocycle can be either detached or pushed away from the surface, opening the pores and allowing the release of encapsulated cargo following a mechanism that resembles the operation of a macroscopic valve.

In a pioneering report, Stoddart, Zink and coworkers demonstrated the controlled encapsulation and release of cargo molecules from MSN using pH-responsive CB[6]-based nanovalves.[98] In this work, a pH jump from neutral to basic values (pH *ca.* 10) promotes the deprotonation of a di-ammonium binding site and consequent dissociation the CB[6] macrocycle from the surface of the MSN, allowing release of the cargo from the pores. However, the basic pH conditions required to open the nanovalves exclude potential biological applications. In a subsequent report, the same research groups further elaborated the structure of the axle molecule to obtain a bi-stable pseudorotaxane that allows cargo release under biologically relevant pH conditions (Figure 4.16).[14]

It should be noted that the nanovalve operation (*i.e.*, wheel translocation) strategy was identical to that reported in the pioneering work of Mock and coworkers, previously commented on.[36] The surface-attached axle, containing two basic aliphatic amino groups and an aniline, characterized by a substantially lower $pK_a$ value, constitute a key element in opening the nanovalve under biologically relevant pH conditions. Under neutral conditions, the CB[6] is bound to the 1,4-diammoniumbutane site, blocking the release of the cargo molecules from pores. At slightly acidic pH values, the aniline protonates and the CB[6] wheel translocate to the stronger binding 1,6-diammoniumhexane site, opening the valve and allowing cargo release. Noteworthily, the $pK_a$ of the aniline group can be easily tuned through the nature of the *para* substituents, which provides an important modular element to adjust the pH at which the valve opens. Besides the employed slightly acidic conditions, the nanovalve can also be efficiently opened at pH > 10 due to the deprotonation of all ammonium groups and consequent dissociation of the pseudorotaxane.

When the MSN just described are functionalized with azobenzene units inside the pores and nanovalves on the surface, gated nanocontainers exhibiting AND logic function can be demonstrated.[99] These MSN release cargo materials only when two stimuli (light and pH) are simultaneously applied, thereby increasing control over the gating phenomena.

CB[6]-1,6-diammoniumhexane pseudorotaxanes installed on the surface of MSN were used as gatekeepers to test different stimulation methods including redox, enzymatic and, in the case of hybrid materials, surface localized thermal activation with oscillating magnetic fields and visible light or near-infrared light stimuli.[100–104]

Photoresponsive nanovalves were also described using CB[7]-cinnamamide pseudorotaxanes mounted on the surface of MSN.[105] UV-light absorption promotes the previously discussed *trans–cis* photoisomerization of the cinnamamide axle, leading to the dissociation of the complex and consequent release of the cargo encapsulated in the pores of the nanoparticles.

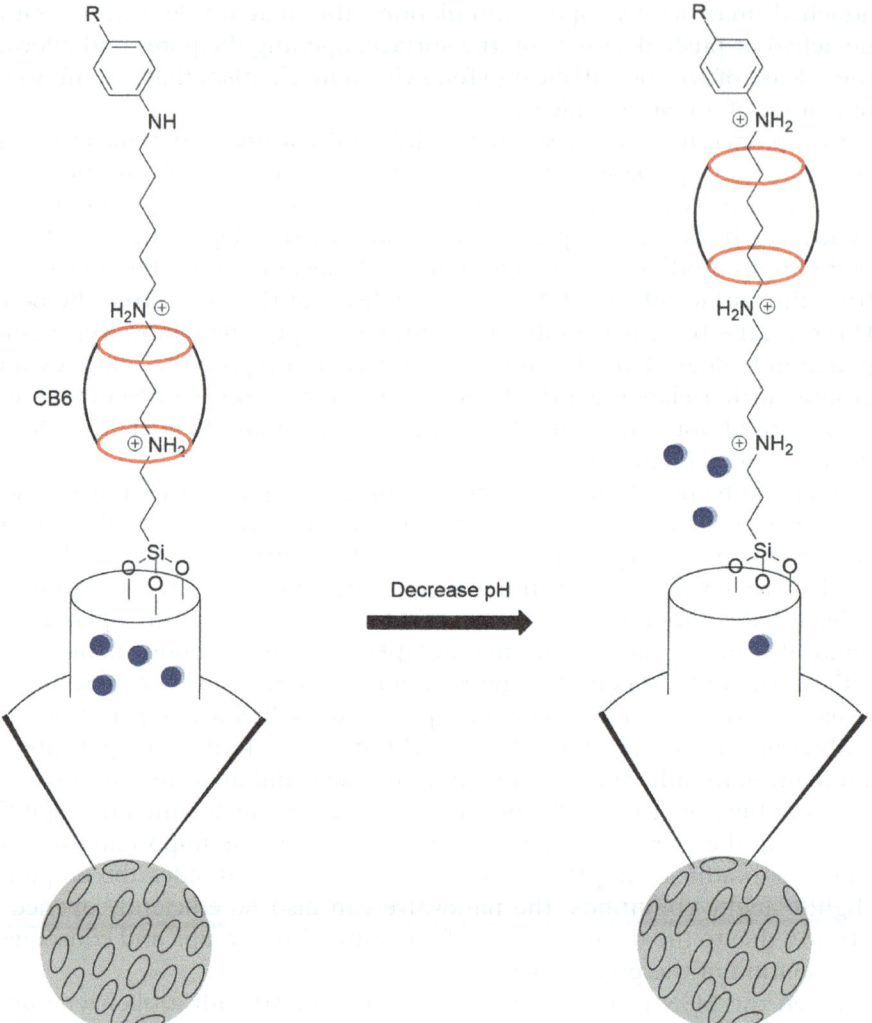

**Figure 4.16**   pH-responsive CB[6]-based nanovalves mounted on the surface of MSN.
The blue dots represent cargo materials encapsulated inside the pores
of the MSN.

In a later report, Zhang and coworkers explored the photochemical
properties of 1:1:1 heteroternary complexes formed from CB[8], bipyridinium and azobenzene derivatives to devise surface-mounted light-
responsive nanovalves on MSN (Figure 4.17).[106]
The strategy was comprised of the covalent functionalization of the MSN
surface with azobenzene derivatives. In the presence of a methylviologen
unit, CB[8] binds to the surface-attached azobenzene molecules through
heteroternary complexation, closing the pores of the MSN. The high

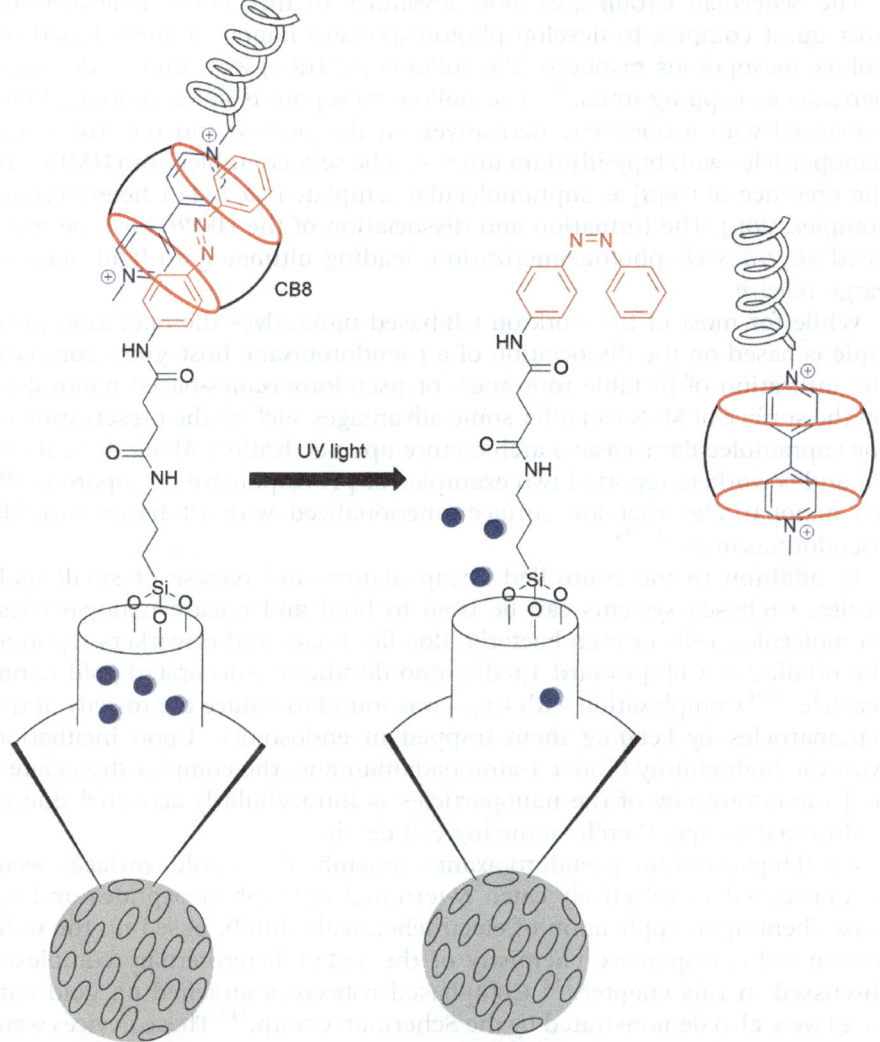

**Figure 4.17**  Photoresponsive nanovalves based on photoswitchable heteroternary complexation of CB[8], azobenzene and 4,4′-bipyridinium derivatives. A self-peptide was attached to the bipyridinium unit for camouflage protection against clearance by macrophages. The blue dots represent cargo materials encapsulated inside the pores of the MSN.

modularity of the system allows the utilization of a self-peptide/bipyridinium conjugate that provides camouflage protection against macrophage clearance. UV-light illumination induces the *trans–cis* photoisomerization of the surface-attached azobenzene guest molecules, leading to the dissociation of the ternary complex and consequent release of cargo materials entrapped inside the pores of the MSN.

The Scherman Group also took advantage of this $1:1:1$ heteroternary host–guest complex to develop photoresponsive nanocontainers based on hollow mesoporous raspberry-like colloids (HMRCs) with iron oxide nanoparticles as capping units.[107] The hollow mesoporous silica colloids, functionalized with azobenzene derivatives on the surface and the iron oxide nanoparticles with bipyridinium units, can be self-assembled into HMRCs in the presence of CB[8] as supramolecular template (*via* $1:1:1$ heteroternary complexation). The formation and dissociation of the HMRCs can be regulated *via trans–cis* photoisomerization, leading ultimately to light-induced cargo release.

While for most of the work on CB-based nanovalves the operation principle is based on the dissociation of a pseudorotaxane host–guest complex, the utilization of bistable rotaxanes- or pseudorotaxanes-based nanovalves on the surface of MSN can offer some advantages such as the preservation of the supramolecular rotaxane architecture upon activation. Along these lines, Fu and coworkers reported two examples of pH-responsive mesoporous silica nanoparticles that are surface-functionalized with CB-based bistable pseudorotaxanes.[108,109]

In addition to the controlled encapsulation and release of small molecules, CB-based systems can be used to bind and release nanoparticles, biomolecules, cells or even bacteria. Rotello, Isaacs and coworkers reported the binding of CB[7] toward 1,6-diammoniumhexane-decorated gold nanoparticles.[110] Complexation with CB[7] was found to reduce the toxicity of the nanoparticles by keeping them trapped in endosomes. Upon incubation with the high-affinity binder 1-aminoadamantane, the complex dissociates, and the cytotoxicity of the nanoparticles is intracellularly activated due to endosomal escape, thereby inducing cell death.

CB[8]-bipyridinium pseudorotaxanes assembled on gold surfaces were demonstrated to selectively catch *N*-terminal tryptophan peptides and release them upon application of electrochemical stimuli, based on the well-known redox responsive chemistry of the $1:1:1$ heteroternary complexes discussed in this chapter.[111] CB[8]-based rotaxanes attached on gold surfaces were also demonstrated by the Scherman Group.[112] These devices were successfully employed to bind and sense dopamine through heteroternary binding. In addition to gold surfaces, the same group also attached CB[8]-bipyridinium catenanes to iron oxide/silica core–shell nanoparticles.[113]

The extraordinary and versatile recognition abilities of CBs allow the development of smart surfaces to attach not only biomolecules but also cells, viruses or bacteria.[114] Jonkheijm and coworkers demonstrated several smart bioactive surfaces based on $1:1:1$ heteroternary CB[8] complexation for reversible surface attachment.[115–117] In an illustrative example, they developed, in collaboration with Ravoo and coworkers, a photoresponsive bioactive surface using the integrin-binding peptide RGD attached to an arylazopyrazole unit. Similarly to azobenzenes, *trans*-arylazopyrazoles bind more tightly to CB[8]-methylviologen complexes than the *cis*-isomers. Experiments with mouse myoblast C2C12 cells showed an increased number of adhered cells on the supramolecular monolayer. Cell imaging showed that the cells can be efficiently released from the surface by UV-light irradiation, which is ascribed to the dissociation of the *cis*-arylazopyrazole peptide conjugate from the 1:1:1 heteroternary complex. Similar photoresponsive self-assembly strategies were also used to bind viruses or bacteria onto surfaces.[116,117]

## 4.3.6 Molecular Information Processing

The examples given of stimuli-responsive CB-based host–guest assemblies involve typically single-input situations. However, Scherman's orthogonal redox- and light-mediated switching of a heteroternary CB[8]-methylviologen-azobenzene complex (see Figure 4.12) has clear potential for the implementation of logic functionality.[75] There are actually some reports in the current literature where the combination of molecular logic and CB chemistry was fruitfully exploited. The Pischel Group developed supramolecular host–guest complexes with guests that show protonation-dependent fluorescence quenching *via* PET as platforms for the demonstration of various simple logic functions such as AND or INHIBIT.[118] The nontrivial XNOR function was demonstrated by the Credi and Tian Groups with a chiral binaphthyl-bipyridinium tweezer and its CB[8] complexes. It was shown that the 1:1 complexation of the terminal bipyridinium moieties alters the helicity and thereby the circular dichroism spectrum of the tweezer guest. On photochemical reduction of the bipyridinium moieties to the radical cations, the complex stoichiometry is altered to 2:1, which has consequences for the helicity of the binaphthyl part. Hence, light and CB[8] as inputs define an XNOR gate whose response is monitored by circular dichroism.

In another work, the Pischel Group showed that the template effect exerted by CB[8] in the [4 + 4] photodimerization of an anthracene guest can be used to interpret a supramolecular keypad lock with light as one input and an adamantane-derived competitor guest as the other input (see Figure 4.18).[119] As a surplus, the response of the system can be followed by fluorescence. Furthermore, the system is fully recyclable by employing features of self-sorting upon the addition of CB[7]. Also, the previously discussed[88] release of memantine from CB[7] by coupling the host–guest equilibrium to the photoinduced formation of a flavylium cation as competitor can be expanded into a release that depends on the operation of a molecular keypad lock function.[120] For this purpose, a chalcone was designed that shows the light-induced transformation into the corresponding flavylium ion only if first the pH is set to 4–5 and then light irradiation is applied. If solutions at neutral pH are irradiated, the fast thermal reconversion of the *cis*-chalcone to the starting *trans*-chalcone puts a halt to the multistep formation of the flavylium ion. In the specific case, this function was combined with the release of a model peptide from the CB[8] cavity, being potentially interesting in controlling the dimerization of peptides[121] by light.

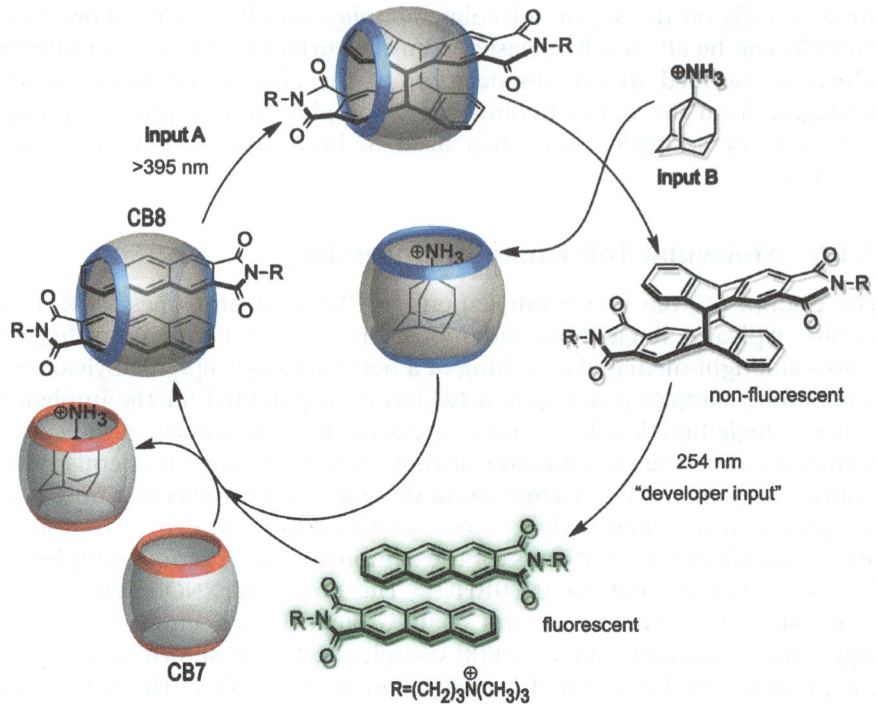

**Figure 4.18**   Supramolecular keypad lock based on CB[8]-templated [4 + 4] photo-dimerization and resetting through self-sorting upon the addition of CB[7].
Reproduced from ref. 119 with permission from the Royal Society of Chemistry.

## 4.4   Conclusion

In this chapter, two principal aspects of the supramolecular chemistry of CB-based (pseudo)rotaxanes were discussed: (1) the programmed synthesis of homo- and heterео(pseudo)rotaxanes, including [2](pseudo)rotaxanes and related higher-order architectures and (2) the stimuli-responsive behavior of such assemblies with potential applications as switches, as sensors, for the release of functional cargo and in biorelevant contexts, as well as for the processing of information at the nanoscale. On the one hand, the use of cucurbiturils for the thermodynamically or kinetically controlled assembly of (pseudo)rotaxanes exploits their rich supramolecular chemistry with special emphasis on their role as prime hosts in self-sorting. On the other hand, the stimuli-responsive facets of CB-based assemblies can be designed in a tailored fashion by devising guests that are addressed by pH changes, redox signals or light, sometimes in an orthogonal manner or by combinations of these inputs. Commonly, upon application of external stimuli, the geometrical and electronic properties of potential guests are altered, leading to changes in binding affinity, mode and stoichiometry. Combining these features of CB hosts and their guests enables the implementation of fascinating artificial devices, literally in a plug-and-play mode and with nearly unmatched functional variety as compared to other commonly used host macrocycles. These trends of modern CB chemistry are showcased in this chapter.

## Acknowledgements

N.B. acknowledges support by the Associated Laboratory for Sustainable Chemistry–Clean Processes and Technologies–LAQV (FCT/MEC fund UID/QUI/50006/2019), and the Portuguese FCT through grants PTDC/QUI-COL/32351/2017 and CEECIND/00466/2017. U.P. is grateful for the generous funding provided by the Spanish Ministerio de Ciencia, Innovación y Universidades (grants CTQ2011-28390 and CTQ2017-89832-P), the Consejería de Economía, Innovación, Ciencia y Empleo, Junta de Andalucía (grants P08-FQM-3685 and P12-FQM-2140), and ERDF.

## References

1. S. Erbas-Cakmak, D. A. Leigh, C. T. McTernan and A. L. Nussbaumer, *Chem. Rev.*, 2015, **115**, 10081–10206.
2. K. I. Assaf and W. M. Nau, *Chem. Soc. Rev.*, 2015, **44**, 394–418.
3. K. Kim, *Chem. Soc. Rev.*, 2002, **31**, 96–107.
4. Y. H. Ko, E. Kim, I. Hwang and K. Kim, *Chem. Commun.*, 2007, 1305–1315.
5. V. N. Vukotic and S. J. Loeb, *Chem. Soc. Rev.*, 2012, **41**, 5896–5906.
6. X. Hou, C. Ke and J. Fraser Stoddart, *Chem. Soc. Rev.*, 2016, **45**, 3766–3780.
7. L. Zhu, M. Zhu and Y. Zhao, *ChemPlusChem*, 2017, **82**, 30–41.
8. E. Pazos, P. Novo, C. Peinador, A. E. Kaifer and M. D. García, *Angew. Chem., Int. Ed.*, 2018, **58**, 403–416.
9. S. J. Barrow, S. Kasera, M. J. Rowland, J. del Barrio and O. A. Scherman, *Chem. Rev.*, 2015, **115**, 12320–12406.
10. E. Masson, X. Ling, R. Joseph, L. Kyeremeh-Mensah and X. Lu, *RSC Adv.*, 2012, **2**, 1213–1247.
11. L. Isaacs, *Acc. Chem. Res.*, 2014, **47**, 2052–2062.
12. A. E. Kaifer, *Acc. Chem. Res.*, 2014, **47**, 2160–2167.
13. K. Kim, W. S. Jeon, J. K. Kang, J. W. Lee, S. Y. Jon, T. Kim and K. Kim, *Angew. Chem., Int. Ed.*, 2003, **42**, 2293–2296.
14. S. Angelos, N. M. Khashab, Y. W. Yang, A. Trabolsi, H. A. Khatib, J. F. Stoddart and J. I. Zink, *J. Am. Chem. Soc.*, 2009, **131**, 12912–12914.
15. G. Celtek, M. Artar, O. A. Scherman and D. Tuncel, *Chem. – Eur. J.*, 2009, **15**, 10360–10363.
16. D. Tuncel, N. Cindir and Ü. Koldemir, *J. Inclusion Phenom. Macrocycl. Chem.*, 2006, **55**, 373–380.

17. E. Masson, X. Lu, X. Ling and D. L. Patched, *Org. Lett.*, 2009, **11**, 3798–3801.

18. P. Branná, M. Rouchal, Z. Prucková, L. Dastychová, R. Lenobel, T. Pospíšil, K. Maláč and R. Vícha, *Chem. - Eur. J.*, 2015, **21**, 11712–11718.

19. V. Ramalingam and A. R. Urbach, *Org. Lett.*, 2011, **13**, 4898–4901.

20. W. Jiang, Q. Wang, I. Linder, F. Klautzsch and C. A. Schalley, *Chem. - Eur. J.*, 2011, **17**, 2344–2348.

21. Z. J. Zhang, H. Y. Zhang, L. Chen and Y. Liu, *J. Org. Chem.*, 2011, **76**, 8270–8276.

22. C. P. Carvalho, Z. Domínguez, C. Domínguez, H. S. El-Sheshtawy, J. P. Da Silva, J. F. Arteaga and U. Pischel, *ChemistryOpen*, 2017, **6**, 288–294.

23. L. Cera and C. A. Schalley, *Chem. Sci.*, 2014, **5**, 2560–2567.

24. W. L. Mock, T. A. Irra, J. P. Wepsiec and T. L. Manimaran, *J. Org. Chem.*, 1983, **48**, 3619–3620.

25. L. Zheng, S. Sonzini, M. Ambarwati, E. Rosta, O. A. Scherman and A. Herrmann, *Angew. Chem., Int. Ed.*, 2015, **54**, 13007–13011.

26. B. C. Pemberton, R. K. Singh, A. C. Johnson, S. Jockusch, J. P. Da Silva, A. Ugrinov, N. J. Turro, D. K. Srivastava and J. Sivaguru, *Chem. Commun.*, 2011, **47**, 6323–6325.

27. D. Tuncel, Ö. Ünal and M. Artar, *Isr. J. Chem.*, 2011, **51**, 525–532.

28. C. Ke, R. a. Smaldone, T. Kikuchi, H. Li, A. P. Davis and J. F. Stoddart, *Angew. Chem., Int. Ed.*, 2013, **52**, 382–387.

29. X. Hou, C. Ke, C. J. Bruns, P. R. McGonigal, R. B. Pettman and J. F. Stoddart, *Nat. Commun.*, 2015, **6**, 6884.

30. J. A. Finbloom, K. Han, C. C. Slack, A. L. Furst and M. B. Francis, *J. Am. Chem. Soc.*, 2017, **139**, 9691–9697.

31. Y. Liu, H. Yang, Z. Wang and X. Zhang, *Chem. - Asian J.*, 2013, **8**, 1626–1632.

32. K. Kim, D. Kim, J. W. Lee, Y. H. Ko and K. Kim, *Chem. Commun.*, 2004, 848–849.

33. Y. Liu, Y. Yu, J. Gao, Z. Wang and X. Zhang, *Angew. Chem., Int. Ed.*, 2010, **49**, 6576–6579.

34. J. del Barrio, P. N. Horton, D. Lairez, G. O. Lloyd, C. Toprakcioglu and O. A. Scherman, *J. Am. Chem. Soc.*, 2013, **135**, 11760–11763.

35. R. Joseph, A. Nkrumah, R. J. Clark and E. Masson, *J. Am. Chem. Soc.*, 2014, **136**, 6602–6607.

36. W. L. Mock and J. Pierpont, *Chem. Commun.*, 1990, 1509–1511.

37. S. I. Jun, J. W. Lee, S. Sakamoto, K. Yamaguchi and K. Kim, *Tetrahedron Lett.*, 2000, **41**, 471–475.

38. J. Wook Lee, K. Kim and K. Kim, *Chem. Commun.*, 2001, 1042–1043.

39. J. W. Lee, S. W. Choi, Y. H. Ko, S. Y. Kim and K. Kim, *Bull. Korean Chem. Soc.*, 2002, **23**, 1347–1350.

40. D. Tuncel, Ö. Özsar, H. B. Tiftik and B. Salih, *Chem. Commun.*, 2007, 1369–1371.

41. D. Tuncel and M. Katterle, *Chem. - Eur. J.*, 2008, **14**, 4110–4116.

42. M. K. Sinha, O. Reany, M. Yefet, M. Botoshansky and E. Keinan, *Chem. – Eur. J.*, 2012, **18**, 5589–5605.
43. T. Ooya, D. Inoue, H. S. Choi, Y. Kobayashi, S. Loethen, D. H. Thompson, Y. H. Ko, K. Kim and N. Yui, *Org. Lett.*, 2006, **8**, 3159–3162.
44. H. Zhang, Q. Wang, M. Liu, X. Ma and H. Tian, *Org. Lett.*, 2009, **11**, 3234–3237.
45. V. Sindelar, S. Silvi and A. E. Kaifer, *Chem. Commun.*, 2006, 2185–2187.
46. V. Sindelar, S. Silvi, S. E. Parker, D. Sobransingh and A. E. Kaifer, *Adv. Funct. Mater.*, 2007, **17**, 694–701.
47. V. Kolman, P. Kulhanek and V. Sindelar, *Chem. – Asian J.*, 2010, **5**, 2386–2392.
48. A. E. Kaifer, W. Li, S. Silvi and V. Sindelar, *Chem. Commun.*, 2012, **48**, 6693–6695.
49. O. Buyukcakir, F. T. Yasar, O. A. Bozdemir, B. Icli and E. U. Akkaya, *Org. Lett.*, 2013, **15**, 1012–1015.
50. Y. Gong, H. Chen, X. Ma and H. Tian, *ChemPhysChem*, 2016, **17**, 1934–1938.
51. A. M. Diniz, N. Basílio, H. Cruz, F. Pina and A. J. Parola, *Faraday Discuss.*, 2015, **185**, 361–379.
52. H.-J. Kim, W. S. Jeon, Y. H. Ko and K. Kim, *Proc. Natl. Acad. Sci. U. S. A.*, 2002, **99**, 5007–5011.
53. H.-J. Kim, J. Heo, W. S. Jeon, E. Lee, J. Kim, S. Sakamoto, K. Yamaguchi and K. Kim, *Angew. Chem., Int. Ed.*, 2001, **40**, 1526–1529.
54. W. S. Jeon, H.-J. Kim, C. Lee and K. Kim, *Chem. Commun.*, 2002, 1828–1829.
55. W. S. Jeon, A. Y. Ziganshina, J. W. Lee, Y. H. Ko, J.-K. Kang, C. Lee and K. Kim, *Angew. Chem., Int. Ed.*, 2003, **42**, 4097–4100.
56. J. W. Lee, I. Hwang, W. S. Jeon, Y. H. Ko, S. Sakamoto, K. Yamaguchi and O. Kim, *Chem. – Asian J.*, 2008, **3**, 1277–1283.
57. A. Trabolsi, M. Hmadeh, N. M. Khashab, D. C. Friedman, M. E. Belowich, N. Humbert, M. Elhabiri, H. A. Khatib, A.-M. Albrecht-Gary and J. F. Stoddart, *New J. Chem.*, 2009, **33**, 254–263.
58. W. S. Jeon, E. Kim, Y. H. Ko, I. Hwang, J. W. Lee, S.-Y. Kim, H.-J. Kim and K. Kim, *Angew. Chem., Int. Ed.*, 2005, **44**, 87–91.
59. Y. H. Ko, I. Hwang, H. Kim, Y. Kim and K. Kim, *Chem. – Asian J.*, 2015, **10**, 154–159.
60. K. Moon, J. Grindstaff, D. Sobransingh and A. E. Kaifer, *Angew. Chem., Int. Ed.*, 2004, **43**, 5496–5499.
61. W. Wang and A. E. Kaifer, *Angew. Chem., Int. Ed.*, 2006, **45**, 7042–7046.
62. D. Sobransingh and A. E. Kaifer, *Org. Lett.*, 2006, **8**, 3247–3250.
63. H. Shi, W.-Q. Sun, R.-L. Lin, C.-H. Liu and J.-X. Liu, *ACS Omega*, 2017, **2**, 4575–4580.
64. A. Y. Ziganshina, Y. H. Ko, W. S. Jeon and K. Kim, *Chem. Commun.*, 2004, 806–807.
65. I. Hwang, A. Y. Ziganshina, Y. H. Ko, G. Yun and K. Kim, *Chem. Commun.*, 2009, 416–418.

66. C. Yang, H. K. Young, N. Selvapalam, Y. Origane, T. Mori, T. Wada, K. Kim and Y. Inoue, *Org. Lett.*, 2007, **9**, 4789–4792.
67. Y. Liu, X.-Y. Li, H.-Y. Zhang, C.-J. Li and F. Ding, *J. Org. Chem.*, 2007, 72, 3640–3645.
68. Z. J. Ding, H. Y. Zhang, L. H. Wang, F. Ding and Y. Liu, *Org. Lett.*, 2011, **13**, 856–859.
69. L. H. Wang, Z. J. Zhang, H. Y. Zhang, H. L. Wu and Y. Liu, *Chin. Chem. Lett.*, 2013, **24**, 949–952.
70. X. Ling and E. Masson, *Org. Lett.*, 2012, **14**, 4866–4869.
71. H. Yin, R. Rosas, D. Gigmes, O. Ouari, R. Wang, A. Kermagoret and D. Bardelang, *Org. Lett.*, 2018, **20**, 3187–3191.
72. S. Choi, S. H. Park, A. Y. Ziganshina, Y. H. Ko, J. W. Lee and K. Kim, *Chem. Commun.*, 2003, 2176.
73. J. Wu and L. Isaacs, *Chem. – Eur. J.*, 2009, **15**, 11675–11680.
74. Y. Kim, Y. H. Ko, M. Jung, N. Selvapalam and K. Kim, *Photochem. Photobiol. Sci.*, 2011, **10**, 1415–1419.
75. F. Tian, D. Jiao, F. Biedermann and O. A. Scherman, *Nat. Commun.*, 2012, **3**, 1207.
76. J. del Barrio, S. T. J. Ryan, P. G. Jambrina, E. Rosta and O. A. Scherman, *J. Am. Chem. Soc.*, 2016, **138**, 5745–5748.
77. H. Cheng, Y. Zhang, C. Xu and Y. Liu, *Sci. Rep.*, 2015, **4**, 4210.
78. L. Zhu, H. Yan, X. J. Wang and Y. Zhao, *J. Org. Chem.*, 2012, 77, 10168–10175.
79. M. Baroncini, C. Gao, V. Carboni, A. Credi, E. Previtera, M. Semeraro, M. Venturi and S. Silvi, *Chem. – Eur. J.*, 2014, **20**, 10737–10744.
80. Z. Miskolczy and L. Biczók, *J. Phys. Chem. B*, 2011, **115**, 12577–12583.
81. Z. Miskolczy and L. Biczók, *Photochem. Photobiol.*, 2012, **88**, 1461–1466.
82. J. R. Nilsson, C. Parente Carvalho, S. Li, J. P. Da Silva, J. Andréasson and U. Pischel, *ChemPhysChem*, 2012, **13**, 3691–3699.
83. F. Pina, M. J. Melo, C. A. T. Laia, A. J. Parola and J. C. Lima, *Chem. Soc. Rev.*, 2012, **41**, 869–908.
84. N. Basílio and F. Pina, *ChemPhysChem*, 2014, **15**, 2295–2302.
85. N. Basílio, L. Cabrita and F. Pina, *J. Agric. Food Chem.*, 2015, **63**, 7624–7629.
86. N. Basílio, V. Petrov and F. Pina, *ChemPlusChem*, 2015, **80**, 1779–1785.
87. N. Basílio, C. A. T. Laia and F. Pina, *J. Phys. Chem. B*, 2015, **119**, 2749–2757.
88. N. Basílio and U. Pischel, *Chem. – Eur. J.*, 2016, **22**, 15208–15211.
89. M. A. Romero, N. Basílio, A. J. Moro, M. Domingues, J. A. González-Delgado, J. F. Arteaga and U. Pischel, *Chem. – Eur. J.*, 2017, **23**, 13105–13111.
90. C. P. Carvalho, V. D. Uzunova, J. P. Da Silva, W. M. Nau and U. Pischel, *Chem. Commun.*, 2011, **47**, 8793–8795.
91. J. Vázquez, M. A. Romero, R. N. Dsouza and U. Pischel, *Chem. Commun.*, 2016, **52**, 6245–6248.

92. S. Sun, W. Gao, F. Liu, J. Fan and X. Peng, *J. Mater. Chem.*, 2010, **20**, 5888–5892.
93. T. K. Monhaphol, S. Andersson and L. Sun, *Chem. – Eur. J.*, 2011, **17**, 11604–11612.
94. E. Y. Chernikova, D. V. Berdnikova, Y. V. Fedorov, O. A. Fedorova, F. Maurel and G. Jonusauskas, *Phys. Chem. Chem. Phys.*, 2017, **19**, 25834–25839.
95. A. M. Brouwer, C. Frochot, F. G. Gatti, D. A. Leigh, L. Mottier, F. Paolucci, S. Roffia and G. W. H. Wurpel, *Science*, 2001, **291**, 2124–2128.
96. V. Balzani, M. Clemente-Leon, A. Credi, B. Ferrer, M. Venturi, A. H. Flood and J. F. Stoddart, *Proc. Natl. Acad. Sci. U. S. A.*, 2006, **103**, 1178–1183.
97. A. Zubillaga, P. Ferreira, A. J. Parola, S. Gago and N. Basílio, *Chem. Commun.*, 2018, **54**, 2743–2746.
98. S. Angelos, Y. W. Yang, K. Patel, J. F. Stoddart and J. I. Zink, *Angew. Chem., Int. Ed.*, 2008, **47**, 2222–2226.
99. S. Angelos, Y. W. Yang, N. M. Khashab, J. F. Stoddart and J. I. Zink, *J. Am. Chem. Soc.*, 2009, **131**, 11344–11346.
100. C. R. Thomas, D. P. Ferris, J. Lee, E. Choi, M. H. Cho, E. S. Kim, J. F. Stoddart, J. Shin, J. Cheon and J. I. Zink, *J. Am. Chem. Soc.*, 2010, **132**, 10623–10625.
101. J. Croissant and J. I. Zink, *J. Am. Chem. Soc.*, 2012, **134**, 7628–7631.
102. D. T. Marquez and J. C. Scaiano, *Langmuir*, 2016, **32**, 13764–13770.
103. J. Liu, X. Du and X. Zhang, *Chem. – Eur. J.*, 2011, **17**, 810–815.
104. M. W. Ambrogio, T. A. Pecorelli, K. Patel, N. M. Khashab, A. Trabolsi, H. A. Khatib, Y. Y. Botros, J. I. Zink and J. F. Stoddart, *Org. Lett.*, 2010, **12**, 3304–3307.
105. Y.-L. Sun, B.-J. Yang, S. X.-A. Zhang and Y.-W. Yang, *Chem. – Eur. J.*, 2012, **18**, 9212–9216.
106. N. Ma, W.-J. Wang, S. Chen, X.-S. Wang, X.-Q. Wang, S.-B. Wang, J.-Y. Zhu, S.-X. Cheng and X.-Z. Zhang, *Chem. Commun.*, 2015, **51**, 12970–12973.
107. C. Hu, K. R. West and O. A. Scherman, *Nanoscale*, 2016, **8**, 7840–7844.
108. T. Chen, N. Yang and J. Fu, *Chem. Commun.*, 2013, **49**, 6555–6557.
109. M. Wang, T. Chen, C. Ding and J. Fu, *Chem. Commun.*, 2014, **50**, 5068–5071.
110. C. Kim, S. S. Agasti, Z. Zhu, L. Isaacs and V. M. Rotello, *Nat. Chem.*, 2010, **2**, 962–966.
111. F. Tian, M. Cziferszky, D. Jiao, K. Wahlström, J. Geng and O. A. Scherman, *Langmuir*, 2011, **27**, 1387–1390.
112. C. Hu, Y. Lan, F. Tian, K. R. West and O. A. Scherman, *Langmuir*, 2014, **30**, 10926–10932.
113. X. Ren, Y. Wu, D. E. Clarke, J. Liu, G. Wu and O. A. Scherman, *Chem. – Asian J.*, 2016, **11**, 2382–2386.
114. M. Wiemann and P. Jonkheijm, *Isr. J. Chem.*, 2018, **58**, 314–325.

115. Q. An, J. Brinkmann, J. Huskens, S. Krabbenborg, J. de Boer and P. Jonkheijm, *Angew. Chem., Int. Ed.*, 2012, **51**, 12233–12237.
116. N. L. Weineisen, C. A. Hommersom, J. Voskuhl, S. Sankaran, A. M. A. Depauw, N. Katsonis, P. Jonkheijm and J. J. L. M. Cornelissen, *Chem. Commun.*, 2017, **53**, 1896–1899.
117. S. Sankaran, J. Van Weerd, J. Voskuhl, M. Karperien and P. Jonkheijm, *Small*, 2015, **11**, 6187–6196.
118. U. Pischel, V. D. Uzunova, P. Remón and W. M. Nau, *Chem. Commun.*, 2010, **46**, 2635–2637.
119. C. P. Carvalho, Z. Domínguez, J. P. Da Silva and U. Pischel, *Chem. Commun.*, 2015, **51**, 2698–2701.
120. M. A. Romero, R. J. Fernandes, A. J. Moro, N. Basílio and U. Pischel, *Chem. Commun.*, 2018, **54**, 13335–13338.
121. H. D. Nguyen, D. T. Dang, J. L. J. van Dongen and L. Brunsveld, *Angew. Chem., Int. Ed.*, 2010, **49**, 895–898.

CHAPTER 5

# Hybrid Supramolecular Assemblies of Cucurbit[n]uril-supported Metal and Other Inorganic Nanoparticles

MHEJABEEN SAYED, SHARMISTHA DUTTA CHOUDHURY AND HARIDAS PAL*

Radiation & Photochemistry Division, Bhabha Atomic Research Centre, Trombay, Mumbai 400085, India
*Email: hpal@barc.gov.in

## 5.1 Introduction

Broadly, the subject domain of supramolecular chemistry is understanding the unique properties and exploring the applications of recognition-guided assemblies of molecular units formed through noncovalent interactions like hydrophobic and hydrophilic interactions, hydrogen bonding, π–π stacking, ion–dipole/dipole–dipole interactions, van der Waals forces, CH–π interactions *etc.* In biology, supramolecular interactions play a dominant role in sustaining life in nature. Extensive studies on artificial supramolecular systems have explored their diverse applications in areas like sensing, drug stabilization, drug delivery, pharmaceuticals, molecular devices, among others.[1-21] In this regard, uses of macrocyclic molecules as receptors (hosts) for different guest dyes and drugs have attracted immense research interest.[6-16,20-25] Most macrocycles have nonpolar cavities suitable to encapsulate

Smart Materials No. 36
Cucurbituril-based Functional Materials
Edited by Dönüs Tuncel
© The Royal Society of Chemistry 2020
Published by the Royal Society of Chemistry, www.rsc.org

hydrophobic molecules, forming host–guest inclusion complexes that can display exotic properties and applications. Although formation constants ($K_f$) of the host–guest complexes formed solely through hydrophobic interactions may not be very high, those involving other noncovalent interactions besides hydrophobic interactions can often have very high $K_f$ values, indicating their large stabilities.[6–16,20–25]

Among different macrocyclic receptors, cucurbit[$n$]uril (CB[$n$]) homologues have attracted considerable interest for about the last one and half decades, due to their ability to form strong host–guest complexes and their satisfactory biocompatibility.[3,6,10–16,23–29] CB[$n$] molecules are cyclic oligomers of glycoluril units joined by pairs of methylene bridges, and common CB[$n$] homologues are CB[5], CB[6], CB[7], CB[8] and CB[10], possessing 5, 6, 7, 8 and 10 glycoluril units, respectively. They are highly symmetric pumpkin-shaped molecules possessing a hydrophobic cavity with two identical portals latched with highly polarizable carbonyl groups.[3,6,10–16,23–32] CB[7] and CB[8] are recognized as the most useful CB[$n$] homologues, as they have adequate cavity sizes to encapsulate many guest molecules. While the relatively smaller CB[7] cavity mostly supports 1:1 inclusion complex formation, the larger CB[8] cavity often leads to 2:1 (guest-to-host) inclusion complex formation encapsulating two guest molecules simultaneously.

The $K_f$ values for the host–guest complexes involving CB[7] and CB[8] are usually quite high, typically in the range of $10^5$–$10^6$ M$^{-1}$. This is mainly due to rigid cavity and highly polarizable portals having large negative charge density that support additional ion–dipole, dipole–dipole and hydrogen bonding interactions for most guest molecules.[3,6,10–16,23–34] With CB[7] and CB[8], in some specific cases involving cationic guests, the $K_f$ values can even go as high as $10^{12}$–$10^{15}$ M$^{-1}$.[3,6,33,34] Important cavity dimensions, along with a schematic of a CB[$n$] cavity and the typical range of the $K_f$ values involving CB[7] and CB[8] hosts, are provided in Figure 5.1.

The CB[$n$] macrocycles are very useful capping agents for various nanomaterials, having potential applications in areas like nanocatalysis, nanomedicines, sensing, plasmonics *etc.* In the following sections, we summarize some of these CB[$n$]-assisted nanomaterials, discussing their structural motifs and application prospects. Considering the scope of the present chapter, however, we restrict our discussion mainly to metal nanoparticles and a few important inorganic nanomaterials. For other kinds of CB[$n$]-assisted nanomaterials, readers are referred to the published review articles that discuss different such systems with diverse perspectives.[10,11,19,35–42]

## 5.2 Cucurbit[$n$]uril-assisted Metal Nanoparticle Assemblies

Noble metal nanoparticles (NPs) have many applications in catalysis, sensing, theranostics, plasmonics and so on.[43–53] Since bare metal NPs undergo quick agglomeration, they have to be stabilized by using suitable capping

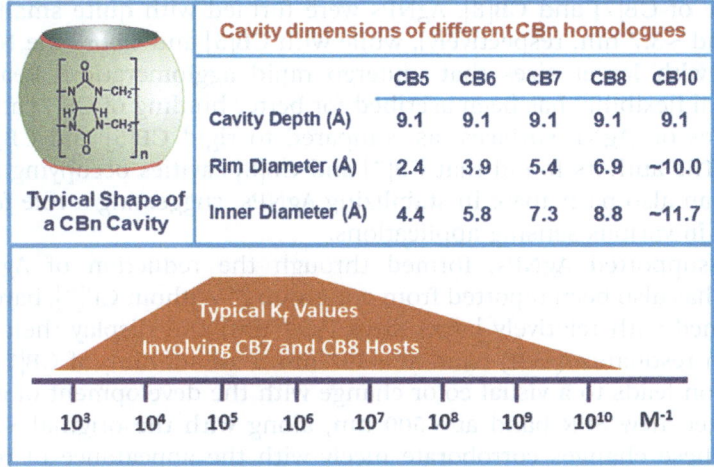

| Cavity dimensions of different CBn homologues | | | | | |
|---|---|---|---|---|---|
| | CB5 | CB6 | CB7 | CB8 | CB10 |
| Cavity Depth (Å) | 9.1 | 9.1 | 9.1 | 9.1 | 9.1 |
| Rim Diameter (Å) | 2.4 | 3.9 | 5.4 | 6.9 | ~10.0 |
| Inner Diameter (Å) | 4.4 | 5.8 | 7.3 | 8.8 | ~11.7 |

Typical Shape of a CBn Cavity

Typical $K_f$ Values Involving CB7 and CB8 Hosts

$10^3$　$10^4$　$10^5$　$10^6$　$10^7$　$10^8$　$10^9$　$10^{10}$　$M^{-1}$

**Figure 5.1**　Schematic of a CB[n] cavity, the list of important cavity dimensions of CB[n] homologues and typical range for $K_f$ values involving CB[7] and CB[8] hosts are shown for quick visualizations.

agents to maintain their suspension in water for higher catalytic activity.[53–57] Preferred capping agents are those that do not bind very strongly with NPs so that the NP surfaces can be easily accessed by the desired reacting species. CB[n] homologues are soft capping agents for metal NPs. Due to the presence of the negative charge density at the portals, CB[n] molecules can bind noncovalently on NP surfaces, providing quite reversible CB[n]-assisted NP assemblies.[45,53,54,58,59] In the following section, we discuss some of the CB[n]-assisted metal NP assemblies, with special emphasis on silver and gold NPs.

### 5.2.1　Supramolecular Assemblies of Silver Nanoparticles Supported by Cucurbit[n]uril

Studies on silver NPs (AgNP) have attracted substantial research interests due to their antibacterial activity and potential biomedical applications.[43–45] A facile, green, one-pot synthesis of well dispersed AgNPs, protected by CB[7], has been reported by Premkumar *et al.*,[45] treating a mixture of AgNO₃ and CB[7] with NaOH, without using any conventional reducing agent. The authors propose that initially formed Ag atoms are first entrapped in the CB[7] cavity and are subsequently reorganized to form relatively small (~5 nm) CB[7]-protected AgNPs. The synthesized AgNP-CB[7] system has been shown to display higher cytotoxicity than bare AgNPs toward human breast adenocarcinoma (MCF-7) and lung bronchoalveolar (NCIH358) cells, suggesting their potential therapeutic applications.

Reduction of AgNO₃ by NaBH₄ in the presence of CB[n] ($n = 5$–8) produces CB[n]-stabilized AgNPs, as reported by Lu and Masson.[59] Interestingly, in the presence of CB[7] and CB[8], AgNPs were formed with quite smaller sizes (~5.3 and ~3.7 nm, respectively), while with CB[5] and CB[6], the NPs were formed with larger sizes that undergo rapid agglomeration. Reasonable structural flexibility has been ascribed for better binding of CB[7] and CB[8] molecules on AgNP surfaces, as compared to rigid CB[5] and CB[6] molecules. The authors found that CB[7] and CB[8] cavities occupying suitable guests can also participate in stabilizing AgNPs, suggesting a role for these systems in various sensing applications.

CB[7]-supported AgNPs, formed through the reduction of $AgNO_3$ by $NaBH_4$, has also been reported from our group.[60] Without CB[7], bare AgNPs are formed with relatively larger sizes (~27 nm) and display their surface plasmon resonance (SPR) band at ~397 nm. Post-addition of CB[7] to this dispersion leads to a visual color change with the development of a largely red-shifted new SPR band at ~500 nm, along with the original ~397 nm band. These changes corroborate nicely with the appearance of relatively smaller AgNPs (~17 nm), suggesting the CB[7]-assisted disintegration of the initially formed larger AgNPs. When $AgNO_3$ was reduced by $NaBH_4$ directly in the presence of CB[7], very small AgNPs (~5 nm) were formed, displaying their SPR band at ~421 nm. As CB[7] portals have a strong cation receptor property,[33,34,61–63] it is proposed that the prelocalization of $Ag^+$ ions around CB[7] portals and the subsequent reduction of these ions lead to the formation of CB[7]-stabilized smaller AgNPs because CB[7] prevents larger particle growth. The SPR absorption spectra and schematics of the AgNP-CB[7] assemblies formed in different cases are shown in Figure 5.2A. Since CB[7] molecules on AgNP surfaces are also capable of recognizing guest molecules using their free portals, synthesized AgNP-CB[7] systems were tested for uptake and release of a model phototherapeutic dye, 5,10,15,20-tetrakis(4-*N*-methylpyridyl) porphyrin (TMPyP).[64–66] The unresolved emission bands of free TMPyP transform into two well resolved emission bands in the presence of AgNP-CB[7], suggesting uptake of TMPyP by AgNP-CB[7] assembly.[65,66] Upon the addition of 1-adamantylamine (ADA), a strong competitive binder for CB[7],[18,31,60,61] the bound TMPyP undergoes stimulus-responsive release, causing reversal of the emission band, as indicated in Figure 5.2B, demonstrating the prospects of such systems in drug delivery and therapeutics.

## 5.2.2 Supramolecular Assemblies of Gold Nanoparticles Supported by Cucurbit[*n*]uril

### 5.2.2.1 *Gold Nanoparticle Assemblies: General Aspects*

Extensive efforts have been made to synthesize gold nanoparticles (AuNPs) with desired sizes because their catalytic activity is largely dependent on particle size.[54,55,58,67–70] Corma *et al.*[58] synthesized CB[*n*]-assisted AuNPs using both the vapor deposition and the chemical reduction methods. Using CB[7], a large fraction of AuNPs were formed with very small size (~1 nm),

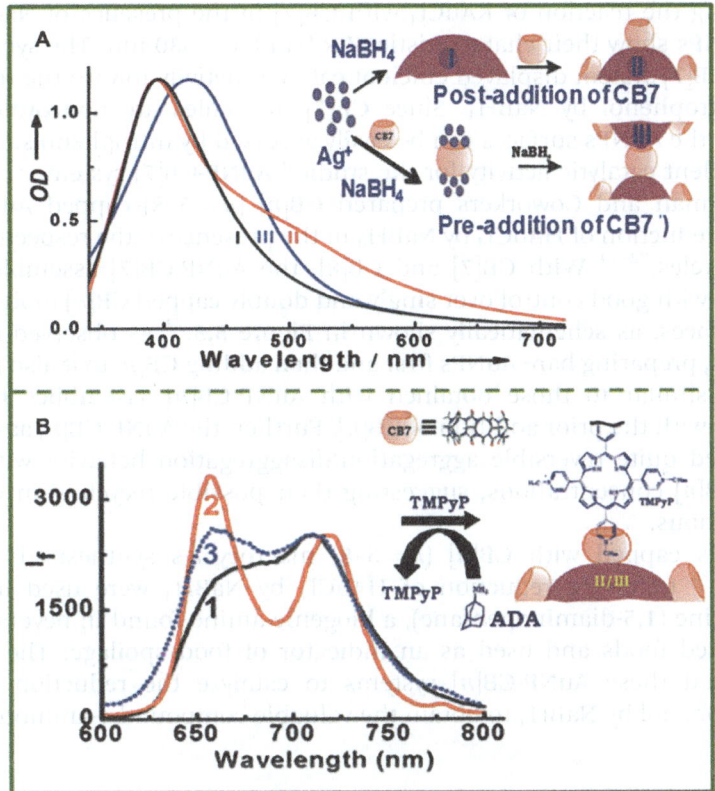

**Figure 5.2**    (A) Absorption spectra for AgNPs formed in the absence (I), under post-addition (II) and under pre-addition (III) of CB[7]. Inset: Schematics of CB[7] supported AgNP formations in different conditions. (B) Fluorescence spectra of TMPyP (~0.5 μM) in the absence (1), in the presence (2) of CB[7] stabilized AgNPs and on addition of ADA (3) to the previous solution. Inset: Schematics of uptake and ADA responsive release of TMPyP using AgNP-CB[7] system.
Reproduced from ref. 60 with permission from the Royal Society of Chemistry.

ascribed to those entrapped in the CB[7] cavities. Since the cavity volume of CB[7] is ~0.279 nm$^3$ and the volume of an Au atom is ~0.013 nm,$^3$ these AuNPs are apparently composed of only few tens of Au atoms. Along with these ultra small AuNPs, a fraction of AuNPs are also formed with sizes of ~4 nm, attributed to externally formed NPs capped by CB[7] molecules. Relatively larger (~6 to 7 nm) AuNPs were always formed on using CB[5] or CB[6], possibly due to their smaller cavity size and rigid structures, not supporting encapsulation of AuNPs inside their cavity and also not preventing the particle growth as efficiently as in the case of CB[7].[58,59]

One-pot synthesis of well dispersed CB[7]-supported AuNPs with relatively small sizes (10–15 nm) has been reported by Premkumar and Geckeler,[71] following the reaction of KAuCl$_4$ with CB[7] in the presence of NaOH, and the AuNPs show their characteristic SPR band at ~530 nm. The synthesized AuNP-CB[7] system displayed efficient catalytic activity toward the reduction of 4-nitrophenol by NaBH$_4$. Since CB[7] molecules are noncovalently attached, the AuNPs surfaces can be easily accessed by nitrophenols, resulting in excellent catalytic activity for the studied AuNP-CB[7] system.

Scherman and Coworkers prepared CB[n] (n = 5–8)–capped AuNPs following reduction of HAuCl$_4$ by NaBH$_4$ in the presence of the respective CB[n] macrocycles.[72–74] With CB[7] and CB[8], the AuNP-CB[7] assemblies were formed with good control over singly and doubly capped CB[n] molecules on NP surfaces, as schematically shown in Figure 5.3.[72] As observed by these authors, preparing bare AuNPs first and then adding CB[n] to it also provides results similar to those obtained with AuNP-CB[n] assemblies prepared directly with the prior addition of CB[n]. Further, the AuNP-CB[n] assemblies displayed quite reversible aggregation/disaggregation behavior with changing CB[n] concentrations, suggesting their possible recycling in different applications.

AuNPs capped with CB[n] (n = 5–8) macrocycles synthesized by Pozo *et al.*[75,76] following reduction of HAuCl$_4$ by NaBH$_4$ were used to detect cadaverine (1,5-diaminopentane), a biogenic amine found in beverages and fermented foods and used as an indicator of food spoilage. The authors also used these AuNP-CB[n] systems to catalyze the reduction of toxic 4-nitrophenol by NaBH$_4$ to obtain the valuable compound 4-aminophenol.[76]

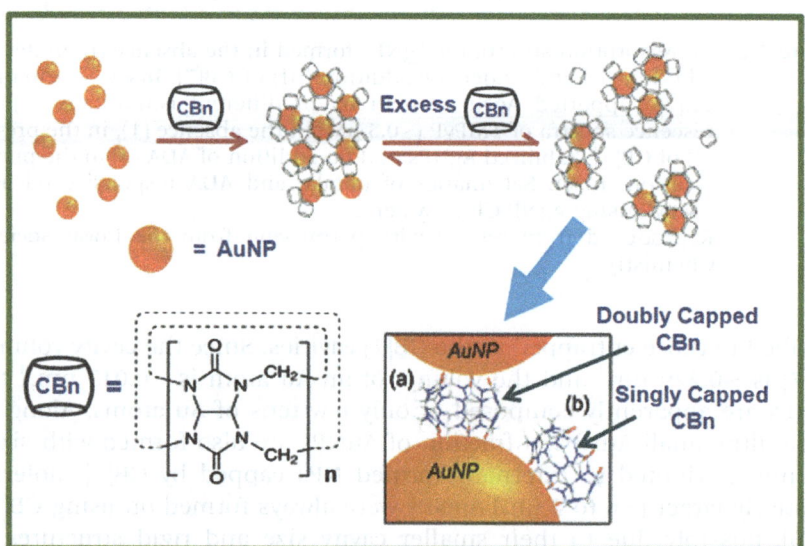

**Figure 5.3**   Schematics of CB[n] assisted AuNP formation with doubly capped (a) and singly capped (b) CB[n] molecules on NP surfaces.
Reproduced from ref. 72 with permission from the Royal Society of Chemistry.

In general, AuNP-CB[n] systems with higher CB[n] homologues were seen to display better catalytic effect than those with smaller CB[n] homologues, attributed to lower surface coverage of NPs by higher CB[n] molecules than lower ones, causing the NP surfaces to be more accessible in the former case than in the latter.

## 5.2.2.2 Cucurbit[n]uril-supported Gold Nanoparticles in Plasmonic Applications

Highly ordered metal NP clusters are important for their applications in device fabrications. Plasmonic coupling between closely placed metal NPs is a topic of extensive research because it can cause enormous enhancement in the optical fields at the gaps between NPs, creating "hot-spots" in the sub-wavelength regions.[36,77–85] Thus, plasmonic NPs arranged in orderly manners can have immense potential in controlling, localizing and enhancing optical signals, providing many new nanophotonic concepts and applications. The use of CB[n] macrocycles to prepare ordered metal NP clusters can offer a number of interesting features. They can act as both spacer and glue, adhering adjacent NPs with fixed interparticle separations of about 0.9 nm, the typical height of the CB[n] cavity. Further, since many of the CB[n] molecules on NP surfaces would have one of their portals free, they can act as a receptor for different guest molecules. Therefore, CB[n]-assisted metal NP clusters can become very useful in the ultrasensitive detection of various analytes using surface-enhanced Raman spectroscopy (SERS), a technique that relies on the localization of analytes at NP surfaces. In this respect, CB[n]-capped NPs are especially useful for those analytes that do not have independent interaction with NP but can be encapsulated by CB[n]s to hold them close to the NP surface, providing a great enhancement of their SERS signals, especially when the analytes are placed at the hot-spots.

Mahajan et al.[79] used dispersed AuNP-CB[n] ($n = 5$–8) systems for SERS studies and observed systematic shifts in the Raman peaks for the vibrational modes of CB[n] with the changing size of the macrocycles, attributed to the size-dependent changes in the steric factors and bond angle strains as the macrocycles bind to NP surfaces. Using these shifts, the authors could characterize and detect different CB[n] homologues with about 10 ppb sensitivity. Similar AuNP-CB[n] systems were also used by Taylor et al.[80] for SERS-based sensing of rhodamine 6G (Rh6G) dye. They observed that AuNP-CB[5] aggregates, especially, provide rigid, fixed and reproducible ~0.9 nm interparticle separations, giving strong and reproducible SERS signals for Rh6G, due to the holding of the dye at the hot-spots of the NP assembly. In the same vein, Tao et al.[81] utilized AuNP-CB[7] assemblies for SERS-based detection of an otherwise inert molecule, ferrocene. These authors adopted a three-step approach: first, preparing the substrate through assembly of AuNPs on amino-functionalized glass; next, forming ferrocene-CB[7] inclusion complexes $(K_f > 10^6 \text{ M}^{-1})$[28] independently and allowing them to get physisorbed on the prepared substrate; and finally, covering the ferrocene-CB[7]–adsorbed substrate with another layer of AuNPs. This procedure ensured that the CB[7]-encapsulated ferrocenes were placed at the hot-spots, as schematically shown in Figure 5.4A. The authors could thus achieve about

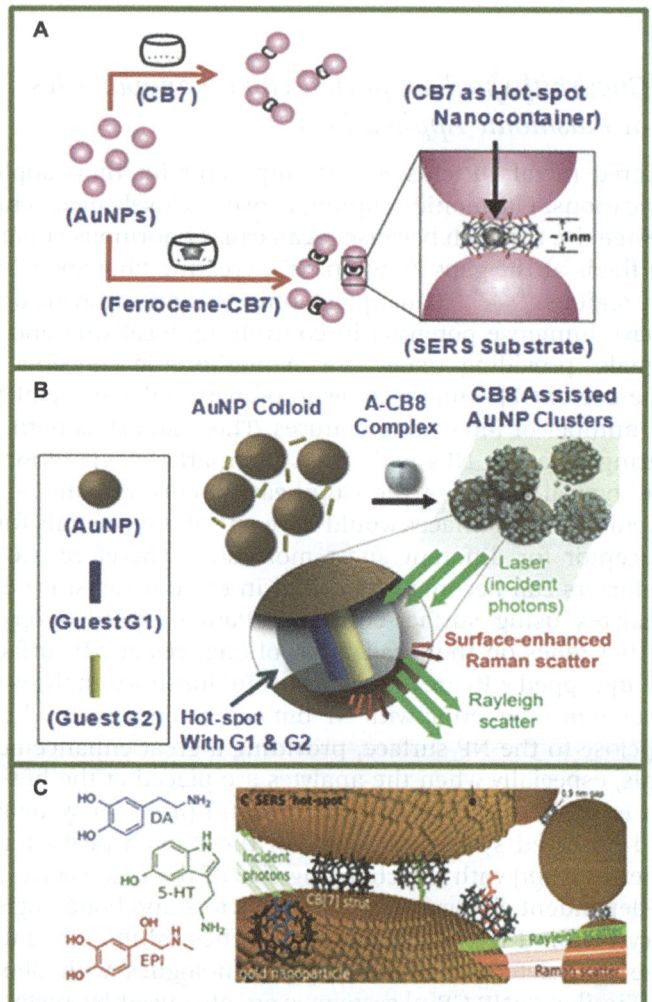

**Figure 5.4**  (A) Schematic of SERS-based ultra sensitive detection of ferrocene using AuNP-CB[7] assembly to hold the analyte at the hot-spot. Reproduced from ref. 81 with permission from the Royal Society of Chemistry. (B) Schematic of SERS-based detection of Guest 2 (G2) in the presence of Guest 1 (G1) using AuNP-CB[7] assembly to hold the G1-CB[8]-G2 ternary complex at the hot-spot. Reproduced from ref. 82 with permission from American Chemical Society, Copyright 2012. (C) Schematic of SERS-based simultaneous fingerprint detection of DA, EPI and 5HT using AuNP-CB[8] assembly to hold the analytes at the hot-spot. Reproduced from ref. 83, https://doi.org/10.1038/srep06785, under the terms of the CC BY 4.0 license, http://creativecommons.org/licenses/by/4.0/.

$10^9$-fold SERS enhancement for ferrocene. With a similar perspective, Kasera et al.[82] used AuNP-CB[8] clusters for their SERS-based detection of a number of analytes in the presence of a known guest, exploiting the ability of CB[8] to support 1:1:1 ternary complex formation involving two guests, G1 and G2. Authors used electron-deficient methyl viologen ($MV^{2+}$) as the known guest, G1, because $MV^{2+}$ shows very strong binding ability with CB[8] ($K_f = 8.5 \times 10^5$ $M^{-1}$). In the absence of G2, only 1:1 G1-CB[8] complexes are formed and placed at the hot-spots, evidenced from their characteristic SERS signals. Once the unknown electron-rich guest, G2, becomes available, both G1 and G2 are cooperatively entrapped into CB[8] cavity, placing the 1:1:1 ternary complex at the hot-spot and producing a very strong SERS signal for the sensitive detection of G2, as schematically shown in Figure 5.4B. Using this method, the authors could detect anthracene, perylene, 2-naphthol, naphthalenediols *etc.* with about 10 pM sensitivity. Since SERS is a technique involving molecular fingerprints, it has the ability to simultaneously detect multiple analytes in a mixture. CB[*n*] macrocycles can complement this ability by holding different analytes simultaneously at the hot-spots of AuNP-CB[*n*] assemblies, giving large SERS enhancements for all the analytes together. This has been demonstrated successfully by Scherman and co-workers[83] by carrying out SERS experiments using AuNP-CB[7] assemblies to detect simultaneously three monoamine neurotransmitters: dopamine (DA), epinephrine (EPI) and serotonin (5HT). The binding constants ($K_f$) of the three monoamines with CB[7] being quite similar ($10^4$–$10^5$ $M^{-1}$), in the presence of excess CB[7], their binding behaviors are not disturbed much even with their simultaneous presence in the solution. Accordingly, all the monoamines can be detected with high sensitivity using their fingerprints in the SERS signals, as schematically shown in Figure 5.4C. Recently, Scherman and coworkers[84] have also investigated CB[7] complexation with single-molecule sensitivity using a nanogap SERS geometry comprised of individual AuNPs on a planar Au surface spaced by a single layer of CB[7] molecules. Characteristic changes were observed in the Raman modes of the host when single guests were included inside its cavity, thus allowing the detection of "filled" and "unfilled" states of CB[7].

Due to their plasmonic properties, one-dimensional (1D) chains of metal NPs with well-defined interparticle separations have immense potential for the development of nanoscale connectors for miniaturized electric and optical circuits. Hüsken et al.[85] have presented an electrokinetic strategy to fabricate 1D AuNP chains by connecting NPs sequentially using CB[7] nanojunctions. By placing AuNP suspension and CB[7] solution in two compartments separated by a nanoporous polycarbonate (PC) membrane and applying a small voltage ($E < 2$ V) between the electrodes in the two compartments, the authors observed the formation of 1D AuNP chains, as characterized by TEM and extinction spectroscopy measurements. It is suggested that CB[7] molecules pass through the pores of the PC membrane into the AuNP compartment and sequentially get attached with the NPs forming the AuNP-CB[7] chains, as schematically shown in Figure 5.5A.

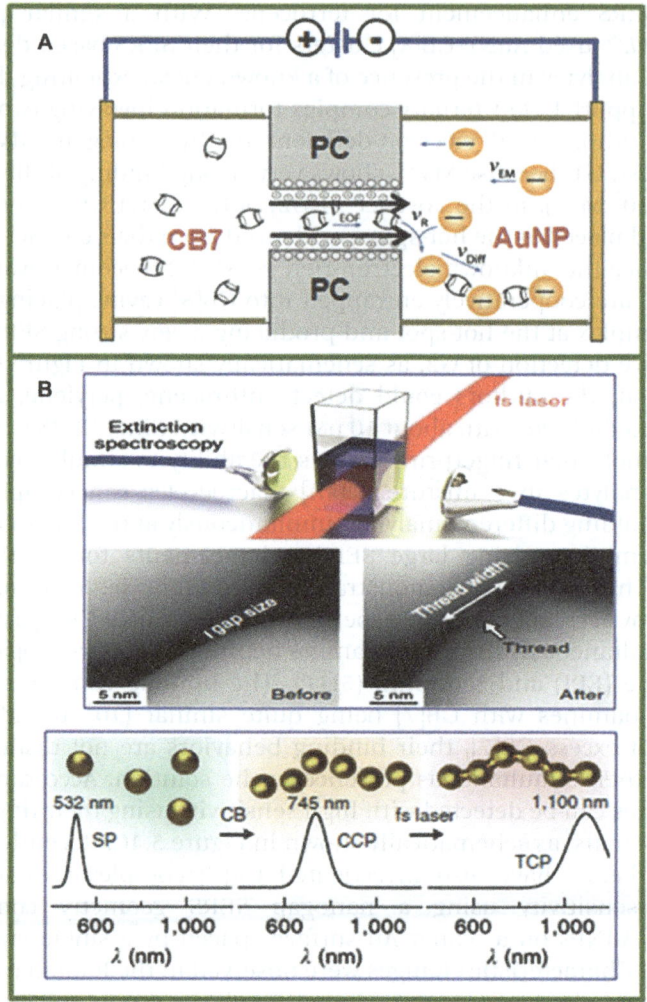

**Figure 5.5**  (A) Schematics of 1D AuNP-CB[7] chain formation through electrokinetic method. Reproduced from ref. 85, http://dx.doi.org/10.1021/nl403224q, under the terms of the CC BY 4.0 license, https://creativecommons.org/licenses/by/4.0. (B) Schematics of plasmon-induced laser threading of AuNPs supported by CB[$n$] and monitoring the process by extinction spectroscopy. Characteristic SPR bands for the isolated AuNPs (single particle, SP), the AuNP-CB[$n$] chains (capacitive chain plasmon, CCP) and the laser-threaded AuNP-chains (threaded chain plasmon, TCP) appear at ~532 nm, ~745 nm and ~1100 nm, respectively. Reproduced from ref. 86, https://doi.org/10.1038/ncomms5568, under the terms of the CC BY 4.0 license, https://creativecommons.org/licenses/by/4.0/.

The precise gap between adjacent NPs in the AuNP-CB[$n$] systems has also been utilized efficiently by Herrmann *et al.*[86] to demonstrate plasmon-induced laser threading to create near-identical, continuous strings of

AuNPs connected by conducting gold threads. In their experiment, AuNPs were first assembled into chains with precise interparticle separations of ~0.9 nm, maintained by CB[7] spacers. Exposing this system to intense femtosecond laser pulses generated an extremely high field at NP junctions, causing the surface Au atoms to diffuse into the gaps due to optical forces and nonthermal melting, forming metal threads connecting adjacent NPs, as shown schematically in Figure 5.5B. It was observed that the laser-induced threading of AuNPs can occur on a large scale in water, leading to a new SPR band in the near-IR region. It has been suggested that this method can be used to create entirely new structures like split rings, chiral strings, *etc.* by properly tailoring the light fields, thus opening up new prospects in photovoltaics, sensing and SERS applications.

Because of their anisotropic shape and the possible tuning of their aspect ratios, significant interest has been paid to gold nanorods (AuNRs) that can display tunable SPR bands over a wider spectral range than isotropic AuNPs. The end-to-end assembly of AuNRs can couple individual SPR bands of the nanorods, resulting in a huge shift in the absorption spectra, which can be controlled by modulating the size of the assembly, having potential applications in sensing and nanoelectronics.[87–89] The formation of end-to-end assembly of AuNRs using supramolecular interactions have been reported by Jones *et al.*[89] The authors first prepared AuNRs with their two ends functionalized with methyl viologen ($MV^{2+}$) moieties. These AuNRs were subsequently treated with poly-ethyleneoxide($(EO)_n$)-based telechelic molecules having naphthol (NP) units at both ends, [NP-$(EO)_n$-NP], in the presence of CB[8]. The end-to-end nanoassembly was thus formed through 1:1:1 complexation, encapsulating both $MV^{2+}$ and NP units simultaneously into the CB[8] cavity, as schematically shown in Figure 5.6. It was further observed that the nanoassembly can be disrupted at will by using ADA as competitive binder, making the system amenable to specific applications.

### 5.2.2.3 Cucurbit[n]uril-assisted Gold Nanoparticles in Health Care

In health care, enzyme-linked immunosorbent assays (ELISA) are important tools where enzyme-modified antibodies are used to detect analytes through enzymatic reactions. In the usual procedure, enzyme-modified antibodies are used in excess for selective binding to a solid surface prepared specifically with immobilized antigens. Allowing antigen-antibody binding to be completed, free enzyme-modified antibodies are removed by washing. Subsequently, the solid surface is treated with a solution containing the substrate whereby a color develops in the solution due to enzyme–substrate reaction, which is used to estimate the substrate concentration. Rica and Velders[90] demonstrated ELISA using AuNP-CB[7] assembly to detect protein,

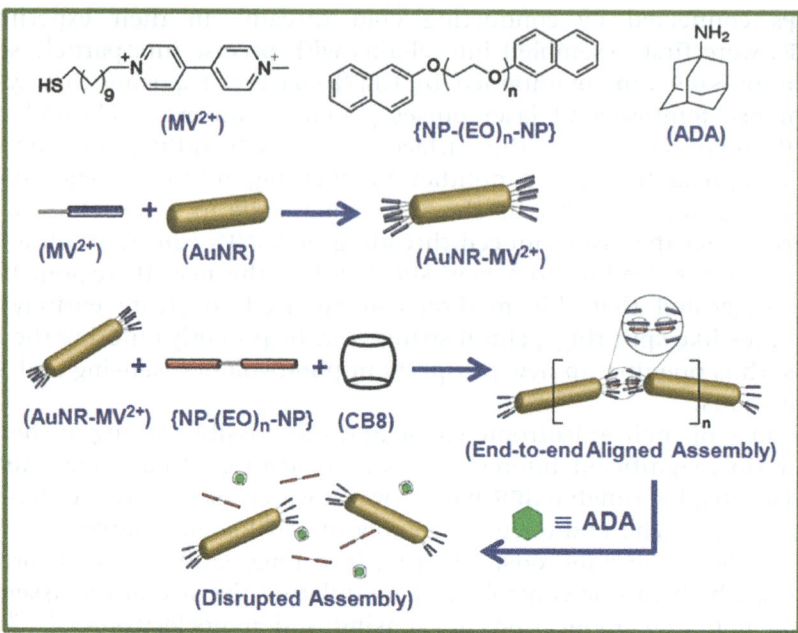

**Figure 5.6**   Schematics of CB[8]-assisted formation of end-to-end nanoassembly of AuNRs and their disintegration by a competitive binder ADA.
Reproduced from ref. 89 with permission from the Royal Society of Chemistry.

following changes in the SPR band due to the aggregation/deaggregation of AuNPs. In a suspension of AuNPs, the presence of CB[7] acts like a glue to aggregate the NPs, which leads to a significant red shift of the SPR band. With the cation receptor property of CB[7],[33,34,61–63] if an enzyme-substrate reaction produces a cationic product that can bind competitively with CB[7], then the nanoaggregates can be disintegrated, which is readily detected by a blue shift of the SPR band. Based on this strategy, the authors had detected mouse IgG protein using a urease-labeled antibody (anti–mouse IgG) as the probe. A solution containing IgG and an excess of urease-labeled antibody was taken such that, after binding of all the IgG, some antibody still remained free in the solution. Subsequently, IgG-bound magnetic beads were added to this solution, so that the free antibodies got bound to IgGs on these beads. The beads were then removed using a magnet, washed thoroughly and mixed with AuNP-CB[7] aggregates. Urea was subsequently added to this mixture, whereby $NH_4^+$ ions were produced through enzymatic reaction to disintegrate the AuNP-CB[7] aggregates. After the magnetic beads were taken out again, the absorption spectrum of the solution was recorded to quantify the enzymatic reaction and thus to estimate the antibody used up by IgG in the solution. The whole procedure is conceptually shown in Figure 5.7. The authors observed that the ratio of the integrated absorbances for 550–700 nm (representing AuNP-CB[7] aggregates) to 490–540 nm (representing

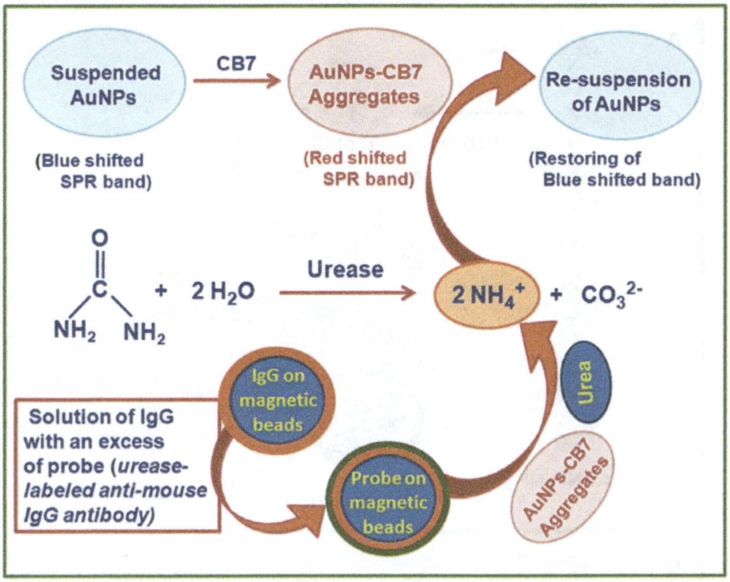

**Figure 5.7**  Schematic of the use of AuNP-CB[7] aggregates in the estimation of mouse IgG protein using urease-labeled antibody (anti–mouse IgG) as the probe for the ELISA method.
The figure is redrawn based on the concept discussed by Rica and Velders in ref. 90.

dispersed AuNPs) regions is very useful in constructing the calibration curve for estimating the IgG concentration in the solution.[90]

The engineering of host–guest systems to respond to suitable triggers inside cells can be utilized for controlled therapeutic applications. Kim *et al.*[91] designed a supramolecular nanohybrid system using 1,6-diamino-hexane (DAH)–terminated AuNP (AuNP-DAH) and CB[7] as complementary units that showed easy uptake by cells. While isolated, AuNP-DAH shows very high cytotoxicity, the AuNP-DAH-CB[7] nanohybrid shows much reduced cytotoxicity. Upon the addition of the competitive binder ADA, which binds stronger with CB[7] than DAH, the nanohybrid suffers the uncapping of CB[7], causing the regeneration of isolated AuNP-DAH and its cytotoxicity inside the cell, as schematically shown in Figure 5.8A.

For therapeutic applications, AuNPs with exotic shapes like nanostars, nanorods, nanoflowers *etc.* are also considered to be important nano-materials as they exhibit SPR bands in the near-infrared (NIR) region (700–1100 nm), which can be used to induce hyperthermia in living organisms by shining with NIR lasers.[92–94] In this regard, gold nanostars (AuNS) have drawn special interests because of their high drug-loading capacity and best photothermal efficiency. While bare AuNSs show very strong aggregation behavior, CB[n] capping resists their aggregation because CB[n] molecules act as rigid spacers to keep AuNSs apart. Quite uniform and stable

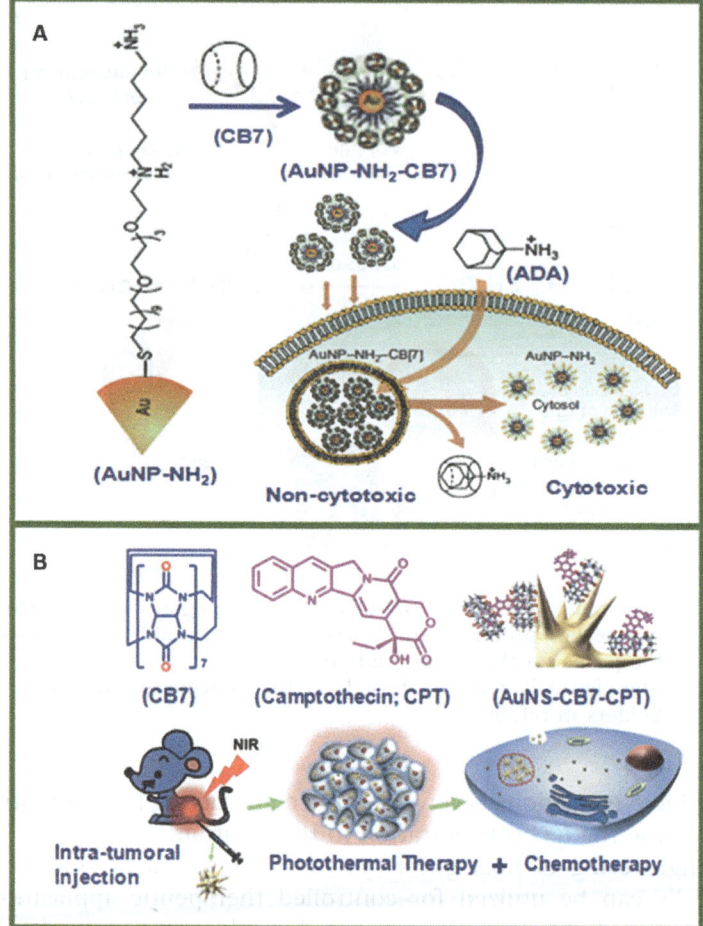

**Figure 5.8** (A) Schematic of the use of AuNP-DAH-CB[7] nanohybrid for controlled therapeutic effect inside cells. Reproduced from ref. 91 with permission from Springer Nature, Copyright 2010. (B) Schematic of the use of AuNS-CB[7] system for cargo loading and NIR light–triggered photothermal release of drug at the target sites for therapeutic application. Reproduced from ref. 94 with permission from American Chemical Society, Copyright 2018.

AuNS-CB[7] dispersion was successfully prepared by Han *et al.*,[93] showing strong NIR absorption and displaying high photothermal efficiency, suitable for therapeutic applications. In the same spirit, Xu *et al.*[94] prepared AuNS-CB[7] assemblies for cargo loading and the NIR light–triggered controlled release of the anticancer drug, camptothecin (CPT), in a tumor, displaying a synergistic cancer treatment through photothermal release of the drug at the target site and induced hyperthermia, to achieve the best chemotherapeutic action, as schematically shown in Figure 5.8B.

### 5.2.3 Cucurbit[*n*]uril-assisted Palladium and Platinum Nanoparticles for Catalytic Applications

A simple one-pot synthesis of palladium nanoparticles (PdNPs) stabilized by CB[6] was reported by Cao *et al.*,[95] treating aqueous $PdCl_2$ and CB[6] (1:1 molar ratio) with ethanolic solution of $NaBH_4$ under rapid addition and constant stirring. TEM microscopy established the formation of well dispersed, nearly spherical CB[6]-stabilized PdNPs with an average size of ∼3 nm. Prepared PdNP-CB[6] system showed extraordinary stability over six months both in suspension and in solid state. In their study, authors also investigated the catalytic activity of the prepared PdNP-CB[6] system toward Suzuki-Miyaura cross-coupling reactions involving different aryl halides and arylboronic acid (*cf.* Scheme 5.1), whereby coupling products were obtained with very good yields under mild reaction conditions. In a subsequent study, Cao *et al.*[96] also synthesized CB[6]-stabilized palladium nanocrystals (PdNCs), having abundant surface defects, which were able to catalyze C–H functionalization of electron-deficient polyfluoroarenes by a coupling reaction with aryl halides (*cf.* Scheme 5.2).

The catalytic activity of PdNC-CB[6] system was found to be quite insensitive to air and moisture and displayed good recyclability with significantly low catalyst loading. CB[6]-stabilized PdNPs were also synthesized by Karami and Naeini,[97] and their catalytic activity was studied for Suzuki reactions to couple various aryl halides with arylboronic acids, both at room temperature and at 40 °C, similarly to the ones reported by Cao *et al.*[95] The synthesis and characterization of uniformly dispersed PdNPs supported by CB[6] and its uses for various hydrogenation reactions have been reported recently by Nandi *et al.*[98] The Pd(II)-CB[6] complex was first prepared by dissolving $PdCl_2$ in DMF and charging it with a proportionate amount of CB[6]. The solvent was subsequently removed, and the dried Pd(II)-CB[6] complex was dispersed in water in an autoclave reactor and pressurized with

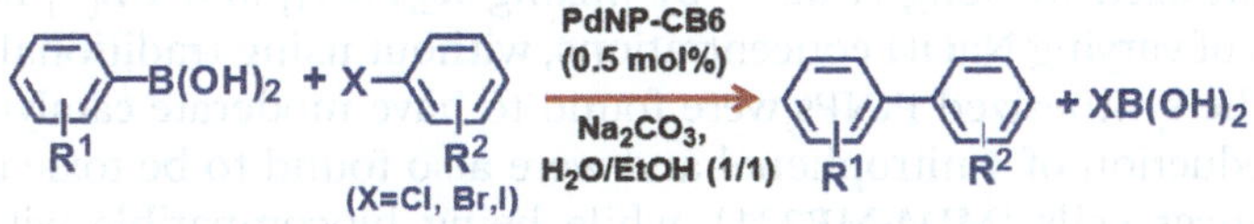

**Scheme 5.1**  General schematic of the Suzuki-Miyaura cross-coupling reactions between substituted aryl halides and arylboronic acids catalyzed by CB[6]-stabilized PdNPs.

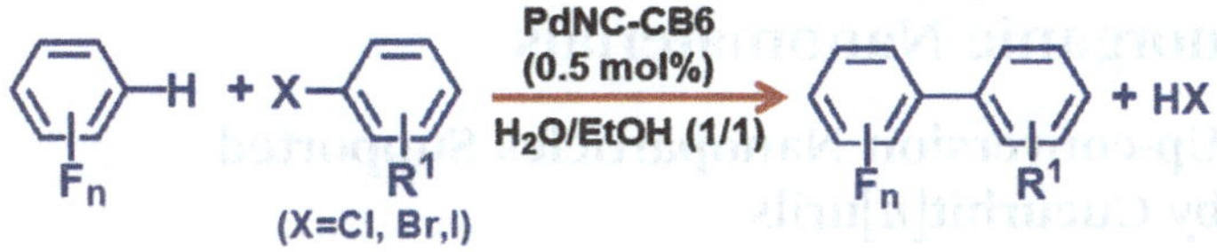

**Scheme 5.2**  The C–H functionalization of electron-deficient polyfluoroarenes by coupling with aryl halides, catalyzed efficiently by CB[6]-stabilized PdNCs.

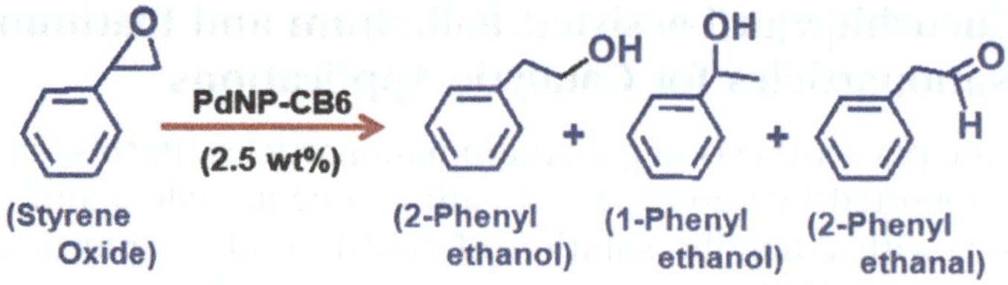

**Scheme 5.3**  Hydrogenation reaction of an epoxide, catalyzed by CB[6]-stabilized PdNPs, under high pressure of H$_2$ gas.

about 10 bar hydrogen; the reduction of Pd(II) to Pd(0) was allowed to occur under 1000 rpm stirring for about 3 h. The black-colored, CB[6]-capped PdNPs thus obtained were dried at 100 °C to obtain the final material. The PdNP-CB[6] catalyst was found to provide excellent conversion yields in the hydrogenation of various epoxides, olefins, carbonyls, nitriles, azo compounds, imines and nitroarenes (*cf.* Scheme 5.3) and could be recycled several times (>10 times) without losing its efficiency.

Platinum-based nanomaterials are widely investigated as electrocatalysts for methanol oxidation, important for developing methanol fuel cells as a green power source. In practice, CO that results from methanol de-composition acts as a strong poison for Pt catalyst deactivation, limiting its performance. Cao *et al.*[99] have reported the fabrication of sub-10-nm platinum nanoparticles (PtNPs) in the presence of CB[6] that can overcome this limi-tation. First, CB[6] and H$_2$PtCl$_6$·6H$_2$O were mixed in water (3 : 1 molar ratio) to obtain a light yellow mixture. This mixture was treated with different reducing agents like ascorbic acid and ethylene glycol in order to obtain well-defined and uniformly dispersed PtNPs of different morphologies. They observed that the CB[6]-PtNPs have enhanced electrocatalytic activity for methanol oxi-dation, better durability and unique poisoning tolerance toward CO, which is believed to be due to the interaction of CO with CB[6] that promotes CO oxidation and reduces its direct bonding with Pt. Size-controlled PtNPs were also synthesized by Song *et al.*[100] by mixing K$_2$[PtCl$_4$] and CB[7] in alkaline solutions of varying NaOH concentrations, without using traditional reducing agents. The synthesized PtNPs were found to have moderate catalytic activity for the reduction of 4-nitrophenol and were also found to be toxic to human breast cancer cells (MDA-MB231), while being biocompatible with normal human embryonic kidney cells (HEK-293).

## 5.3  Cucurbit[*n*]uril-supported Assemblies of Other Inorganic Nanomaterials

### 5.3.1  Up-conversion Nanoparticles Supported by Cucurbit[*n*]urils

Lanthanide-doped up-converting nanoparticles (UCNPs) with constitution as β-NaYF$_4$:Yb,X/NaYF$_4$ (X = Er,Tm) that convert NIR light to visible wavelengths have emerged as a new class of optical probes that overcome many limitations of conventional fluorescent dyes.[101–105] Functionalization of the hydrophobic surfaces of UCNPs and their bioconjugation, however, remains quite challenging. In this regard, CB[n] macrocycles have played an important role for improving the properties of UCNPs. Very recently, Sun *et al.*[101] have shown that CB[7] can undergo efficient ligand exchange with oleic acid (OA) molecules, usually present on the surface of conventionally synthesized UCNPs, to form water-soluble CB[7]-coated UCNPs. The CB[7]-UCNPs additionally provide a number of macrocyclic cavities for binding of biomolecules, as demonstrated using an adamantyl conjugated IgG protein (IgG-ADA), where the ADA end of the conjugate binds strongly with CB[7], as schematically shown in Figure 5.9A. The cytotoxicity of

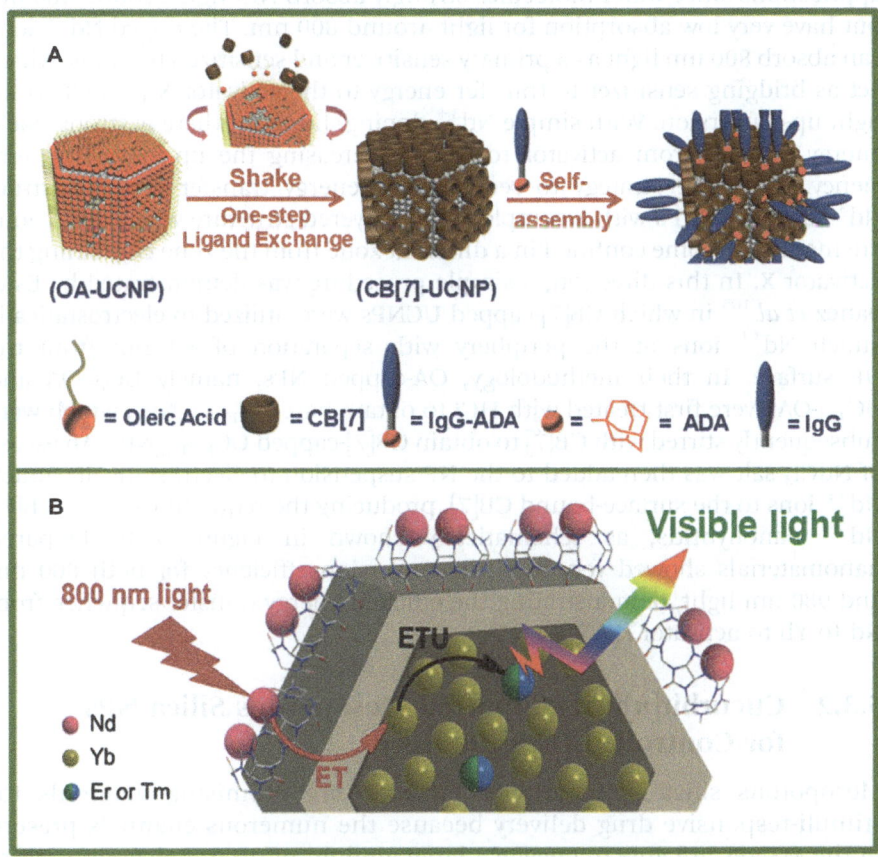

**Figure 5.9** (A) Schematic of ligand exchange on UCNPs with CB[7] and subsequent bioconjugation. Reproduced from ref. 101 with permission from the Royal Society of Chemistry. (B) Schematic of the $UC_{Er(Tm)}$NPs covered with $Nd^{3+}$ ions attached to surface-bound CB[7] molecules. Reproduced from ref. 105 with permission from the Royal Society of Chemistry.

CB[$n$]-UCNP@IgG-ADA was tested by MTT assay using HeLa cell line, and the NPs were found to be biocompatible. Immunofluorescence imaging of E-cadherin IgG bound CB[$n$]-UCNPs revealed that the prepared NPs were able to specifically target the E-cadherin transmembrane protein of the HeLa cells. In another study, CB[$n$]-capped UCNPs were found to form exclusion complexes with dyes like methylene blue and pyronin, *via* their free carbonyl portal, thus making it possible to load a high concentration of the dyes close to the UCNPs, consequently leading to efficient energy transfer from the UCNPs to the dyes, which was also useful for singlet oxygen generation with NIR excitation.[102]

Over the past decade, UCNPs with doped $Nd^{3+}$ have been investigated to achieve up-conversion with 800 nm light, rather than with the usual 980 nm light, as obtained with $Yb^{3+}$ dopant. This is especially required for biological applications since water molecules strongly absorb NIR light around 980 nm but have very low absorption for light around 800 nm. The doped $Nd^{3+}$ ions can absorb 800 nm light as a primary sensitizer and sensitize $Yb^{3+}$ ions, which act as bridging sensitizer to transfer energy to the activator X ($=$ Er,Tm) for light up-conversion. With simple $Nd^{3+}$ doping, however, there is strong back-energy transfer from activator to $Nd^{3+}$, decreasing the up-conversion efficiency.[104,105] One strategy to reduce back energy transfer is to construct $Nd^{3+}$-doped UCNPs with a complex onion-layered structure where $Nd^{3+}$ ions are made to become confined in a different zone from the zone containing the activator X. In this direction, a simple procedure was demonstrated by Estebanez *et al.*[105] in which CB[7]-capped UCNPs were utilized to electrostatically attach $Nd^{3+}$ ions at the periphery with separation of ~1 nm from the NP surface. In their methodology, OA-capped NPs, namely $UC_{Er}$-OA and $UC_{Tm}$-OA, were first treated with HCl to obtain bare $UC_{Er(Tm)}$NPs, which were subsequently stirred with CB[7] to obtain CB[7]-capped $UC_{Er(Tm)}$NPs. An excess of $NdCl_3$ salt was then added to the NP suspension to electrostatically attach $Nd^{3+}$ ions to the surface-bound CB[7], producing the required $UC_{Er(Tm)}$-CB[7]-$Nd^{3+}$ nanohybrids, as schematically shown in Figure 5.9B. Prepared nanomaterials showed very high up-conversion efficiency for both 800 nm and 980 nm light, demonstrating the efficient energy transfer sequence from Nd to Yb to activator (Er/Tm).

## 5.3.2 Cucurbit[$n$]uril-supported Mesoporous Silica NPs for Controlled Drug Delivery

Mesoporous silica nanoparticles (MSNPs) are promising materials for stimuli-responsive drug delivery because the numerous channels present in the MSNPs are able to encage a large amount of drug molecules within, and their pore openings can be used for the controlled release of drugs.[106–108] However, like UCNPs, the MSNPs are also nondispersive in water, which limits their effective uses. Making MSNPs dispersed in physiological media for their biological stimuli-responsive uses has been a longtime major challenge. Liu *et al.*[108] have skillfully utilized the noncovalent interaction of CB[7] to prepare water-dispersive, biocompatible, $Fe_3O_4$-embedded magnetic MSNPs useful for the controlled binding and enzyme-responsive release of drugs. The surfaces of magnetic MSNPs were first tethered with 1,4-butanediamine (BD) stalks, while drug molecules (calcein) were entrapped in the pores of MSNPs. At neutral pH, since BD is protonated, CB[7] macrocycles get strongly bound with BD stalks and thus act as the lids for the pores of MSNPs, blocking drug release. In this

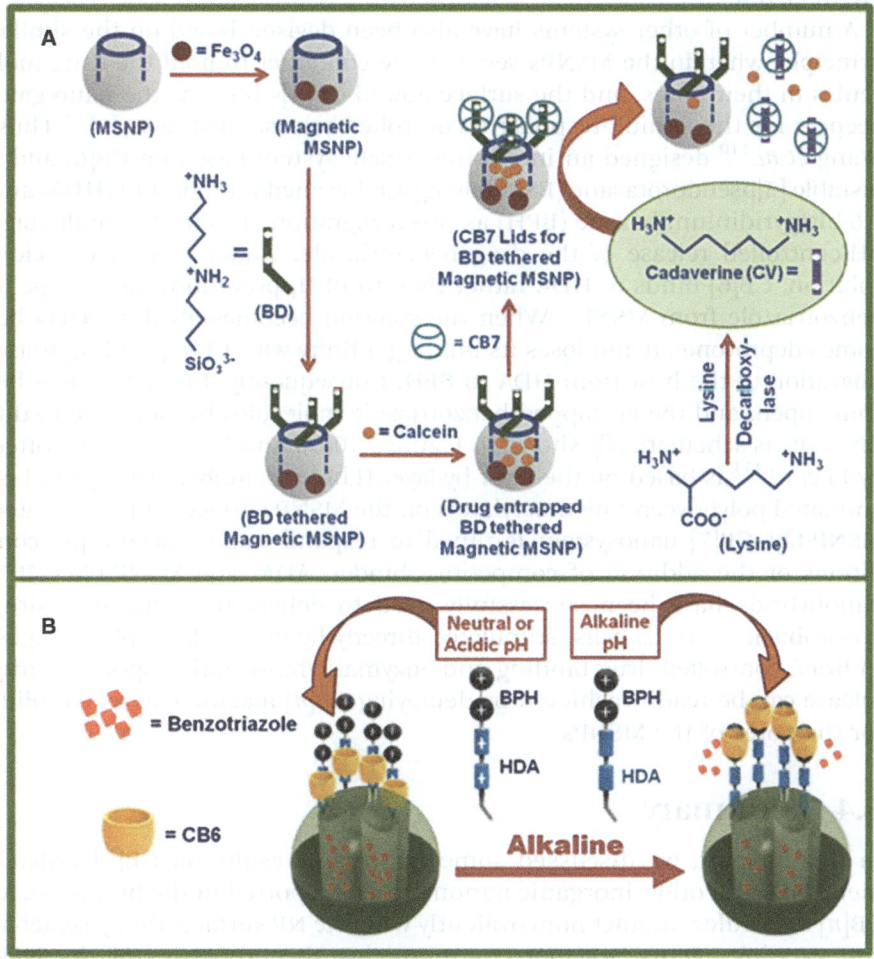

**Figure 5.10** (A) Schematics of the MSNP-BD-CB[7] system used for controlled binding and enzyme-responsive release of drug using CB[7] to act as nanolid for the pores of MSNPs. The figure is redrawn based on the concept discussed by Liu *et al.* in ref. 108. (B) Schematics of the MSNP-BSR-CB[6] system used for holding and controlled release of drugs using CB[6] molecules to act as nano gatekeepers for the pores of the MSNPs. The figure is redrawn based on the concept discussed by Wang *et al.* in ref. 110.

situation, the amino acid lysine, which has a very low binding affinity for CB[7], cannot disturb the bound CB[7] molecules on MSNP surfaces. In the presence of the enzyme lysine decarboxylase, however, decarboxylation of lysine occurs, producing cadaverine (CV), which exists in protonated form at neutral pH. Since protonated CV has much stronger binding with CB[7] than protonated BD, the enzymatic reaction triggers displacement of CB[7] from BD stalks to become bound with CV, causing the opening of the pores of MSNPs, thus releasing the entrapped drugs, as schematically shown in Figure 5.10A.

A number of other systems have also been devised based on the similar principle, wherein the MSNPs serve as the container to hold the drug molecules in their pores, and the surface-bound CB[*n*]s serve as the nano gatekeepers for the stimuli-responsive controlled release of drugs.[109,110] Thus, Wang *et al.*[110] designed an interesting MSNP system based on CB[6] and a bistable [2]pseudorotaxane (BSR) having 1,6-hexanediammonium (HDA) and 1,6-bis(pyridinium)hexane (BPH) as two recognition sites for the multistage pH-controlled release of the drug benzotriazole. Under neutral or acidic solution, CB[6] binds to HDA rather than to BPH, preventing the escape of benzotriazole from MSNPs. When the solution becomes alkaline, HDA becomes deprotonated and loses its binding affinity with CB[6], leading to the migration of the host from HDA to BPH. Consequently, the nanovalves become open, and the entrapped benzotriazole molecules become free to diffuse out, as schematically shown in Figure 5.10(b). Another strategy reported by Li *et al.*[111] is based on the layer-by-layer (LbL) assembly of CB[7] and bisaminated poly(glycerol methacrylate)s on the MSNP surface. This integrated MSNP-LbL-CB[7] nanosystem is tuned to respond under specific pH conditions or the addition of competitive binder, ADA. The MSNP-LbL-CB[7] nanohybrids have been successfully used to deliver the anticancer drug doxorobucin to HeLa cells, stimulated directly by intracellular pH changes. In brief, controlled drug binding and enzymatic/biostimuli-responsive drug release can be readily achieved by deploying CB[*n*] macrocycles as nanolids for the pores of the MSNPs.

## 5.4  Summary

In this chapter, we discussed some important results on CB[*n*]-assisted metal NPs and other inorganic nanomaterials reported in the literature. As CB[*n*] molecules interact noncovalently with the NP surface, they just act as soft capping agents, without compromising the properties of the NPs. Thus, CB[*n*]-assisted metal NPs and other inorganic nanomaterials have been investigated quite extensively in regard to their synthesis, characteristics and potential applications in diverse areas like sensing, plasmonics, drug delivery, health care and so on. Considering the scope of the present chapter, we have tried to discuss some of these aspects involving the metal NPs and some inorganic nanomaterial systems that are supported by CB[*n*] macrocycles. We hope that the presentation made in this chapter will draw the interest of researchers to explore many more aspects of CB[*n*]-supported nanomaterials for their useful applications and benefits to mankind.

## Acknowledgements

We thank all the colleagues, collaborators and authors of the published papers whose contributions have helped us in writing this chapter. We are also thankful to our home institute, Bhabha Atomic Research Centre, for its generous support extended in all our scientific endeavors.

## References

1. G. V. Oshovsky, D. N. Reinhoudt and W. Verboom, *Angew. Chem., Int. Ed.*, 2007, **46**, 2366.
2. H. Dodziuk, *Introduction to Supramolecular Chemistry*, Kluwer Academic Publishers, Dordrecht, Boston, 2002.
3. K. Kim, N. Selvapalam, Y. H. Ko, K. M. Park, D. Kim and J. Kim, *Chem. Soc. Rev.*, 2007, **36**, 267.
4. B. Rybtchinski, *ACS Nano*, 2011, **5**, 6791.
5. Y. H. Ko, E. Kim, I. Hwang and K. Kim, *Chem. Commun.*, 2007, 1305.
6. M. Sayed and H. Pal, *J. Mat. Chem. C*, 2016, **4**, 2685.
7. L. C. Palmer and S. I. Stupp, *Acc. Chem. Res.*, 2008, **41**, 1674.
8. Y. Zhao, F. Sakai, L. Su, Y. Liu, K. Wei, G. Chen and M. Jiang, *Adv. Mater.*, 2013, **25**, 5215.
9. K. Liu, Y. Kang, Z. Wang and X. Zhang, *Adv. Mater.*, 2013, **25**, 5530.
10. X. Ma and Y. Zhao, *Chem. Rev.*, 2015, **115**, 7794.
11. X. J. Loh, *Mater. Horiz.*, 2014, **1**, 185.
12. G. Schaeffer, O. Fuhr, D. Fenske and J.-M. Lehn, *Chem. – Eur. J.*, 2014, **20**, 179.
13. K.-J. Chen, M. A. Garcia, H. Wang and H.-R. Tseng, *Supramolecular Nanoparticles for Molecular Diagnostics and Therapeutics. Supramolecular Chemistry*, John Wiley & Sons, New York, 2012.
14. J. M. Zayed, N. Nouvel, U. Rauwald and O. A. Scherman, *Chem. Soc. Rev.*, 2010, **39**, 2806.
15. H.-J. Schneider, *Applications of Supramolecular Chemistry for 21st Century Technology*, Taylor & Francis, Boca Raton, FL, 2012.
16. H. Pal, *Sci. Adv. Today*, 2016, **2**, 25275.
17. M. E. Brewster and T. Loftsson, *Adv. Drug Deliverery Rev.*, 2007, **59**, 645.
18. G. Hettiarachchi, D. Nguyen, J. Wu, D. Lucas, D. Ma, L. Isaacs and V. Briken, *PLoS One*, 2010, **5**, e10514.
19. L. Wang, L.-L. Li, Y.-S. Fan and H. Wang, *Adv. Mater.*, 2013, **25**, 3888.
20. R. Villalonga, R. Cao and A. Fragoso, *Chem. Rev.*, 2007, **107**, 3088.
21. J. Mohanty, N. Thakur, S. Dutta Choudhury, N. Barooah, H. Pal and A. C. Bhasikuttan, *J. Phys. Chem. B*, 2012, **116**, 130.
22. H. Dube, M. R. Ams and J. Rebek, Jr., *J. Am. Chem. Soc.*, 2010, **132**, 9984.

23. R. N. Dsouza, U. Pischel and W. M. Nau, *Chem. Rev.*, 2011, **111**, 7941.

24. M. Sayed, F. Biedermann, V. D. Uzunova, K. I. Assaf, A. C. Bhasikuttan, H. Pal, W. M. Nau and J. Mohanty, *Chem. - Euro. J.*, 2015, **21**, 691.

25. K. I. Assaf and W. M. Nau, *Chem. Soc. Rev.*, 2015, **44**, 394.

26. E. I. Cucolea, C. Tablet, H.-J. Buschmann and L. Mutihac, *J. Inclusion Phenom. Macrocyclic Chem.*, 2015, **83**, 103.

27. M. Sayed, M. Sundararajan, J. Mohanty, A. C. Bhasikuttan and H. Pal, *J. Phys. Chem. B*, 2015, **119**, 3046.

28. J. W. Lee, S. Samal, N. Selvapalam, H.-J. Kim and K. Kim, *Acc. Chem. Res.*, 2003, **36**, 621.

29. J. Lagona, P. Mukhopadhyay, S. Chakrabarti and L. Isaacs, *Angew. Chem., Int. Ed.*, 2005, **44**, 4844.

30. A. C. Bhasikuttan, H. Pal and J. Mohanty, *Chem. Commun.*, 2011, **47**, 9959.

31. W. Ong and A. E. Kaifer, *J. Org. Chem.*, 2004, **69**, 1383.

32. J. Mohanty, S. Dutta Choudhury, H. P. Upadhyaya, A. C. Bhasikuttan and H. Pal, *Chem. - Euro. J.*, 2009, **15**, 5215.

33. H.-J. Buschmann and E. Schollmeyer, *J. Inclusion Phenom. Mol. Recognit. Chem.*, 1997, **29**, 167.

34. M. V. Rekharsky, T. Mori, C. Yang, Y. H. Ko, N. Selvapalam, H. Kim, D. Sobransingh, A. E. Kaifer, S. Liu, L. Isaacs, W. Chen, S. Moghaddam, M. K. Gilson, K. Kim and Y. Inoue, *Proc. Natl. Acad. Sci. U. S. A.*, 2007, **104**, 20737.

35. I. Ghosh and W. M. Nau, *Adv. Drug Delivery Rev.*, 2012, **64**, 764.

36. S. Gürbüz, M. Idrisa and D. Tuncel, *Org. Biomol. Chem.*, 2015, **13**, 330.

37. X. Ma and Y. Zhao, *Chem. Rev.*, 2015, **115**, 7794.

38. H. Yang, B. Yuan, X. Zhang and O. A. Scherman, *Acc. Chem. Res.*, 2014, **47**, 2106.

39. L. Isaacs, *Acc. Chem. Res.*, 2014, **47**, 2052.

40. J. Du, P. Zhang, X. Zhao and Y. Wang, *Sci. Rep.*, 2017, **7**, 6064.

41. H. Jung, K. M. Park, J.-A. Yang, E. J. Oh, D.-W. Lee, K. Park, S. H. Ryu, S. K. Hahn and K. Kim, *Biomaterials*, 2011, **32**, 7687.

42. G. Yun, Z. Hassan, J. Lee, J. Kim, N.-S. Lee, N. H. Kim, K. Baek, I. Hwang, C. G. Park and K. Kim, *Angew. Chem., Int. Ed.*, 2014, **53**, 6414.

43. M. Rai, A. Yadav and A. Gade, *Biotechnol. Adv.*, 2009, **27**, 76.

44. M. Liong, B. France, K. A. Bradley and J. I. Zink, *Adv. Mater.*, 2009, **21**, 1684.

45. T. Premkumar, Y. Lee and K. E. Geckeler, *Chem. - Eur. J.*, 2010, **16**, 11563.

46. K. Saha, S. S. Agasti, C. Kim, X. Li and V. M. Rotello, *Chem. Rev.*, 2012, **112**, 2739.

47. N. Elahi, M. Kamali and M. H. Baghersad, *Talanta*, 2018, **184**, 537.

48. R. M. Tripathi, A. Shrivastav and B. R. Shrivastav, *Artif. Cells, Nanomed., Biotechnol.*, 2015, **43**, 311.

49. K. Haume, S. Rosa, S. Grellet, M. A. Śmiałek, K. T. Butterworth, A. V. Solov'yov, K. M. Prise, J. Golding and N. J. Mason, *Cancer Nano*, 2016, **7**, 8.

50. M. Sengani, A. M. Grumezescu and V. D. Rajeswar, *OpenNano*, 2017, **2**, 37.
51. S. Linic, U. Aslam, C. Boerigter and M. Morabito, *Nat. Mater.*, 2015, **14**, 567.
52. V. Montes-García, J. Pérez-Juste, I. Pastoriza-Santos and L. M. Liz-Marzán, *Chem. – Euro. J.*, 2014, **20**, 10874.
53. M.-C. Daniel and D. Astruc, *Chem. Rev.*, 2004, **104**, 293.
54. A. Wolf and F. Schüth, *Appl. Catal., A*, 2002, **226**, 1.
55. B. J. Hornstein and R. G. Finke, *Chem. Mater.*, 2003, **15**, 899.
56. S. Praharaj, S. K. Ghosh, S. Nath, S. Kundu, S. Panigrahi, S. Basu and T. Pal, *J. Phys. Chem. B*, 2005, **109**, 13166.
57. G. Budroni and A. Corma, *Angew. Chem., Int. Ed.*, 2006, **45**, 3328.
58. A. Corma, H. García, P. Montes-Navajas, A. Primo, J. Calvino and S. Trasobares, *Chem. – Eur. J.*, 2007, **13**, 6359.
59. X. Lu and E. Masson, *Langmuir*, 2011, **27**, 3051.
60. N. Barooah, A. C. Bhasikuttan, V. Sudarsan, S. Dutta Choudhury, H. Pal and J. Mohanty, *Chem. Commun.*, 2011, **47**, 9182.
61. S. Dutta Choudhury, J. Mohanty, H. Pal and A. C. Bhasikuttan, *J. Am. Chem. Soc.*, 2010, **132**, 1395.
62. M. Shaikh, J. Mohanty, A. C. Bhasikuttan, V. D. Uzunova, W. M. Nau and H. Pal, *Chem. Commun.*, 2008, 3681.
63. S. J. Barrow, S. Kasera, M. J. Rowland, J. del Barrio and O. A. Scherman, *Chem. Rev.*, 2015, **115**, 12320.
64. L. Kaestner, M. Cesson, K. Kassab, T. Christensen, P. D. Edminson, M. J. Cook, I. Chambrier and G. Jori, *Photochem. Photobiol. Sci.*, 2003, **2**, 660.
65. J. Mohanty, A. C. Bhasikuttan, S. Dutta Choudhury and H. Pal, *J. Phys. Chem. B*, 2008, **112**, 10782.
66. A. C. Bhasikuttan, S. Dutta Choudhury, H. Pal and J. Mohanty, *Isr. J. Chem.*, 2011, **51**, 634.
67. S. Carrettin, J. Guzman and A. Corma, *Angew. Chem., Int. Ed.*, 2005, **44**, 2242.
68. A. Corma and P. Serna, *Science*, 2006, **313**, 332.
69. A. Abad, C. Almela, A. Corma and H. García, *Tetrahedron*, 2006, **62**, 6666.
70. A. Abad, P. Concepcion, A. Corma and H. García, *Angew. Chem., Int. Ed.*, 2005, **44**, 4066.
71. T. Premkumar and K. E. Geckeler, *Chem. – Asian J.*, 2010, **5**, 2468.
72. T.-C. Lee and O. A. Scherman, *Chem. Commun.*, 2010, **46**, 2438.
73. T.-C. Lee and O. A. Scherman, *Chem. – Euro. J.*, 2012, **18**, 1628.
74. R. W. Taylor, R. J. Coulston, F. Biedermann, S. Mahajan, J. J. Baumberg and O. A. Scherman, *Nano Lett.*, 2013, **13**, 5985.
75. M. del Pozo, E. Casero and C. Quintana, *Microchim. Acta*, 2017, **184**, 2107.
76. M. del Pozo, E. Blanco, P. Hernández, J. A. Casas and C. Quintana, *J. Nanopart. Res.*, 2018, **20**, 121.

77. M. Schnell, A. García-Etxarri, A. J. Huber, K. Crozier, J. Aizpurua and R. Hillenbrand, *Nat. Photonics*, 2009, **3**, 287.

78. L. Polavarapu, J. Pérez-Juste, Q.-H. Xu and L. M. Liz-Marzán, *J. Mater. Chem. C*, 2014, **2**, 7460.

79. S. Mahajan, T.-C. Lee, F. Biedermann, J. T. Hugall, J. J. Baumberga and O. A. Scherman, *Phys. Chem. Chem. Phys.*, 2010, **12**, 10429.

80. R. W. Taylor, T.-C. Lee, O. A. Scherman, R. Esteban, J. Aizpurua, F. M. Huang, J. J. Baumberg and S. Mahajan, *ACS Nano*, 2011, **5**, 3878.

81. C. Tao, Q. An, W. Zhu, H. Yang, W. Li, C. Lin, D. Xu and G. Li, *Chem. Commun.*, 2011, **47**, 9867.

82. S. Kasera, F. Biedermann, J. J. Baumberg, O. A. Scherman and S. Mahajan, *Nano Lett.*, 2012, **12**, 5924.

83. S. Kasera, L. O. Herrmann, J. del Barrio, J. J. Baumberg and O. A. Scherman, *Sci. Rep.*, 2014, **4**, 6785.

84. D. O. Sigle, S. Kasera, L. O. Herrmann, A. Palma, B. de Nijs, F. Benz, S. Mahajan, J. J. Baumberg and O. A. Scherman, *J. Phys. Chem. Lett.*, 2016, **7**, 704.

85. N. Hüsken, R. W. Taylor, D. Zigah, J.-C. Taveau, O. Lambert, O. A. Scherman, J. J. Baumberg and A. Kuhn, *Nano Lett.*, 2013, **13**, 6016.

86. L. O. Herrmann, V. K. Valev, C. Tserkezis, J. S. Barnard, S. Kasera, O. A. Scherman, J. Aizpurua and J. J. Baumberg, *Nat. Commun.*, 2014, **5**, 4568.

87. G. Lu, L. Hou, T. Zhang, W. Li, J. Liu, P. Perriat and Q. Gong, *J. Phys. Chem. C*, 2011, **115**, 22877.

88. L. Xu, H. Kuang, L. Wang and C. Xu, *J. Mater. Chem.*, 2011, **21**, 16759.

89. S. T. Jones, J. M. Zayed and O. A. Scherman, *Nanoscale*, 2013, **5**, 5299.

90. R. de la Rica and A. H. Velders, *Small*, 2011, **7**, 66.

91. C. Kim, S. S. Agasti, Z. Zhu, L. Isaacs and V. M. Rotello, *Nat. Chem.*, 2010, **2**, 962.

92. W. X. Niu, Y. A. A. Chua, W. Q. Zhang, H. J. Huang and X. M. Lu, *J. Am. Chem. Soc.*, 2015, **137**, 10460.

93. Y. Han, X. Yang, Y. Liu, Q. Ai, S. Liu, C. Sun and F. Liang, *Sci. Rep.*, 2016, **6**, 22239.

94. P. Xu, Q. Feng, X. Yang, S. Liu, C. Xu, L. Huang, M. Chen, F. Liang and Y. Cheng, *Bioconjugate Chem.*, 2018, **29**, 2855.

95. M. Cao, J. Lin, H. Yang and R. Cao, *Chem. Commun.*, 2010, **46**, 5088.

96. M. Cao, D. Wu, W. Su and R. Cao, *J. Catal.*, 2015, **321**, 62.

97. K. Karami and N. H. Naeini, *Appl. Organomet. Chem.*, 2015, **29**, 33.

98. S. Nandi, P. Patel, A. Jakhar, N. H. Khan, A. V. Biradar, R. I. Kureshy and H. C. Bajaj, *ChemistrySelect*, 2017, **2**, 9911.

99. M. Cao, D. Wu, S. Gao and R. Cao, *Chem. – Eur. J.*, 2012, **18**, 12978.

100. S. G. Song, T. Premkumar and C. S. Song, *NANO*, 2018, **13**, 1850007.

101. Y. Sun, W. Zhang, B. Wang, X. Xu, J. Chou, O. Shimoni, A. T. Ung and D. Jin, *Chem. Commun.*, 2018, **54**, 3851.

102. L. Francés-Soriano, M. González-Béjar and J. Pérez-Prieto, *Nanoscale*, 2015, **7**, 5140.

103. M. Y. Hossan, A. Hor, Q. Luu, S. J. Smith, P. S. May and M. T. Berry, *J. Phys. Chem. C*, 2017, **121**, 16592.
104. X. Li, F. Zhang and D. Zhao, *Chem. Soc. Rev.*, 2015, **44**, 1346.
105. N. Estebanez, J. Ferrera-González, L. Francés-Soriano, R. Arenal, M. González-Béjar and J. Pérez-Prieto, *Nanoscale*, 2018, **10**, 12297.
106. P. Yang, Z. Quan, L. Lu, S. Huang and J. Lin, *Biomaterials*, 2008, **29**, 692.
107. A. Bernardos, E. Aznar, M. D. Marcos, R. Martínez-Máñez, F. Sancenón, J. Soto, J. M. Barat and P. Amorós, *Angew. Chem., Int. Ed.*, 2009, **48**, 5884.
108. J. Liu, X. Du and X. Zhang, *Chem. – Eur. J.*, 2011, **17**, 810.
109. S. Angelos, Y.-W. Yang, N. M. Khashab, J. F. Stoddart and J. I. Zink, *J. Am. Chem. Soc.*, 2009, **131**, 11344.
110. M. D. Wang, T. Chen, C. D. Ding and J. J. Fu, *Chem. Commun.*, 2014, **50**, 5068.
111. Q.-L. Li, Y. Sun, Y.-L. Sun, J. Wen, Y. Zhou, Q.-M. Bing, L. D. Isaacs, Y. Jin, H. Gao and Y.-W. Yang, *Chem. Mater.*, 2014, **26**, 6418.

CHAPTER 6

# Cucurbituril-assisted Supramolecular Polymeric Hydrogels

AISAN KHALIGH[a,b] AND DÖNÜS TUNCEL*[a,b]

[a] Department of Chemistry, Bilkent University, Ankara 06800, Turkey;
[b] UNAM – National Nanotechnology Research Center, Institute of Materials Science and Nanotechnology, Bilkent University, Ankara 06800, Turkey
*Email: dtuncel@fen.bilkent.edu.tr

## 6.1 Hydrogels

Hydrogels that consist of three-dimensional networks of hydrophilic polymers are a highly important class of materials.[1] This porous structure can retain a large amount of water through surface tension or capillary effect, which takes the gel form and offers the possibility for the storage of hydrophilic biomolecules.[2,3] Hydrogels are often classified into two major groups on the basis of the chemical or physical nature of their cross-linking (Figure 6.1): synthetic hydrogels (covalent approach) and supramolecular hydrogels (noncovalent approach).[3,4]

Synthetic hydrogels are prepared by cross-linking appropriate oligomers or polymers *via* nonreversible permanent covalent bonds. Since their discovery by Wichterle and Lim in 1960, these hydrogels have been of great interest for biomedical applications.[5,6] However, most of the covalently cross-linked polymeric hydrogels are brittle, poorly transparent and without self-healing ability once the cross-linked network is broken.[7] These deficiencies, which limit their biomedical applications, could be solved by

Smart Materials No. 36
Cucurbituril-based Functional Materials
Edited by Dönüs Tuncel
© The Royal Society of Chemistry 2020
Published by the Royal Society of Chemistry, www.rsc.org

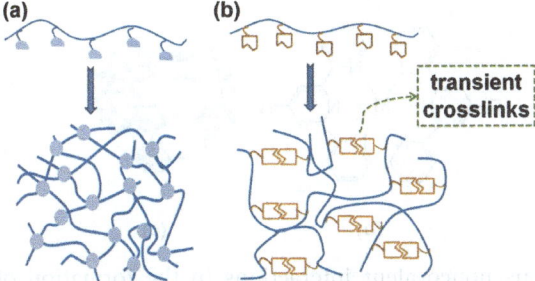

**Figure 6.1**  Schematic representation of the (a) covalently cross-linked synthetic hydrogels (static hydrogels) and (b) noncovalently cross-linked supramolecular hydrogels (reversible hydrogels).
Reproduced from ref. 4 with permission from the Royal Society of Chemistry.

utilizing dynamic noncovalent interactions as the structural cross-links between polymer chains. The noncovalent approach provides reversible structures that can be used for the fabrication of self-healing materials with tunable mechanical properties for expected applications.[2,3,8]

## 6.2  Supramolecular Polymeric Hydrogels

Supramolecular polymeric hydrogels (SPHs), a novel class of noncovalently cross-linked polymer materials, have received considerable attention over the last two decades due to their significant mechanical properties gained from the use of polymer chains, as well as their stimuli responsivity and processability inherent to the supramolecular units used in cross-linking.[2] The supramolecular cross-linking of polymer chains by various noncovalent interactions (Figure 6.2), such as hydrogen bonding, metal–ligand coordination, electrostatic interaction, and host–guest interactions, decreases structural flexibility and results in the formation of three-dimensional networks.[3] These aqueous polymeric networks have excellent biocompatibility, biodegradability and biostability, and their mechanical properties can be tuned by polymer loading and cross-link density. They exploit specific, directional and reversible noncovalent interactions that are responsive to various external stimuli, including physical (temperature, light, voltage, magnetic field) and chemical (pH, ionic strength, redox agent, glucose, enzyme, competitive host/guest) parameters of the surrounding environment.[2,3,8–11] Generally, these external stimuli have a significant impact on the cross-linkages, and they can cause swelling or dissociation of the cross-linked hydrogel networks. SPHs can effectively change their physicochemical properties upon exposure to external stimuli. Besides, they can be easily modified with numerous functional components to achieve desired functions such as bioactive properties, optoelectronic properties, as well as self-healing and shape-memory abilities.[3] Due to these unique properties, the

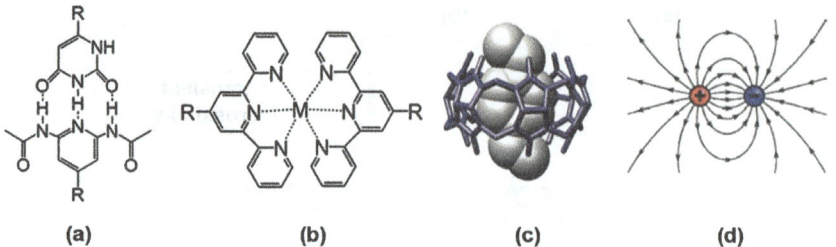

**(a)**                    **(b)**                    **(c)**                    **(d)**

**Figure 6.2**   Various noncovalent interactions in the formation of supramolecular
hydrogels: (a) hydrogen bonding, (b) metal–ligand coordination, (c)
host–guest recognition and (d) electrostatic interaction.
Reproduced from ref. 3 with permission from the Royal Society of
Chemistry.

SPHs show great potential for a wide variety of biomedical applications such
as drug delivery, wound dressing and healing, tissue engineering scaffolds
and diagnostic devices.[12–16]

Host–guest recognition is the most used noncovalent interaction for the
preparation of SPHs. Among the different types of macrocyclic hosts,
cyclodextrins (CD) and CB[n] have gained particular attention in supramolecular hydrogel chemistry.[3,17–20] These macrocyclic hosts are nontoxic and
biocompatible. They can form inclusion complexes in aqueous media by
hosting a guest molecule within their cavity.[21] Herein, water molecules are
complexed in the cavity of host molecules *via* limited hydrogen bonding and
weak interactions. These high-energy water molecules are released by guest
complexation, the leading method in positive energy contribution to guest
inclusion.[22]

CDs consist of glucopyranose units with a hydrophilic outer surface and a
hydrophobic central cavity. In the past decade, CD-based supramolecular
polymeric hydrogels and their biomedical applications in drug delivery and
tissue engineering have been widely investigated. They mainly bind to
neutral and anionic guest molecules. Binding in CDs occurs through
hydrophobic interaction between the guest molecule and the inner cavity;
however, these macrocyclic hosts have low binding affinity for their guests
$(K_{eq} \sim 10^4$ M$^{-1})$, which results in poor mechanical properties.[23] In contrast,
CB[n]s have significantly higher affinity for binding guest molecules, as high
as $10^{15}$ M$^{-1}$.[2,4]

## 6.3   CB[n] as Supramolecular Host

CB[n]s ($n = 5$–8 and 10) are a family of macrocyclic host molecules made of
glycoluril monomers linked by methylene bridges. These macrocyclic hosts
are composed of two hydrophilic portals decorated with partially negatively
charged carbonyl groups and a hydrophobic cavity. CB[n]s are usually synthesized by heating a mixture of glycoluril and formaldehyde in 1 : 2 ratio

under acidic conditions for several hours. This results in a mixture of CB[$n$] homologues ($n = 5$–8) and glycoluril oligomers, which can be isolated. All CB[$n$] homologues have the same height (9.1 Å); however, they differ in the sizes of their cavities (ranging from 4.4 to 8.8 Å), which is determined by the number of glycoluril units in the macrocycle. The water solubility of CB[$n$] varies in an odd–even fashion: CB[5] and CB[7] are highly water soluble ($S_{H_2O}$: 20–30 mM), whereas CB[6] and CB[8] exhibit low water solubility of 0.018 mM and <0.01 mM, respectively.[4,24,25]

CB[$n$]s are efficient host molecules in molecular recognition and have an exceptionally high affinity for positively charged compounds attributed to their carbonyl portals. These macrocyclic hosts, with rigid structure and the ability for complexation of a wide variety of guest molecules through the hydrophobic effect, ion–dipole and hydrogen bond interaction, have been considered perfect model systems for the study of molecular recognition and the building blocks of simple molecular machines.[20,22] The solubility of CB[$n$]s in organic solvents is less than $10^{-5}$ M, and therefore the host–guest chemistry of CB[$n$]s has mostly been studied in aqueous media.[4]

## 6.3.1 Molecular Recognition Properties of CB[$n$] Hosts

As previously described, CB[$n$] homologues are efficient host molecules in molecular recognition; however, they have different molecular recognition properties due to differences in the number of glycoluril units and therefore in their varying cavity and portal sizes.

CB[5] and its homologues (dimethylglycoluril and cyclohexanoglycoluril) can accommodate inside their cavity various gas molecules including krypton, xenon, nitrogen, oxygen, argon, nitrous oxide, nitric oxide, carbon monoxide, carbon dioxide, methane, ethylene and ethane, as well as small solvent molecules such as methanol and acetonitrile. Moreover, they can bind simultaneously two cations (alkali metals, alkali-earth, $NH_4^+$, $Pb^{2+}$, $Cd^{2+}$) through electrostatic interactions with both of the carbonyl portals.[4]

CB[6] can bind alkali-metal, alkaline-earth, transition metal, lanthanide cations and several gas molecules, as well as positively charged and neutral organic guests, due to its larger inner cavity. CB[6] is known to be able to host alkyl ammonium ions due to the partial occupation of the cavity by hydrophobic oligomethylene chains and strong ion–dipole interactions of ammonium cation with one of the carbonyl portals. In particular, CB[6] can tightly bind protonated polyamines such as 1,6-diaminohexane or spermine through strong 1 : 1 host–guest interactions.[4]

CB[7] has significant binding affinities for positively charged amphiphilic guests such as adamantine (AD), *p*-xylylene, ferrocene and trimethylsilyl derivatives bearing one or two amino groups, as well as viologen derivatives. Some of the CB[7]@guest molecule complexes display equilibrium association constants equal and even higher than that of avidinbiotin interaction ($K_{eq} = 10^{15}$ M$^{-1}$), which represents the strongest noncovalent interactions for a synthetic system.[4]

Isaacs *et al.*, in 2005, reported an equilibrium association constant of $10^{12}$ $M^{-1}$ for the rimantadine@CB[7] complex.[26] In the same year, Kim, Kaifer, Inoue *et al.*, reported the remarkably high affinity of $K_{eq} = 5.0 \times 10^{15}$ $M^{-1}$ for 1-(2 aminoethylamino)adamantane@CB[7] complex.[27] Kim, Inoue, Gilson *et al.*, in 2007 reported a $K_{eq} = 3.0 \times 10^{15}$ $M^{-1}$ for the host–guest pair of 1,10 bis(trimethylammoniomethyl)ferrocene@CB[7].[28] Inoue, Gilson *et al.* in 2011 reported the host–guest complexes of bicyclo[2.2.2]octane and adamantine with CB[7], each having a $K_{eq}$ value above $10^{15}$ $M^{-1}$. During the formation of these complexes, the enthalpy and entropy values increased due to the strong 1 : 1 host–guest interaction and dehydration of CB[7], respectively.[29]

By increasing the cavity size of the supramolecular host from CB[5] to CB[8], the size of the guests for 1 : 1 complexes increases. CB[8] forms strong 1 : 1 complexes with positively charged and relatively large guests such as adamantane derivatives, cyclen, cyclam macrocyles (also their doubly charged Cu and Zn complexes) and long alkylammonium aliphatic chains.[4]

Interestingly, CB[8] can simultaneously accommodate two guests, either two different guests to form a 1 : 1 : 1 heteroternary complex or two of the same guest to form a 1 : 2 homoternary complex. Herein, the dynamic supramolecular interactions have been used as a reversible linking motif to bind two separate entities, leading to the formation of hetro- or homoternary CB[8]-mediated complexes in an aqueous solution.[2,4,11,22] CB[8] is highly symmetric, and its encapsulation complexes can be readily characterized with $^1$H-NMR spectroscopy. The encapsulated aromatic units can be easily attached to different molecular or macromolecular backbones, which enable the design of various monomers.[30]

# 6.4 CB[8]-assisted Supramolecular Polymeric Hydrogels

As previously mentioned, by considering the advantages of the rich host–guest chemistry of CB[8] and its recognition properties, this supramolecular host has attracted widespread attention for the construction of stimuli-responsive nanostructures such as supramolecular polymeric hydrogels.

## 6.4.1 Supramolecular Polymeric Hydrogels Based on CB[8]-mediated Heteroternary Complexation

SPHs formed *via* molecular recognition of CB[8] for specific chemical motifs pendant from polymer chains have been widely studied since 2010. Herein, CB[8]-mediated host–guest heteroternary complexation (1 : 1 : 1) is a powerful and effective way for the preparation of dynamic and stimuli-responsive SPHs (Figure 6.3). These heteroternary complexes are formed between CB[8], an electron-poor first guest (such as methyl viologen) and an electron-rich second guest (such as naphthol, dibenzylfuran or pyrene) through the

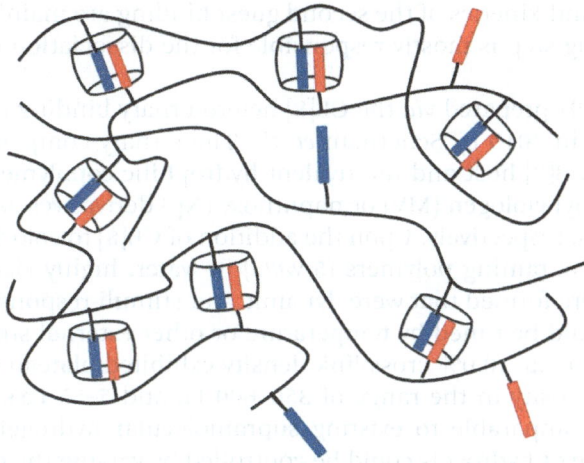

**Figure 6.3** Schematic representation of the heteroternary CB[8]-based polymeric hydrogel network.
Reproduced from ref. 4 with permission from the Royal Society of Chemistry.

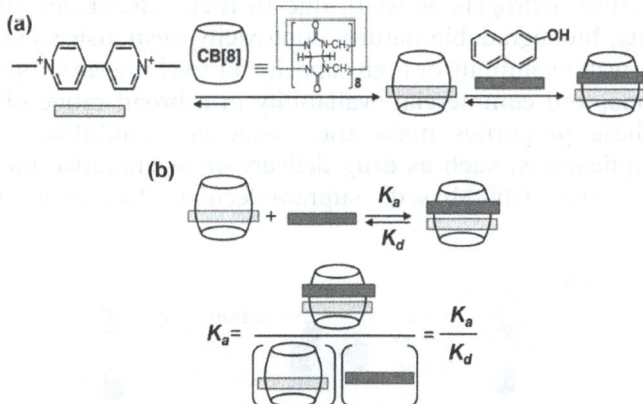

**Figure 6.4** Schematic representation of the (a) two-step, three-component binding of CB[8] in water; (b) thermodynamic and kinetic parameters of second-guest binding.
Reproduced from ref. 2 with permission from American Chemical Society, Copyright 2010.

multiple noncovalent interactions including electrostatic and hydrophobic effects, which leads in remarkably high-equilibrium binding affinities ($K_{eq}$ up to $10^{14}$ $M^{-2}$). Herein, CB[8] acts as a dynamic interlink between two guests.[11,31] As can be seen in Figure 6.4, the CB[8]-mediated heteroternary complexes form in a stepwise binding process: first, the electron-poor guest enters ($K_{eq1}$), followed by the electron-rich guest ($K_{eq2}$). While two $K_a$ values are related to this binding scheme (overall $K$ in $M^{-2}$), the dynamic equilibrium and kinetics of the second guest binding are mainly considered, as this binding step is mostly responsible for the dissociation of the ternary complex.[8,30,32]

The first SPH prepared *via* the CB[8] heteroternary binding motif in water was reported in 2010 by Scherman *et al.*[2] The ternary complex was formed between the CB[8] host and multivalent hydrophilic copolymers containing pendant methyl viologen (MV) or naphthoxy (Np) derivatives as the first and second guests, respectively. Upon the addition of CB[8] to colorless solutions of the guest-containing polymers (5 wt%) in water, highly viscous and colored SPHs were formed that were dynamic and stimuli-responsive, and their properties could be tuned by temperature or other external stimuli. Herein, hydrogels with 2.5–10.0% cross-link density exhibited plateau modulus and zero-shear viscosity in the range of 350–600 Pa and 5–55 Pa s, respectively, which were comparable to existing supramolecular hydrogels. The cross-linking degree of hydrogels could be controlled by varying the concentration of CB[8]. These hydrogels demonstrated thermal reversibility and facile modulation of their microstructures (such as pore sizes) with further additions of CB[8] and thermal treatment (Figure 6.5).[2]

Polysaccharides (hyaluronic acid, carboxymethylcellulose and chitosan) are among the mostly used biopolymers for fabrication of hydrogels (supramolecular hydrogels as well), due to their advantages such as biocompatibility, biodegradable nature, nontoxicity, responsiveness to enzymatic activity, water solubility or high capacity for swelling *via* simple chemical modifications, and commercial availability in a broad range of molecular weights. These properties make them excellent candidates for a wide range of applications, such as drug delivery or regenerative medicine.[33,34] Accordingly, the CB[8]-based supramolecular hydrogels containing

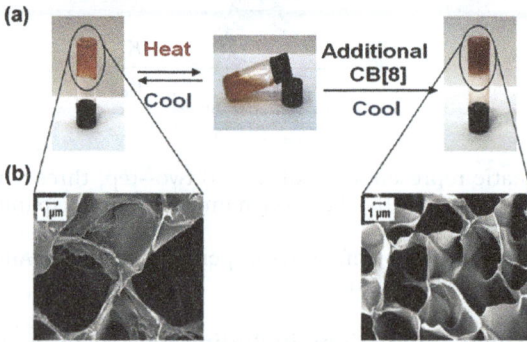

**Figure 6.5**   (a) Photographs of reversible sol–gel conversion of supramolecular hydrogel by varying temperature; (b) SEM images showing network reversibility in which further addition of CB[8] to a hydrogel with low cross-link density and then thermal treatment lead to an increase in cross-link density.
Reproduced from ref. 2 with permission from American Chemical Society, Copyright 2010.

polysaccharides are expected to be more biocompatible and smart due to the excellent properties of CB[8] and polysaccharide derivatives. Therefore, the previously reported CB[8]-SPHs mostly contain polysaccharides in their network. Here are some examples.

Using the host–guest binding ability of CB[8] with MV and Np derivatives, ultra-high water content supramolecular polymeric hydrogel (UW-SPH) with minimal polymer content (0.5 wt%) was synthesized. For this purpose, high molecular weight hydroxyethyl cellulose (HEC, $Mn = 1.3$ MDa) and poly(vinyl alcohol) (PVA, $Mn = 1.5$ MDa) were first functionalized with MV- and NP-binding moieties, respectively. By mixing MV-HEC with CB[8] and Np-PVA in an aqueous solution, a lightly colored transparent hydrogel was formed immediately (Figure 6.6). It was found that these hydrogels had highly tunable mechanical properties that could be easily controlled by modifying the component loadings. This feature allows for rapid self-healing of the materials after deformation by damage. The prepared UW-SPH exhibited stimuli-responsiveness to a multitude of external stimuli including temperature, chemical potential and competing guests. They also showed excellent shear-thinning properties (due to the low polymer content and noncovalent cross-linking). Generally, the simple preparation method even at biologically relevant temperatures, availability from inexpensive renewable resources, tunable mechanical properties and biocompatibility highlight this UW-SPH as an exciting candidate for use in various biomedical and industrial applications.[8]

UW-SPHs were later applied for the controlled release of bovine serum albumin (BSA, $Mw = 67\,000$ g mol$^{-1}$) and lysozyme ($Mw = 14\,000$ g mol$^{-1}$), as

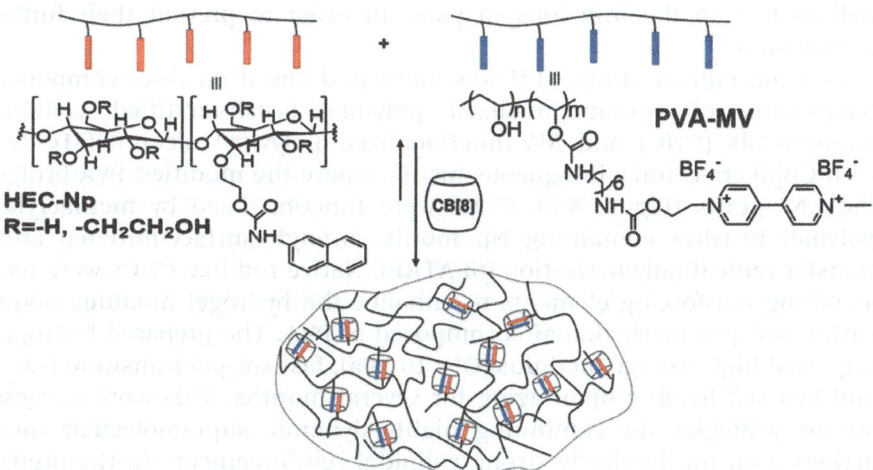

**Figure 6.6**  Schematic representation of the preparation of supramolecular hydrogel through strong host–guest interactions of CB[8] with MV-HEC and Np-PVA in water.
Reproduced from ref. 8 with permission from American Chemical Society, Copyright 2012.

model proteins. The main aim of the work was to study the effect of the protein molecular weight and polymer loadings of the hydrogels on the protein release rate. The hydrogels containing two different polymer loadings (0.5 and 1.5 wt%) were prepared. Results showed that BSA released more slowly than lysozyme and that the UW-SPH with 1.5 wt% polymer content showed a sustained release of BSA over the course of 160 days. By increasing the polymer content of hydrogel, the mesh size of the network decreased leading to the slower release of protein. This feature makes these materials potential candidates for sustained therapeutic application such as biomedical delivery devices.[35]

In another study, an antimicrobial SPH was formed through the ternary 1:1:1 host–guest interaction of CB[8] with a strong siderophore pendant from a polymer backbone and then utilized as consolidant for archaeological wood conservation. Guar/boronic acid functionalized viologen derivative (guar/MV-BA) was served as first guest polymer and a 50:50 mixture of catechol- and naphthol-functionalized chitosan (PolyCatNap) as second guest polymer. The chitosan polymer was cross-linked with the MV-BA guar upon the addition of CB[8]. Beside the natural antibacterial properties of chitosan, MV-BA remarkably enhanced the biocidal activity of the network. Formation and disassembly of the ternary complex was mediated by heat and modulating the water content. The obtained system not only exhibited all the properties of such SPHs but also showed biological degradation (due to MV-BA, chitosan) and structural stability (due to guar, chitosan, CB[8] cross-links), as well as iron saturation (due to catechol) to chelate and trap $Fe^{3+}$ ions. Therefore, the designed CB[8]-SPH could be pulled into the porous wooden artifacts to strengthen them and provide biocidal protection as well as to trap the iron ions in place in order to prevent their further degradation.[36]

In a subsequent study, SPH was fabricated based on three-component recognition of Np-functionalized polymer brush-modified cellulose nanocrystals (CNC) and MV-functionalized polyvinyl alcohol (MV-PVA) with CB[8] cross-links in aqueous media, where the modified PVA bridges the CNC-grafts (Figure 6.7). CNCs were functionalized by methacrylate polymer brushes containing Np motifs through surface-initiated atom transfer radical polymerization (SI-ATRP). Native rod-like CNCs were used as strong reinforcing elements to enhance the hydrogel modulus bound within soft polymeric domains composed of PVA. The prepared hydrogels imparted high storage modulus ($G' < 10$ kPa), fast sol–gel transition ($<6$ s), and fast self-healing upon aging for several months. This work suggests strong strategies for combining highly dynamic supramolecular interactions with mechanically strong colloidal reinforcements in the preparation of the next generation of advanced dynamic materials from renewable resources.[10]

The investigations on CB[8]-SPH *via* heteroternary complex formation were continued in the reporting of the simple method to fabricate monodisperse SPH beads using droplet-based microfluidics. The droplet-based

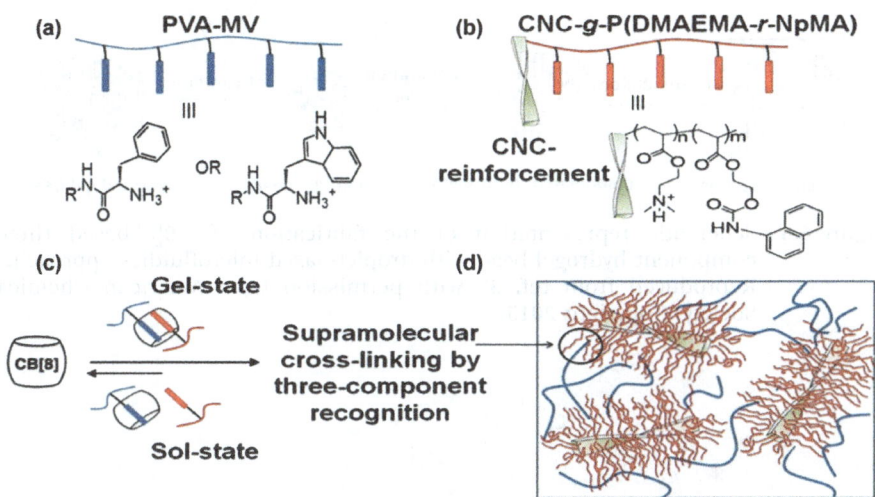

**Figure 6.7** Schematic representation of the (a) PVA containing the methyl viologen functionality (PVA-MV); (b) CNCs containing copolymer grafts of protonated dimethylaminoethyl methacrylate and naphthyl methacrylate repeat units (CNC-g p(DMAEMA-r-NpMA)); (c) CB[8] host; (d) supramolecular hydrogel network based on three-component recognition of CB[8] with PVA-MV and CNC-g-p(DMAEMA-r NpM).
Reproduced from ref. 10 with permission from John Wiley & Sons, © 2014 WILEY-VCH Verlag GmbH & Co. KGaA, Weinheim.

microfluidics approach has rapidly developed in the last decade due to the numerous advantages including low fabrication costs, small sample volumes, reduced analysis times, improved sensitivity, high-throughput analysis, enhanced operational flexibility and facile automation. This technology has been employed with great promise in chemical reactions, molecular detection, imaging, drug delivery, diagnostics, cell biology and other fields. Recently, droplet-based microfluidics has been applied for the preparation of hydrogel beads using monodispersed microdroplets as a template for particle synthesis. This approach provides the uniform loading of cargo inside the hydrogel particle. In the study conducted by Abell *et al.*, monodisperse SPH beads were prepared by the droplet-based microfluidics method through the 1:1:1 ternary host–guest interaction between CB[8] and a mixture of naphthyl-functionalized poly(hydroxyethyl cellulose) copolymer (HEC$_{Np}$) and methyl viologen-functionalized poly((vinylbenzyl)-trimethylammonium chloride) copolymer (SAM$_{MV}$). As can be seen in Figure 6.8, the formation of hydrogel beads is a stepwise process: at first, microdroplets containing a dilute mixture of SAM$_{MV}$, HEC$_{Np}$ and CB[8] were generated in a microfluidic device. The droplets were then collected on a microscope slide, and their water was evaporated at room temperature, thereby concentrating its contents and increasing the CB[8] cross-linking within the droplet. Finally, upon adding a little amount of deionized water to the dried particles (rehydration), monodisperse hydrogel beads were formed.

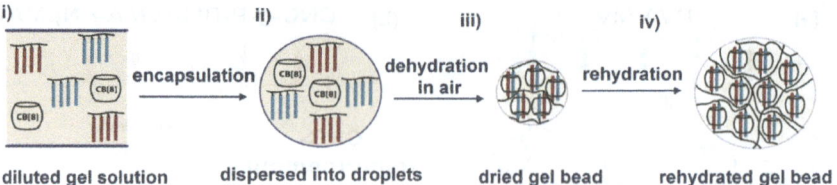

Figure 6.8    Schematic representation of the fabrication of CB[8]-based three-component hydrogel bead with droplet-based microfluidics approach. Reproduced from ref. 37 with permission from American Chemical Society, Copyright 2015.

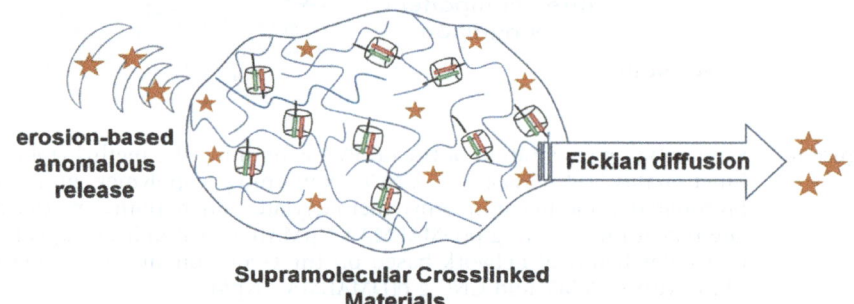

Figure 6.9    Schematic representation of the release mechanism of a model drug from the prepared hydrogel network. Reproduced from ref. 39 with permission from Elsevier, Copyright 2014.

The size of the hydrogel could be modified by varying the polymer loading or the dimensions of the microfluidic flow focus junction.[37]

The SPH beads were then applied as a carrier for the controlled release of fluorescein-tagged dextran (FITC). It was found that dependent on the relative content of the polymers and CB[8] in the hydrogel formulation, both the sustained and the fast release of cargo could be achieved. Moreover, the rapid release of cargo (minutes) triggered was observed upon the addition of the competing guest, 1-adamantylamine, likely due to the disruption of the supramolecular cross-links.[37]

It has been found that the mechanical strength and viscoelastic properties of CB[8]-SPHs are typically affected by polymer physics and its interactions, the dynamics of cross-linking, the type of second guest and its dissociation activation energy. Therein, the effect of the second guest molecule on the viscoelastic properties of the CB[8]-SPH formed *via* heteroternary complexation was investigated using different second guests (naphthole (Np), di benzylfuran (DBF) and pyrene (Pyr)), which have identical thermodynamic equilibrium constants in the formation of heteroternary complexes with CB[8]- and MV-bearing polymers. The SPH networks were prepared *via* host–guest recognition of CB[8] with MV-bearing styrene and second guest–bearing poly(dimethylacrylamide) in aqueous media (Figure 6.9). It was found that the second guests had completely different dissociation activation energies ($E_{ad}$ values for Np 30 kJ mol$^{-1}$, DBF 54 kJ mol$^{-1}$ and Pyr 89 kJ mol$^{-1}$) despite their similar associative rates. These different second guests resulted in the physically cross-linked hydrogels with similar strength (plateau modulus), concentration and structure but different network dynamics. Based on the oscillatory rheology results, Pyr-bearing hydrogels and Np-bearing hydrogels were the most and the least elastic samples, respectively, due to their different dissociation activation energies ($E_{ad}$) of the second guests. In fact, a second guest with a higher dissociation activation energy leads to the formation of stronger hydrogels with longer cross-link lifetime.[38]

In a follow-up study, the effect of the network dynamics on *in vitro* release of Rhodamine-B (Rh-B, a common water-soluble imaging agent used in microscopy) by the previously mentioned SPHs, was investigated. Hydrogel networks involving second guests with lower $E_{ad}$ showed higher release constants because these networks with rapid cross-link dynamics allow quicker diffusive rates of water and Rh-B in and out of the hydrogel. Herein, easy water diffusion into the hydrogel network facilitates erosion of hydrogel through swelling mechanisms. Therefore, the release mechanism is the combination of erosion-based and Fickian diffusional releases (Figure 6.9). Fickian diffusion refers to the solute transport process in which the polymer relaxation time ($t_r$) is much higher than the characteristic solvent diffusion time ($t_d$). When $t_r \approx t_d$, the macroscopic drug release becomes anomalous or non-Fickian. This study demonstrated that the cross-link dynamic is an important factor in determining the release mechanism of therapeutic cargo from a hydrogel.[39]

## 6.4.2 Supramolecular Polymeric Hydrogels Based on CB[8]-mediated Homoternary Complexation

CB[8]-mediated 1 : 2 homoternary complexation as another effective motif for preparation of dynamic SPHs has been studied since 2013. Using this binding motif, the SPH network is simplified from a three-component to a two-component system (CB host and two of the same guest molecules). Similar to the heteroternary complexation, 1 : 2 homoternary complexes likely form in a stepwise binding process: one of the guest molecules enters first ($K_{eq1}$), followed by the other guest molecule ($K_{eq2}$).

Several guest motifs have been used for CB[8] homoternary complexation. The self-assembly of polymer chains through naturally occurring and nontoxic units is very attractive in developing the robust supramolecular polymeric hydrogels with minimal toxicity for biomedical applications. Aromatic amino acid monomers of phenylalanine and tryptophan are two such molecular units. They bind within the CB[8] cavity in a 2 : 1 fashion through the multiple noncovalent interactions acting synergistically, resulting in remarkably high binding equilibrium constants ($K_{eq}$ up to $10^{11}$ M$^{-2}$). Amino acids are different from MV and Np guest moieties because they do not always yield a visible charge transfer complex, and they are naturally occurring amino acids with no significant toxicity profile. Therefore, hydrogels derived from these amino acids have a reduced toxicity profile. Furthermore, the amino acid recognition motifs can be attached to larger and natural polymers such as cellulose derivatives to further improve water content and biocompatibility for use in drug delivery, 3D cell culture and regenerative medicine.[11,40]

Scherman *et al.* in 2013 reported the first 1 : 2 homoternary CB-SPH system based on the complex formation between CB[8] host and synthetic styrenic polymers containing pendant phenylalanine or tryptophan aromatic amino acids (Figure 6.10). It was found that phenylalanine@CB[8] hydrogels were much stronger than tryptophan@CB[8] and that the SPH properties could be tuned simply by controlling the CB[8] concentration. This study led to the improvement of biomedical systems for drug delivery as the hydrogel formed without any potentially toxic guest moieties for CB[8]-based cross-linking. Nontoxicity and the stimuli-responsive nature of the prepared SPH, as well as easy synthesis of the various components, make this system well suited for a variety of important biomedical applications.[11]

The investigations on the SPH formation *via* 1 : 2 host–guest recognition of CB[8] with amino acids were continued by developing a facile method for the functionalization of polysaccharides with amino acid. As previously described, polysaccharides are one of the most widely used biopolymers for the fabrication of SPHs. However, the functionalization of polysaccharides for the introduction of dynamic and noncovalent binding moieties is a difficult process due to their high viscose mixtures, large degrees of steric hindrance and strong hydrogen bonding networks that disables reactive functionalities such as hydroxyl groups. In this work, a facile and mild two-step method was reported for the chemical functionalization of polysaccharides hyaluronic acid (HA), carboxymethyl cellulose (CMC), hydroxyethyl cellulose (HEC) and guar with the dipeptide, cysteine-phenylalanine (CF), which is a good guest for CB[8]. As can be seen in Figure 6.11, methacrylate-functionalized polysaccharides were synthesized (HA-MA, HEC-MA and CMC-MA) and then functionalized with pendant dipeptide CF through the *in situ* reduction of

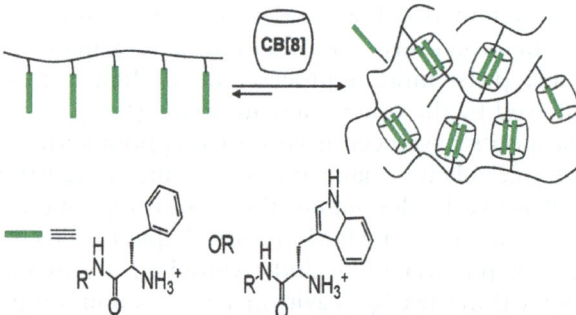

**Figure 6.10**   Schematic representation of the hydrogel formation based on 1 : 2 homoternary complexation of CB[8] and amino acid–bearing styrenic polymers.
Reproduced from ref. 11 with permission from the Royal Society of Chemistry.

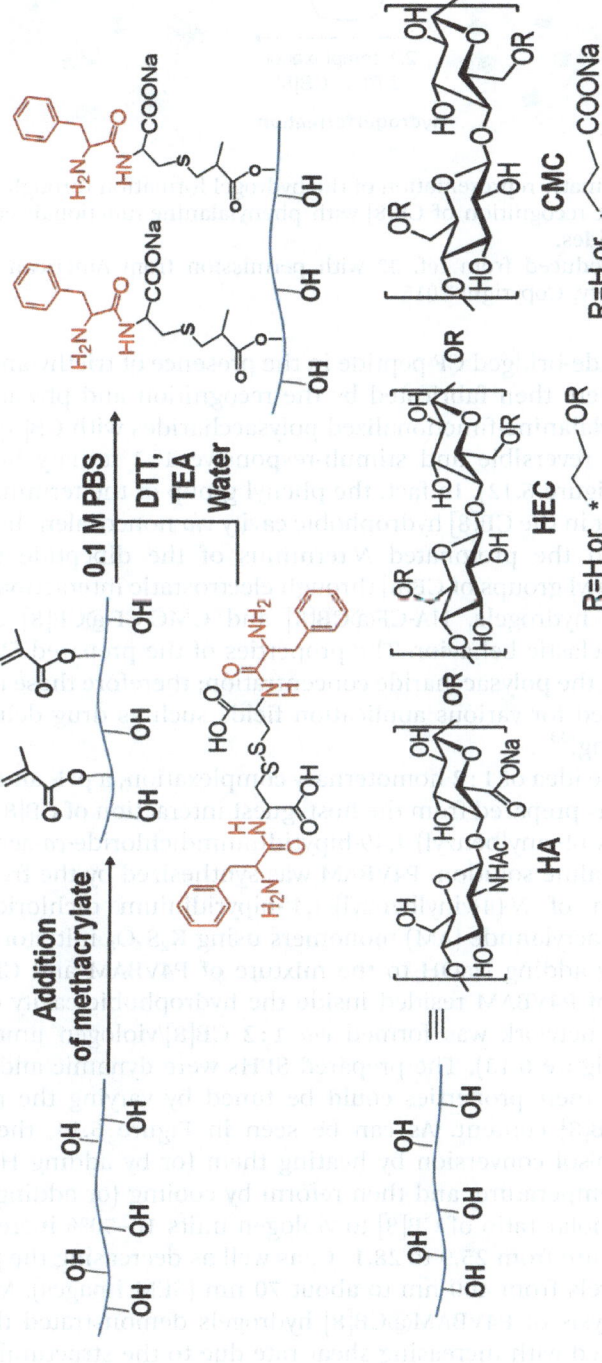

**Figure 6.11** Schematic representation of the two-step method for functionalization of polysaccharides with the dipeptide, cysteine-phenylalanine.
Reproduced from ref. 33 with permission from American Chemical Society, Copyright 2015.

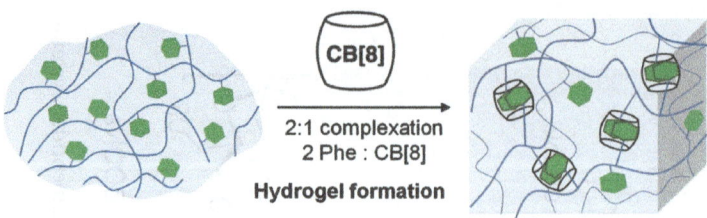

**Figure 6.12** Schematic representation of the hydrogel formation through 1 : 2 host–guest recognition of CB[8] with phenylalanine-functionalized polysaccharides.
Reproduced from ref. 33 with permission from American Chemical Society, Copyright 2015.

precursor disulfide-bridged CF peptide in the presence of triethylamine. The SPH networks were then fabricated by the recognition and physical cross-linking of phenylalanine-functionalized polysaccharides with CB[8] in water through strong, reversible and stimuli-responsive 1 : 2 ternary host–guest complexation (Figure 6.12). In fact, the phenyl group of the terminal amino acid was resident in the CB[8] hydrophobic cavity *via* noncovalent host–guest interactions, and the protonated *N*-terminus of the dipeptide unit was bonded to carbonyl groups of CB[8] through electrostatic interaction. Among the synthesized hydrogels, HA-CF@CB[8] and CMC-CF@CB[8] exhibited impressive viscoelastic behavior. The properties of the prepared SPHs were tuned by varying the polysaccharide concentration; therefore these materials can be engineered for various application fields such as drug delivery and tissue engineering.[33]

Using the same idea of 1 : 2 homoternary complexation, a pH- and thermosensitive SPH was prepared from the host–guest interaction of CB[8] and MV units of poly(*N*-(4-vinylbenzyl)-4,49-bipyridiniumdichloride-*co*-acrylamide) (P4VBAM) in alkaline solution. P4VBAM was synthesized by the free radical copolymerization of *N*-(4-vinylbenzyl)-4,4'-bipyridinium dichloride (4VB) monomers with acrylamide (AM) monomers using $K_2S_2O_8$ initiator in water (Figure 6.13). By adding NaOH to the mixture of P4VBAM and CB[8], two bipyridyl units of P4VBAM resided inside the hydrophobic cavity of CB[8], and a hydrogel network was formed *via* 1 : 2 CB[8]/viologen units binary complexation (Figure 6.13). The prepared SPHs were dynamic and stimuli-responsive, and their properties could be tuned by varying the pH, temperature and CB[8] content. As can be seen in Figure 6.14, these SPHs undergo a gel-to-sol conversion by heating them (or by adding HCl), flow by raising the temperature, and then reform by cooling (or adding NaOH). Increasing the molar ratio of CB[8] to viologen units 10–50% increased the gel–sol temperature from 25.9 to 28.1 °C, as well as decreasing the pore size within the xerogels from 400 nm to about 70 nm (SEM images). Moreover, rheological analysis of P4VBAM@CB[8] hydrogels demonstrated that their viscosity decreased with increasing shear rate due to the structural damage of the hydrogels, and this demonstrated that the prepared P4VBAM@CB[8]

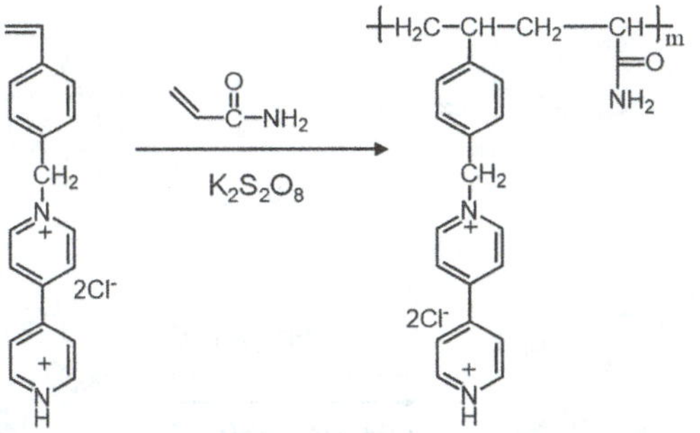

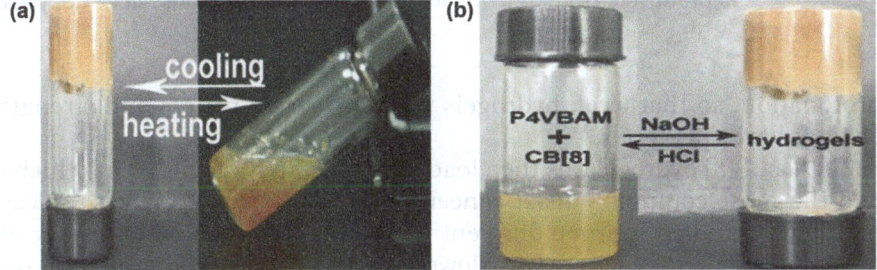

**Figure 6.13** Synthesis reaction of P4VBAM copolymer.
Reproduced from ref. 9 with permission from the Royal Society of Chemistry.

**Figure 6.14** Reversible gel–sol conversion of P4VBAM@CB[8] hydrogels by varying (a) temperature and (b) pH.
Reproduced from ref. 9 with permission from the Royal Society of Chemistry.

hydrogels are viscoelastic, bearing both elastic and viscous responses to mechanical disruption due to the dynamic nature of the ternary complexation.[9]

A second example of the pH and thermo-responsive SPH was reported through 1 : 2 host–guest recognition of CB[8] with modified chitosan, *N*-(4-diethylaminobenzyl) chitosan (EBCS), in aqueous acidic solution. Upon the addition of CB[8], the hydrophobic cavity of this host accommodated the aromatic ring of an EBCS guest, and therefore the host–guest interactions, most probably ion–dipole interactions between the protonated diethyl aminobenzyl groups with CB[8] portals, led to the formation of the novel SPH with thermo- and pH-sensitive properties (Figure 6.15). Upon heating, this SPH to 50 °C, it transformed into the solution due to the weakening of the noncovalent host–guest interactions; therefore the benzene ring moved out from the CB[8] cavity, and the hydrogel network was gradually degraded. EBCS@CB[8] hydrogel could be formed upon cooling to room temperature. Herein, solution pH was critical factor for hydrogel stability and strength due to the importance of the diethyl aminobenzyl group

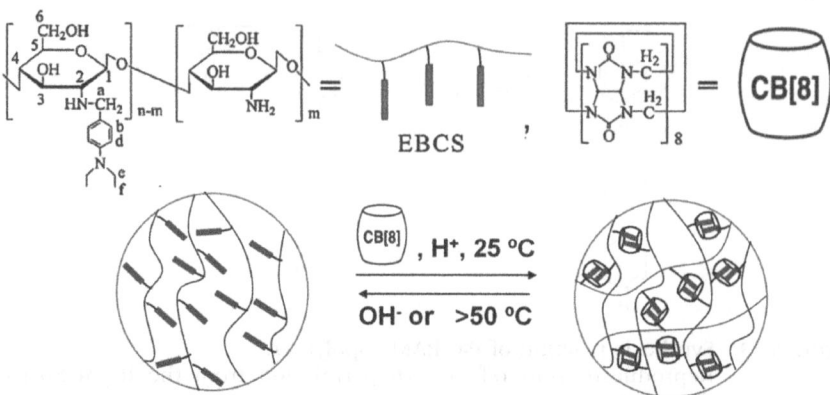

**Figure 6.15**  Schematic representation of the formation of EBCS@CB[8] supramolecular hydrogel network in acidic solution.
Reproduced from ref. 41 with permission from Elsevier, Copyright
2013.

protonation, so that weak hydrogels were formed at pH 6.8 and strengthened at lower pH levels.[41]

EBCS@CB[8] hydrogel was then loaded with 5-fluorouracil (5-Fu), which is
one the most commonly used cancer drugs, and then its *in vitro* release
profile was studied in three different pH values (1.2, 4 and 6.8). In low pH
media, the drug release rate was slower, probably due to the formation of a
strong EBCS@CB[8] network in the strong acidic solutions. On the other
hand, at a higher pH value (6.8), the drug released quickly because the
hydrogel network was gradually disassembled. Therefore, the ease of using
this EBCS@CB[8] hydrogel in biomedical applications was quite restricted.[41]

Using such a homoternary complexation approach, a novel self-healing
SPH was synthesized by Cao and his coworkers in 2016, through the
host–guest recognition of CB[8] with naphthaline groups of poly($N^1,N^1,N^2,N^2$-
tetramethyl-$N^1$-(naphthalen-2-ylmethyl)-$N^2$-(4-vinylbenzyl) ethane-1,2 diami-
nium-*co-N,N* dimethylacrylamide) (PTNVE/DMA) (Figure 6.16). Enhanced π–π
interaction of two naphthaline units with CB[8], as well as the dipole–dipole
interaction between the ammonium cations of PTNVE/DMA and carbonyl
groups of CB[8], led to the formation of stable complexes with a self-healing
property. The self-healing property of the PTNVE/DMA@CB[8] hydrogel was
affected by varying the molar ratio of CB[8] to naphthaline units. Additionally, adding a competitive guest such as 1-adamantanamine hydrochloride
(ADA) into hydrogels showed that the formation and degradation of supramolecular hydrogels are reversible and further indicated that the main mo-
tive force of the gelation is the host–guest interaction between CB[8] and
naphthaline groups in the side chains of PTNVE/DMA. These self-healing
PTNVE/DMA@CB[8] hydrogels, with a highly ordered cavity microstructure
and facile synthesis method, can be used in several fields such as biomedical
materials.[42]

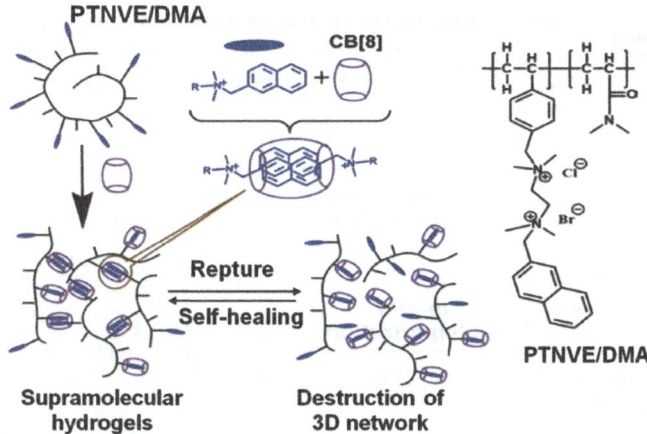

**Figure 6.16** Schematic representation of the formation of self-healing PTNVE/ DMA@CB[8] hydrogels.
Reproduced from ref. 42 with permission from Taylor & Francis, Copyright 2016.

## 6.4.3 Newly Developed CB[8]-assisted Supramolecular Polymeric Hydrogels with Improved Mechanical Properties

Most of the previously reported CB[8]-SPHs have been synthesized *via* the dynamic host–guest complexation of CB[8] with the mixtures of two auxiliary functionalized polymers in aqueous media. However, these supramolecular networks have moderate mechanical characteristics, and a higher degree of dynamic cross-linking is needed to improve them.[43,44] Unfortunately, this strategy has been restricted due to the limited aqueous solubility of CB[8], the high viscosity of the polymer solution, as well as the limited potential for scale-up and manipulation.

Scherman and coworkers in 2017 reported a facile and modular synthetic approach to prepare stimuli-responsive CB[8]-SPH based on *in situ* polymerization of vinyl monomer solution (acrylamide-, acrylate- and imidazolium-based hydrophilic monomers) in the presence of a CB[8] host and polymerizable guest molecule (Figure 6.17). 1-Benzyl-3-vinylimidazolium bromide was applied as a polymerizable guest molecule to form 1 : 2 complexes with the CB[8] host, and the CB[8] ternary conjugate is then served as a dynamic cross-linker to build the hydrogel network after photo-initiated polymerization at room temperature. Predetermined amounts of CB[8], vinyl monomers and 1-benzyl-3-vinylimidazolium bromide were dissolved in deionized water under a nitrogen gas atmosphere, and the solution was then exposed to UV irradiation (4.8 mW cm$^{-2}$) for 6 h in the presence of photo-initiator.[43]

Using such an *in situ* polymerization approach, higher loading of CB[8] as well as overall monomer concentration can be accessed compared with previously reported CB[8] hydrogels. This leads to the fabrication of hydrogel

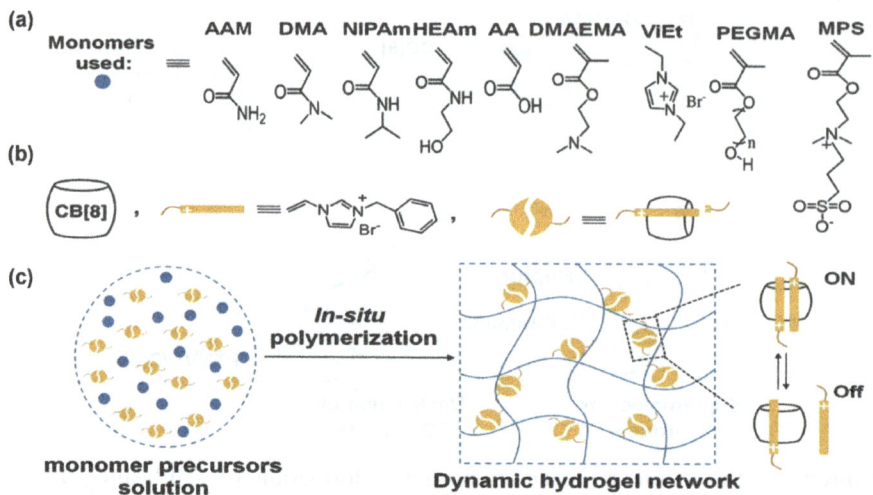

**Figure 6.17**   Schematic representation of (a) the dynamic supramolecular hydrogel
network generated *via* (b) photo-initiated *in situ* polymerization of vinyl
monomers in the presence of CB[8] and (c) a polymerizable guest
molecule.
Reproduced from ref. 43, https://doi.org/10.1002/pola.28667, under the
terms of the CC BY 4.0 licence, https://creativecommons.org/licenses/
by/4.0/.

networks with higher degrees of noncovalent cross-linking and to further results in outstanding properties such as improved mechanical strength, self-
healing, energy dissipation, toughness and shear-thinning. Furthermore, a
wide variety of monomers and guest molecules can be utilized in this approach, leading to a library of hydrogel networks with a range of desirable
properties. The photo-initiated *in situ* polymerization approach can be implemented in large scale, and the prepared CB[8]-SPHs have shown great
promise in a number of important application fields such as wearable and
self-healable electronic devices, sensors and structural biomaterials.[43–45]

Using the same idea, a novel wound dressing material based on the CB[8]-
SPH was fabricated by Zhang *et al.* in 2017 through the *in situ* polymerization
of a supramonomer. As is well-known, frequent wound dressing changes can
cause the individual anxiety and stress, as well as additional trauma to newly
formed tissue. Hydrogels are one of the modern wound dressings that have
most of the desirable properties of an ideal dressing. However, hydrogel
wound dressings often adhere to the wound, and therefore their removal is
painful and time-consuming and can damage the newly formed tissue.
Herein, CB[8]-SPH was developed as a wound dressing material that can
accelerate wound healing and be easily removed from a wound. For this, in
an aqueous solution containing CB[8], tripeptide Phe-Gly-Gly ester derivative
(FGG-EA) and acryl amid (AAm), the 1 : 2 host–guest recognition of CB[8] and
FGG-EA yielded a cross-linkable supramonomer with two terminal acrylate

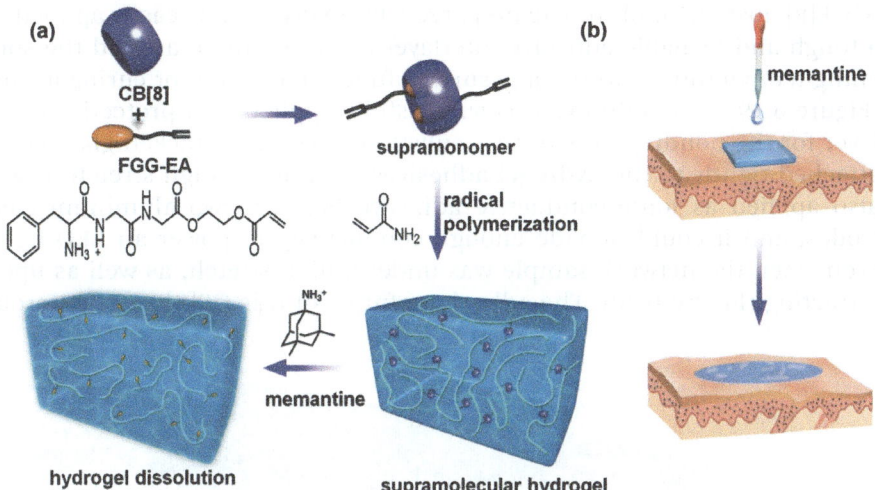

**Figure 6.18**  Schematic representation of the (a) supramolecular hydrogel network prepared from photo-initiated *in situ* polymerization of acrylamide and FGG-EA@CB[8] supramonomer; (b) its dissolution process upon memantine irrigation.
Reproduced from ref. 12 with permission from American Chemical Society, Copyright 2017.

moieties, which further copolymerized with AAm through the photo-initiated *in situ* copolymerization and formed the transparent CB[8]-SPH (Figure 6.18), which was nontoxic and self-healable with desirable mechanical properties. The prepared CB[8]-SPH dissolved within 2 min after exposure to memantine (a potential wound irrigant and a competing guest molecule for CB[8]); therefore it can relieve pain, reduce wound healing time and be easily removed from a wound. Furthermore, the ofloxacin-loaded CB[8]-SPH showed an acceptable release profile within 1 h, which demonstrated that the formed SPH has the ability to load and deliver therapeutic agents such as antimicrobials to prevent infection and promote wound healing.[12]

Using the same *in situ* polymerization approach, CB[8]-SPHs as dynamic and healable adhesives for various nonporous and porous substrates were fabricated in 2018. As is well-known, gel networks and elastomeric materials have excellent adhesion properties to solid substrates; however, such an adhesion strategy is not usable in high water content systems such as those used in structural biomaterials and wearable electronic devices. Herein, CB[8]-SPH networks can be developed to solve these problems. The incorporation of CB[8] molecular recognition into the adhesives not only improves the interfacial toughness between disparate substrates but also endues unique properties such as reversibility and self-healable adhesion.[46]

To this, the CB[8]-SPH networks were fabricated *via* photo-initiated *in situ* polymerization of acryl amid monomer (AAm, 94 mol%) and 1-benzyl-3-vinylimidazolium bromide monomer (5 mol%) in the presence of CB[8] host (2.5 mol%). The prepared CB[8]-SPHs can be easily applied as a tough and healable adhesive interlayer by curing them around the softening temperature, without any surface functionalization or curing agents (Figure 6.19). The adhesive was successfully applied as a protective interlayer for the bonding of two glass substrates, whereby cracked glass chips attached tightly to the hydrogel adhesives even under high stretch. It was also applied as ionic conductive adhesive between two aluminum electrodes, and it could provide enough conductivity to power an LED light, even when the network sample was under a high stretch, as well as upon retracting (Figure 6.20). The adhesion affinity of this CB[8]-SPH to porous

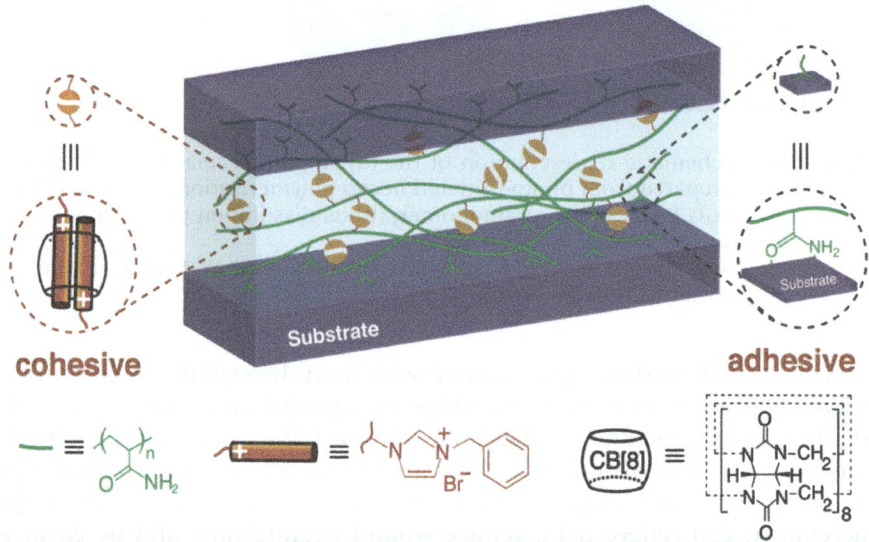

**Figure 6.19**   Schematic representation of the CB[8]-based polymeric hydrogel as an adhesive interlayer between two substrates.
Reproduced from ref. 46 with permission from John Wiley & Sons, © 2018 WILEY-VCH Verlag GmbH & Co. KGaA, Weinheim.

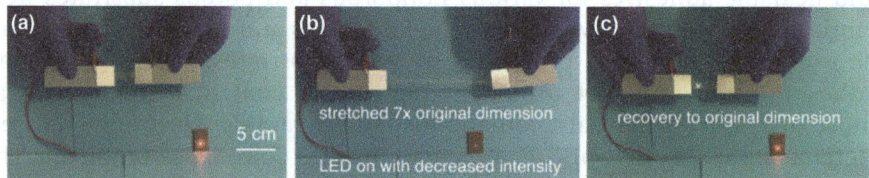

**Figure 6.20**   Photographs of the (a) tough bonding of two aluminum electrodes with the CB[8]-based hydrogel network, which provides enough conductivity to power an LED, (b) even when the network sample is under a high stretch (c) and upon retracting.
Reproduced from ref. 46 with permission from John Wiley & Sons, © 2018 WILEY-VCH Verlag GmbH & Co. KGaA, Weinheim.

substrates such as wood (Sitka spruce) and bone (pig phalange samples) was further evaluated. The wood joints were tightly bonded by hydrogel interlayer, and they could easily hold a weight up to several kilograms. Meantime, a fracture in the wood (away from the adhesion layer) was observed prior to any mechanical failure within the bonded lap joints. Furthermore, two bone pieces (pig phalange samples) were strongly bonded together with the CB[8]-SPH adhesive, whereby the interfacial adhesion stress reached over 2 MPa. This study demonstrated the high potential of such CB[8]-SPH adhesives in many application fields such as stretchable and wearable electronics, wood conservation, restoration of important historical artifacts and tissue/bone regeneration.[46]

## 6.5 Functionalized CB[6]-assisted Supramolecular Polymeric Hydrogels

As previously described (Section 6.3.1), CB[6] can make exceptionally stable complexes with aminoalkanes, especially 1,6-diaminohexane (DAH) *via* the strong host–guest interactions. Therefore, another applicable method for preparation of supramolecular polymeric hydrogels has been introduced based on the two-component system of CB[6]@alkylammonium ion host–guest pairs. Herein, hyaluronic acid (HA) has been the most used as polymer backbone. HA is a polysaccharide that naturally presents in the human body. Functionalized CB[6]-SPHs are formed *via* the 1 : 1 host–guest interactions of CB[6]-grafted HA and aminoalkane-grafted HA. The most interesting feature of such hydrogels is that free alkylamonnium moieties of the network can be further modularly functionalized with divers multifunctional tags-attached CB[6]. The functionalized CB[6]-SPHs have been known as excellent candidates for many biomedical applications such as drug delivery and tissue engineering.

Taking advantage of the strong host–guest interaction of alkylammonium to CB[6] under physiological conditions, Kim and his coworkers reported the fabrication of biocompatible and robust CB[6]-SPHs through simple mixing of CB[6]-conjugated hyaluronic acid (CB[6]-HA) and diaminohexane-conjugated HA (DAH-HA). CB[6]-HA was synthesized by attaching (allyloxy)$_{12}$CB[6] to a thiol-functionalized HA through a thiol–ene click reaction. The high affinity of (DAH) units for CB[6] in the presence of cells led to the *in situ* formation of cell-captured CB[6]/DAH-HA hydrogels after 2 min. The remaining DAH-HA moieties of the hydrogel network could be modularly modified with fluorescent marker tagged-CB[6] (Figure 6.21). The CB[6]/DAH-HA hydrogels with 2 wt% polymer loading showed remarkably high storage modulus (2.4 ± 0.2 kPa) and acceptable biocompatibility. Furthermore, they were soft enough to use as an extracellular matrix for cell proliferation and differentiation without any cytotoxicity. The enzymatic degradability of the hydrogel network was confirmed with hyaluronidase (HAase) treatment, which showed that the enzymatic degradation occurred

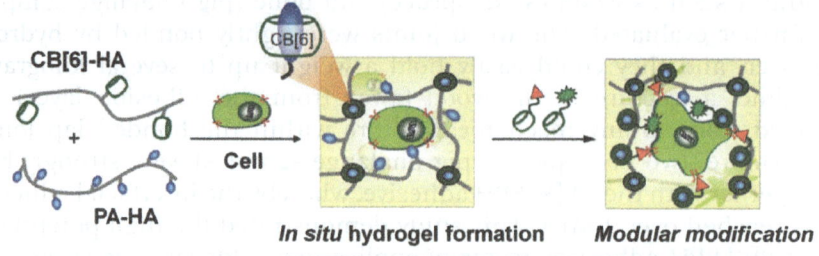

**Figure 6.21** Schematic representation of the *in situ* formation of CB[6]/DAH-HA hydrogel and its modular modification.
Reproduced from ref. 13 with permission from American Chemical Society, Copyright 2012.

within 24 h after the treatment process. Subsequently, the *in situ* formation of CB[6]/DAH-HA hydrogel under the skin of mice was confirmed by sequentially injecting CB[6]-HA and DAH-HA solutions. The hydrogel was formed after a few minutes and preserved its shape for more than 2 weeks. The formed hydrogel was modularly modified *in vivo* by the simple injection of fluorescein isothiocyanate-CB[6] (FITC-CB[6]) and emitted fluorescence for up to 11 days. Generally, these unique characteristics demonstrate the potential of CB[6]/DAH-HA hydrogels as a biocompatible 3D artificial extracellular matrix (ECM) for various biomedical applications such as tissue engineering and drug delivery.[13]

The prepared CB[6]/DAH-HA hydrogel was later developed as an ECM for the spatiotemporally controlled chondrogenesis of human mesenchymal stem cells (hMSCs).[47] Mesenchymal stem cells (MSCs) are one of the most studied stem cells for tissue engineering applications due to their ability to self-replicate and differentiate into various cell types such as adipocytes, osteocytes, chondrocytes, myocytes, cardiomyocytes and neurons, as well as their long life span, facile proliferation *ex vivo*, tumor tropism, and prolonged transgene expression.[48] Herein, one of the main aims is to keep the cell population at the delivered sites and to deliver sufficient growth factors for effective tissue regeneration.[47] MSCs can be controlled by external stimuli such as biologically active molecules (small molecules, peptides and proteins) and the physical properties of ECM. Supramolecular hydrogels are promising candidates to use as artificial ECMs for the differentiation of MSCs to specific cells and tissues.[48] In the present study, CB[6]-HA was synthesized *via* thiol-ene click photoreaction of monoallyoxy-functionalized CB[6] with thiol-functionalized HA, and then it was mixed with DAH-HA in the presence of hMSCs and transforming growth factors-β3 (TGF-β3), resulting in the *in situ* formation of hMSCs-encapsulated monoCB[6]/PA-HA hydrogels with controlled cross-linking density and physical properties. TGF-β3 was also encapsulated in the hydrogel network, and free DAH moieties were modularly modified with dexamethasone-CB[6] (Dexa-CB[6]) *via* the strong host–guest interaction of CB[6] and DAH. It was observed that more than 95% of the hMSCs in these hydrogels survived and proliferated, even after 10 days' incubation. The modularly modified monoCB[6]/DAH-HA hydrogels with FITC-CB[6] emitted florescence for more than 10 days. Spatial control of the HMSCs was achieved by modulating cross-link density of the hydrogel *via* controlling the functionalization amount of HA with mono-allyloxy-CB[6]. Moreover, the differentiation of the hMSCs was controlled by various release profiles of TGF-β3 and dexamethasone (Dexa). The effective chondrogenic differentiation of hMSCs encapsulated in the monoCB[6]/DAH-HA hydrogel with TGF-β3 and Dexa-CB[6] was confirmed by bio-chemical glycosaminoglycan content analysis, real-time quantitative PCR, histological and immunohistochemical analyses.[47]

In a subsequent study, CB[6]/HA hydrogel was developed as an artificial ECM for the long-term MSC therapy and tissue engineering. MSCs were first engineered with recombinant adenoviral (rAd) vectors carrying enhanced green fluorescent protein (EGFP) or IL-12M-producing genes (EGFP/rAd or IL-12M/rAd vectors) after making a complex with $Fe^{3+}$ (Figure 6.22(a)). The CB[6]/HA hydrogels were then prepared *via in situ* hydrogel formation between monoCB[6]-HA and DAH-HA in the presence of engineered MSC (eMSC) and modularly modified with dexa-$CB_{12}$[6] (Figure 6.22(b)). All-*trans*-retinoic acid (RA) drug was also introduced to hydrogels *via* the modification of the HA backbone in DAH-HA. The combination of Dexa and RA cues in the hydrogel network produced a synergistic effect on transgene expression. EGFP was well expressed by the eMSCs in Dexa-monoCB[6]/RA-DAH-HA

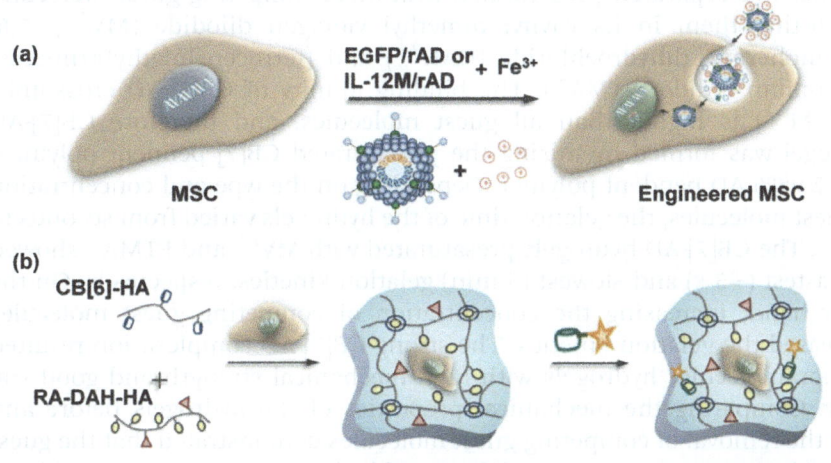

**(a)**

MSC    EGFP/rAD or IL-12M/rAD + $Fe^{3+}$    Engineered MSC

**(b)**

CB[6]-HA

RA-DAH-HA

*In situ* **Hydrogel Formation    Modular Post-Modification**

**Figure 6.22** Schematic representation of the (a) transduction of mesenchymal stem cells (MSC); (b) *in situ* formation of Dexa-monoCB[6]/RA-DAH-HA hydrogel in the presence of eMSC and its modular modification. Reproduced from ref. 48 with permission from John Wiley & Sons, Copyright © 2014 WILEY-VCH Verlag GmbH & Co. KGaA, Weinheim.

hydrogels in mice for more than 60 days, exhibiting the biological activity of eMSCs. Moreover, IL-12M was constantly produced in the hydrogels by eMSCs/IL-12M, inhibiting tumor growth and remarkably improving the survival period. This study confirmed the ability of Dexa-monoCB[6]/RA-DAH-HA hydrogels to use as a platform scaffold for tissue engineering and the treatment of cancer, as well as other intractable diseases such as stroke and central nervous system disorders.[48]

## 6.6 Injectable and Printable CB-assisted Supramolecular Polymeric Hydrogels with Controllable Gelation Kinetics

The effective control of gelation kinetics of CB-SPHs without any side effect on the structure and properties of these materials is one of the major challenges in hydrogel technology, especially in the field of injectable and printable hydrogels. Herein, a method based on the competing guest molecules was introduced to control the gelation kinetics of CB[7]-adamantane (CB[7]-AD) cross-linked supramolecular hydrogels. The formation of CB[7]-AD pair is fast due to the strong host–guest interaction ($K_{eq}$ up to $10^{14}$ $M^{-1}$), and this process can be hindered by presaturation of CB[7] with a weak competing guest molecule before its interaction with AD, which further results in deceleration of gelation kinetics of the CB[7]-AD cross-linked SPHs. To this end, two *N,N*-dimethyl acrylamide–based polymers with CB[7] and AD pendants were first synthesized, and before mixing, the CB[7] pendant polymer was separately presaturated with three competing guest molecules by hosting them in its cavity: dimethyl viologen diiodide ($MV^{2+}$), 1,6-diaminohexane dihydrochloride ($DAH^{2+}$) and (ferrocenylmethyl)trimethyl ammonium iodide ($FTMA^+$). The binding affinity of CB[7]-AD cross-links ($10^{14}$ $M^{-1}$) is higher than all guest molecules, and therefore CB[7]-AD hydrogel was formed by mixing the presaturated CB[7] pendent polymers with 2 wt% AD pendent polymer. Depending on the type and concentration of guest molecules, the gelation time of the hydrogels varied from seconds to hours. The CB[7]-AD hydrogels presaturated with $MV^{2+}$ and $FTMA^+$ showed the fastest (~5 s) and slowest (5 min) gelation kinetics, respectively. On the other hand, increasing the concentration of competing guest molecules decreased the gelation kinetics. The strong CB[7]-AD complexation resulted in supramolecular hydrogels with high mechanical strength and good stability. Comparing the mechanical properties of the hydrogels before and after the removal of competing guest molecules demonstrated that the guest molecules were not a structural factor of hydrogels and that they did not have any significant effect on the mechanical property of the hydrogels.[49]

Furthermore, the ability of the developed process to control the gelation kinetics of CB[7]-AD cross-linked hydrogels was examined for injection and printing applications (Figure 6.23). Fluorescein isothiocyanate (FITC)-labeled hydrogels were used for better observation. $DAH^{2+}$ was selected as

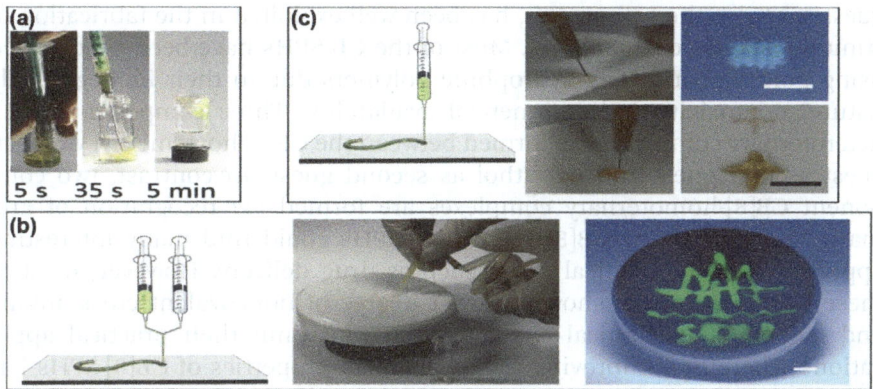

**Figure 6.23** Photographs of the (a) injection of CB[7]-AD cross-linked hydrogels treated with DAH$^{2+}$; (b) printing of CB[7]-AD cross-linked hydrogel treated with DAH$^{2+}$; (c) printing of CB[7]-AD cross-linked hydrogel treated with FTMA$^+$.
Reproduced from ref. 49, https://doi.org/10.1038/srep20722, under the terms of the CC BY 4.0 licence, https://creativecommons.org/licenses/by/4.0/.

the model competing guest molecule due to its proper gelation kinetics (5 min). Injection of CB[7]-AD cross-linked hydrogels was done by vortex mixing of FITC labeled–CB[7] pendant polymer treated with DAH$^{2+}$ and AD pendent polymer. The solution was then injected into water by syringe and gelled in water within 5 min (Figure 6.23(a)). Printing of CB[7]-AD cross-linked hydrogel treated with DAH$^{2+}$ was also tested by mixing the polymer solutions during printing. The mixture gelled soon after its printing on the substrate (Figure 6.23(b)). Moreover, printing the CB[7]-AD cross-linked hydrogel presaturated with FTMA$^+$ (with a slow gelation time of 5 h) was also done by mixing the polymer solutions before the printing process. The mixture was left for 2 h to achieve the proper mechanical strength, and then it was printed on pristine substrate. After curing for 24 h, the hydrogel patterns could be transferred (Figure 6.23(c)).[49]

## 6.7 Summary and Outlook

In this chapter, we provided an overview of the fabrication, properties and applications of CB-SPHs formed through the host–guest interactions of CB homologues. The supramolecular cross-linking of hydrophilic polymer chains through dynamic noncovalent interactions results in the fabrication of reversible 3D SPH networks with stimuli-responsivity, processability and tunable mechanical properties. These CB-SPHs can be modified with various functional components to achieve the desired functions such as bioactivity and optoelectronic properties, as well as self-healing and shape-memory abilities.

Among CB homologues, the ability of CB[8] to form ternary complexes through the accommodation of two guest molecules, either two different guests or twice the same guest, has been well exploited in the fabrication of stimuli-responsive CB[8]-SPHs. Most of the CB-SPHs have been synthesized using polysaccharides as hydrophilic polymers due to their biodegradable nature, nontoxicity and commercial availability. Three component CB[8]-heteroternary complexes are formed between the CB[8] host, methyl viologen guest as first guest and naphthol as second guest. In contrast, two component CB[8]-homoternary complexes are formed *via* recognition of aromatic amino acids by CB[8]. The CB[8]-SPHs could find many interesting applications in biomedical fields such as drug delivery. However, most of these CB[8]-SPHs have shown a lower degree of noncovalent cross-linking and moderate mechanical characteristics that limit their practical applications. Therefore, improving the mechanical properties of CB[8]-SPHs has become increasingly necessary. Recently, *in situ* polymerization of vinyl monomer in the presence of the CB[8] host and polymerizable guest molecule has been used as a facile and simple synthesis method for the fabrication of CB[8]-SPHs with a higher degree of noncovalent cross-linking. This approach can be implemented in large scale. Furthermore, various kinds of monomers and guest molecules can be utilized in this method that enlarge the use of CB[8]-SPHs with enhanced mechanical and biological properties in a number of important fields such as wearable and self-healable electronic devices, sensors and structural biomaterials.

CB[6] and CB[7] have been also used as supramolecular hosts for the construction of CB-SPHs through 1 : 1 host–guest interactions. Cross-linking of polymer chains with strong CB[6]@alkylammonium host–guest pair is an effective motif for the fabrication of dynamic functionalized CB[6]-SPHs. Free alkylammonium moieties of the hydrogel networks can be further modularly functionalized with divers multifunctional tags-attached CB[6]. These CB[6]-SPHs have been known as excellent candidates for many biomedical applications such as drug delivery and tissue engineering. Furthermore, another strong host–guest pair, CB[7]-adamantane, has been used in the development of injectable and printable CB[7]-SPHs through the control of the gelation kinetics of prepared hydrogels. This is achieved by presaturation of CB[7] with a weak competing guest molecule before its interaction with AD.

In summary, considerable progress has been made in developing CB-SPHs for the desired applications. However, it is quite important to develop new flexible approaches that could provide a high degree of dynamic cross-linking in the synthesis of CB-SPHs, as well as improved mechanical and physicochemical properties.

# Acknowledgements

We gratefully acknowledge the financial support of the Scientific and Technological Research Council of Turkey (TUBITAK) grant number 215Z035 and Bikent University, UNAM-National Nanotechnology Research Center.

# References

1. T. R. Hoare and D. S. Kohane, *Polymer*, 2008, **49**, 1993.
2. E. A. Appel, F. Biedermann, U. Rauwald, S. T. Jones, J. M. Zayed and O. A. Scherman, *J. Am. Chem. Soc.*, 2010, **132**, 14251.
3. R. Dong, Y. Pang, Y. Su and X. Zhu, *Biomater. Sci.*, 2015, **3**, 937.
4. E. A. Appel, J. del Barrio, X. J. Loh and O. A. Scherman, *Chem. Soc. Rev.*, 2012, **41**, 6195.
5. H. Zhang, L. Bré, T. Zhao, B. Newland, M. Da Costa and W. Wang, *J. Mater. Chem. B*, 2014, **2**, 4067.
6. Y. Dong, W. U. Hassan, R. Kennedy, U. Greiser, A. Pandit, Y. Garcia and W. Wang, *Acta Biomater.*, 2014, **10**, 2076.
7. N. Peppas, Y. Huang, M. Torres-Lugo, J. Ward and J. Zhang, *Annu. Rev. Biomed. Eng.*, 2000, **2**, 9.
8. E. A. Appel, X. J. Loh, S. T. Jones, F. Biedermann, C. A. Dreiss and O. A. Scherman, *J. Am. Chem. Soc.*, 2012, **134**, 11767.
9. H. Yang, H. Chen and Y. Tan, *RSC Adv.*, 2013, **3**, 3031.
10. J. R. McKee, E. A. Appel, J. Seitsonen, E. Kontturi, O. A. Scherman and O. Ikkala, *Adv. Funct. Mater.*, 2014, **24**, 2706.
11. M. J. Rowland, E. A. Appel, R. J. Coulston and O. A. Scherman, *J. Mater. Chem. B*, 2013, **1**, 2904.
12. W. Xu, Q. Song, J.-F. Xu, M. J. Serpe and X. Zhang, *ACS Appl. Mater. Interfaces*, 2017, **9**, 11368.
13. K. M. Park, J.-A. Yang, H. Jung, J. Yeom, J. S. Park, K.-H. Park, A. S. Hoffman, S. K. Hahn and K. Kim, *ACS Nano*, 2012, **6**, 2960.
14. M. C. I. M. Amin, N. Ahmad, M. Pandey, M. M. Abeer and N. Mohamad, *Expert Opin. Drug Delivery*, 2015, **12**, 1149.
15. W. Wang, H. Wang, C. Ren, J. Wang, M. Tan, J. Shen, Z. Yang, P. G. Wang and L. Wang, *Carbohydr. Res.*, 2011, **346**, 1013.
16. B. Hu, C. Owh, P. L. Chee, W. R. Leow, X. Liu, Y.-L. Wu, P. Guo, X. J. Loh and X. Chen, *Chem. Soc. Rev.*, 2018, **47**, 6917.
17. J. Li, in *Inclusion Polymers*, ed. G. Wenz, Springer, Berlin, Heidelberg, 2009, vol. 222, pp. 175.
18. G. Chen and M. Jiang, *Chem. Soc. Rev.*, 2011, **40**, 2254.
19. X.-L. Ni, X. Xiao, H. Cong, L.-L. Liang, K. Cheng, X.-J. Cheng, N.-N. Ji, Q.-J. Zhu, S.-F. Xue and Z. Tao, *Chem. Soc. Rev.*, 2013, **42**, 9480.
20. S. Gürbüz, M. Idris and D. Tuncel, *Org. Biomol. Chem.*, 2015, **13**, 330.
21. N. Al-Dubaili, K. El-Tarabily and N. I. Saleh, *Sci. Rep.*, 2018, **8**, 2839.
22. M. Wiemann and P. Jonkheijm, *Isr. J. Chem.*, 2018, **58**, 314.
23. T. Loftsson and D. Duchêne, *Int. J. Pharm.*, 2007, **329**, 1.
24. M. S. A. Khan, Ph.D. Dissertation, Masaryk University, Faculty of Science, 2012.
25. J. W. Lee, S. Samal, N. Selvapalam, H.-J. Kim and K. Kim, *Acc. Chem. Res.*, 2003, **36**, 621.
26. S. Liu, C. Ruspic, P. Mukhopadhyay, S. Chakrabarti, P. Y. Zavalij and L. Isaacs, *J. Am. Chem. Soc.*, 2005, **127**, 15959.

27. W. S. Jeon, K. Moon, S. H. Park, H. Chun, Y. H. Ko, J. Y. Lee, E. S. Lee, S. Samal, N. Selvapalam and M. V. Rekharsky, *J. Am. Chem. Soc.*, 2005, **127**, 12984.

28. M. V. Rekharsky, T. Mori, C. Yang, Y. H. Ko, N. Selvapalam, H. Kim, D. Sobransingh, A. E. Kaifer, S. Liu and L. Isaacs, *Proc. Natl. Acad. Sci. U. S. A.*, 2007, **104**, 20737.

29. S. Moghaddam, C. Yang, M. Rekharsky, Y. H. Ko, K. Kim, Y. Inoue and M. K. Gilson, *J. Am. Chem. Soc.*, 2011, **133**, 3570.

30. J. Tian, L. Zhang, H. Wang, D.-W. Zhang and Z.-T. Li, *Supramol. Chem.*, 2016, **28**, 769.

31. F. Biedermann and O. A. Scherman, *J. Phys. Chem. B*, 2012, **116**, 2842.

32. U. Rauwald, F. Biedermann, S. P. Deroo, C. V. Robinson and O. A. Scherman, *J. Phys. Chem. B*, 2010, **114**, 8606.

33. M. J. Rowland, M. Atgie, D. Hoogland and O. A. Scherman, *Biomacromolecules*, 2015, **16**, 2436.

34. D. Pasqui, M. De Cagna and R. Barbucci, *Polymers*, 2012, **4**, 1517.

35. E. A. Appel, X. J. Loh, S. T. Jones, C. A. Dreiss and O. A. Scherman, *Biomaterials*, 2012, **33**, 4646.

36. Z. Walsh, E.-R. Janeček, J. T. Hodgkinson, J. Sedlmair, A. Koutsioubas, D. R. Spring, M. Welch, C. J. Hirschmugl, C. Toprakcioglu and J. R. Nitschke, *Proc. Natl. Acad. Sci. U. S. A.*, 2014, **111**, 17743.

37. X. Xu, E. A. Appel, X. Liu, R. M. Parker, O. A. Scherman and C. Abell, *Biomacromolecules*, 2015, **16**, 2743.

38. E. A. Appel, R. A. Forster, A. Koutsioubas, C. Toprakcioglu and O. A. Scherman, *Angew. Chem.*, 2014, **126**, 10202.

39. E. A. Appel, R. A. Forster, M. J. Rowland and O. A. Scherman, *Biomaterials*, 2014, **35**, 9897.

40. L. M. Heitmann, A. B. Taylor, P. J. Hart and A. R. Urbach, *J. Am. Chem. Soc.*, 2006, **128**, 12574.

41. Y. Lin, L. Li and G. Li, *Carbohydr. Res.*, 2013, **92**, 429.

42. J. Cao, L. Meng, S. Zheng, Z. Li, J. Jiang and X. Lv, *Int. J. Polym. Mater. Polym.*, 2016, **65**, 537.

43. J. Liu, C. Soo Yun Tan, Y. Lan and O. A. Scherman, *J. Polym. Sci., Part A*, 2017, **55**, 3105.

44. J. Liu, C. S. Y. Tan, Z. Yu, N. Li, C. Abell and O. A. Scherman, *Adv. Mater.*, 2017, **29**, 1605325.

45. J. Liu, C. S. Y. Tan, Z. Yu, Y. Lan, C. Abell and O. A. Scherman, *Adv. Mater.*, 2017, **29**, 1604951.

46. J. Liu and O. A. Scherman, *Adv. Funct. Mater.*, 2018, **28**, 1800848.

47. H. Jung, J. S. Park, J. Yeom, N. Selvapalam, K. M. Park, K. Oh, J.-A. Yang, K. H. Park, S. K. Hahn and K. Kim, *Biomacromolecules*, 2014, **15**, 707.

48. J. Yeom, S. J. Kim, H. Jung, H. Namkoong, J. Yang, B. W. Hwang, K. Oh, K. Kim, Y. C. Sung and S. K. Hahn, *Adv. Healthcare Mater.*, 2015, **4**, 237.

49. H. Chen, S. Hou, H. Ma, X. Li and Y. Tan, *Sci. Rep.*, 2016, **6**, 20722.

CHAPTER 7

# Cucurbituril Containing Supramolecular Nanomaterials

REHAN KHAN[a,b] AND DÖNÜS TUNCEL*[a,b]

[a] Department of Chemistry, Bilkent University, Ankara 06800, Turkey;
[b] UNAM – National Nanotechnology Research Center, Institute of Materials Science and Nanotechnology, Bilkent University, Ankara 06800, Turkey
*Email: dtuncel@fen.bilkent.edu.tr

## 7.1 Introduction

Smart functional nanostructures that are adaptive and responsive to different external stimuli (light, heat, pH, competitive guests, redox *etc.*) could be assembled in aqueous milieus by making use of supramolecular chemistry, which basically relies on reversible noncovalent interactions, including hydrogen bonding, $\pi$–$\pi$ stacking, van der Waals forces or hydrophobic interactions.[1,2] Host–guest inclusion complex formation is particularly appealing as high selectivity between the host and guest molecules provides dynamic but strong interactions and offers a number of possibilities in the design of supramolecular nanostructures with desirable topological diversity and programmable functions for specific applications.[3–7] The macrocycles most commonly used as host molecules are cyclodextrins,[8–10] pillarenes,[11] calixarenes[12] and cucurbiturils,[13–17] and they can accommodate guest molecules in their cavities on the basis of shape and size complementariness. Among them, cucurbit[$n$]urils (CB[$n$]) are a relatively new class of macrocycles with versatile recognition properties and an ability to accommodate different organic guest molecules in aqueous solution with exceptionally high binding constants.[13–17]

Smart Materials No. 36
Cucurbituril-based Functional Materials
Edited by Dönüs Tuncel
© The Royal Society of Chemistry 2020
Published by the Royal Society of Chemistry, www.rsc.org

In recent years, CB[n]-containing functional nanomaterials have been receiving increasing attention due to their versatile applications in areas including but not limited to theranostics, photonics, sensing and catalysis.[18-22] CB[n]s can be employed in the preparation of these materials by making use of their host–guest complexation abilities with a variety of guests[14-17,23-46] or by the conjugation of their functionalized derivatives to platforms having appropriate functional groups that will allow them to form nanomaterials.[18,19,47-49]

This chapter provides an overview of the preparation and application of CB[n]-containing supramolecular nanomaterials. In Section 7.2, we discuss supramolecular nanostructures (*e.g.*, nanoparticles, micelles, vesicles and capsules) constructed through the host–guest chemistry of cucurbiturils, and in Section 7.3, we discuss nanostructures formed through the conjugation of functionalized CBs to appropriate platforms.

## 7.2  CB Containing Functional Nanostructures: Through Host–Guest Chemistry of CBs

Properly functionalized amphiphilic molecules consisting of hydrophobic and hydrophilic groups can self-assemble into various nanostructures in aqueous media.[5,7,50] Depending on the nature of an amphiphile, whether it is a small molecule or a polymeric material, and on the ratio of hydrophilic to hydrophobic parts, the nanostructure could take the form of nano-particles, micelles or higher-ordered aggregates such as vesicles.

These structures are also characterized by a packing parameter $P$, which is expressed mathematically as $P = V/AL$, where $V$ is the volume, $L$ the length of the hydrophobic part of the amphiphile and $A$ the hydrophobic–hydrophilic interfacial area. As the value of $P$ increases, spherical ($P = 1/3$), cylindrical ($P = 1/2$) and planar structures ($P = 1$) are formed.[50a]

Amphiphilic molecules obtained through host–guest interactions of host macrocyclic molecules, such as calixarene, cyclodextrin and cucurbituril with suitable guest molecules called supra-amphiphiles, proved to be a promising building block for designing a new generation of smart delivery systems because of their dynamic nature that allows their reversibility as well as their biocompatibility and versatility.[5,7]

Lately, by taking advantage of the abundant host–guest chemistry of CB homologues, a variety of functional and smart supramolecular aggregates have been prepared. Among the CB homologues, CB[7] for its good water solubility and CB[8] for its large cavity have been utilized extensively for these purposes. CB[8] is capable of simultaneously accommodating two guests inside its hydrophobic cavity, forming a ternary complex with methyl viologen (MV) and an electron-rich second guest, including naphthalene, pyrene and tryptophan (see Scheme 7.1).[23-28] The charge transfer (CT) interaction between the electron deficient–MV and the electron-rich second

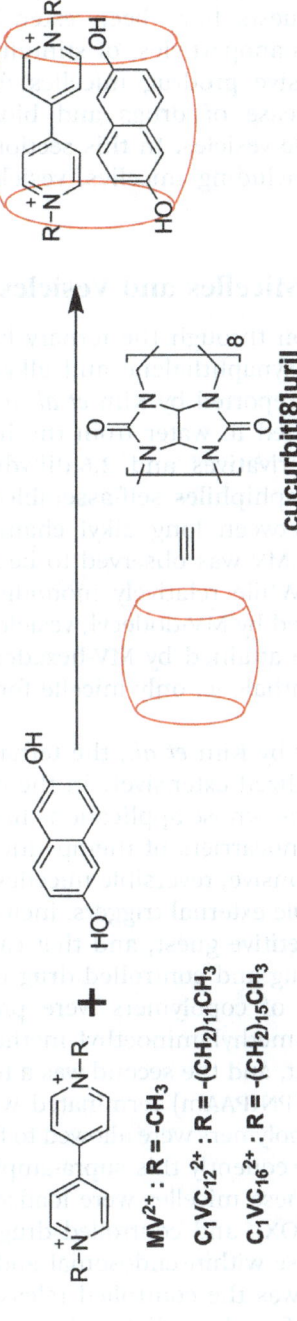

**Scheme 7.1** Molecular structures of CB[8], 2,6-dihydroxynaphthalene and alkylated methyl viologen (MV) and their ternary complex formation.[23]

guests, as well as the hydrophobic interaction, facilitates the formation of the ternary complex.

In this regard, the ability of CB[8] to form ternary complexes with suitably sized and functionalized guests have been extensively explored in the preparation of single-chain nanoparticles, of stimuli-responsive, reversible micelles and of pH-responsive prodrug micelles for the encapsulation, delivery and controlled release of drugs and bioactive nanostructures based on peptide amphiphile vesicles. In this section, the CB-assisted formation of nanostructures including micelles, vesicles, nanoparticles and colloidosomes is discussed.

## 7.2.1   Supramolecular Micelles and Vesicles

Micelle and vesicle formation through the ternary host–guest–guest interaction of CB[8], 2,6-dihydroxynaphthalene and alkylated methyl viologen (MV) (Scheme 7.1) was first reported by Kim *et al.* in 2002.[23] In this work, first supra-amphiphiles formed in water from the heteroternary complexation of CB[8] with MV derivatives and 2,6-dihydroxynaphthalene, and subsequently these supra-amphiphiles self-assembled into vesicles due to hydrophobic interactions between long alkyl chains. The length of the hydrophobic alkyl chains of MV was observed to be affecting the size and dispersion of the vesicles. While relatively monodisperse vesicles with a 20-nm diameter were obtained by MV-dodecyl, vesicles ranging in diameter from 20 nm to 1.2 μm were attained by MV-hexadecyl. In the absence of CB[8] and 2,6-dihydroxynaphthalene, only micelle formation of the MV derivative was observed.

Since the landmark paper by Kim *et al.*, the ternary complex formation ability of CB[8] has been utilized extensively in the construction of smart, self-assembled nanostructures, whose applications have been demonstrated in the stimuli-responsive nanocarriers of therapeutic cargos. For instance, supramolecular stimuli-responsive, reversible micelles[29] have been reported that are responsive to multiple external triggers, including temperature, pH and the addition of a competitive guest, and that can be used for the encapsulation of anticancer drug and controlled drug release.[30] To construct this nanocarrier, two types of copolymers were prepared; one of them was a pH-responsive poly(dimethylaminoethyl methacrylate) (PDMAEMA)-terminated naphthalene guest, and the second was a temperature-responsive poly(*N*-isopropylacrylamide) (PNIPAAm) terminated with a methyl viologen guest (Figure 7.1). These two polymers were allowed to form a ternary complex with CB[8] in water, and subsequently this supra-amphiphile self-assembled in water to form micelles. These micelles were loaded with the chemotherapeutic drug doxorubicin (DOX), and controlled drug release was achieved through a pH-triggered release within endosomal and lysosomal vesicles at around pH 4. Also possible was the controlled release of drugs *via* remote heating methods, such as infrared irradiation because of the temperature-responsiveness nature of DOX.

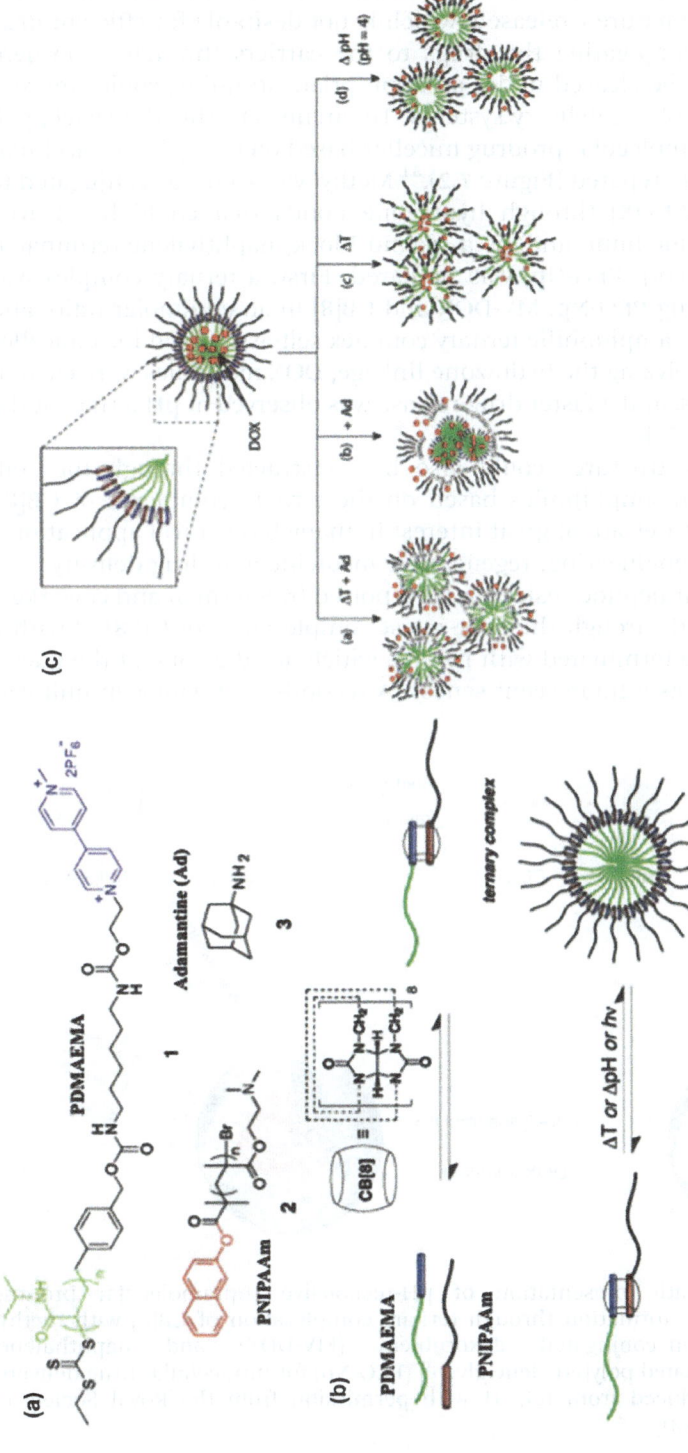

**Figure 7.1** (a) Chemical structures of poly(*N*-isopropylacrylamide) (PNIPAAm), terminated with methyl viologen guest (**1**), poly(dimethyl-aminoethyl methacrylate) (PDMAEMA), terminated with naphthalene guest (**2**) adamantaneamine (Ad) (**3**); (b) supra-amphiphile formation through ternary complex of CB[8] and **1** and **2**; (c) subsequent assembly into a micellar superstructure, hierarchical self-assembly of the supramolecular entity under different conditions and its subsequent mode of drug release after being exposed to different triggers.
Reproduced from ref. 30 with permission from American Chemical Society, Copyright 2012.

When drugs are loaded into the nanocarriers by noncovalent interactions, they could be prematurely released, which is not desirable for efficient drug delivery. Thus, conjugating the drugs to the carriers through a covalent bond, which can be cleaved under an appropriate stimulus, could improve the effectiveness of the delivery systems. To circumvent this drawback, pH-responsive supramolecular prodrug micelles based on CB[8] for intracellular drug delivery was prepared (Figure 7.2).[31] Methyl viologen was conjugated to doxorubicin (MV-DOX) through hydrazone bonds that could be cleaved under an acidic medium, and, as a second block, naphthalene-terminated poly(ethylene glycol) (PEG-Np) was prepared. First, a ternary complex was obtained by mixing PEO-Np, MV-DOX and CB[8] in an equimolar ratio, and subsequently this amphiphilic ternary complex self-assembled into micelles in water. By hydrolyzing the hydrazone linkage, DOX molecules were cleaved from the micelles, and a faster drug release was observed at pH 5 than at the physiological pH 7.4.

Bioactive nanostructures could also be constructed through the self-assembly of supra-amphiphiles based on the ternary complexes of CB[8]. These nanostructures are of great interest in many biomedical applications, including tissue engineering, regenerative medicine and drug delivery.

Supramolecular peptide vesicles were reported by Sherman and coworkers that was prepared through the host–guest complexation of CB[8][32a] with a peptide sequence terminated with pyrene, which acted as one of the guests for CB[8] as well as a fluorescent sensor. A second-guest, viologen unit was

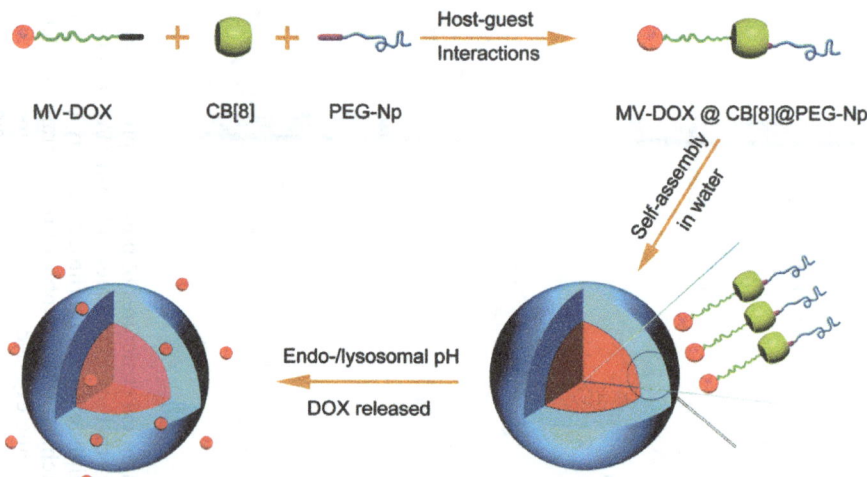

**Figure 7.2**   Schematic presentation of pH-responsive supramolecular prodrug micelle formation through ternary complexation of CB[8] with methyl viologen–conjugated doxorubicin (MV-DOX) and naphthalene-terminated poly(ethylene glycol) (PEO-Np) for intracellular drug delivery. Reproduced from ref. 31 with permission from the Royal Society of Chemistry.

conjugated to a long hydrophobic linker. Vesicle formation was achieved by the self-assembly of suprapeptide amphiphiles resulting from the ternary complex of pyrene- and viologen-containing blocks (Figure 7.3). These vesicles exhibit a stimuli-responsive behavior and can undergo disassembly in the presence of competitive guests, such as 2,6-dihydroxynaphthalene and 1-adamantylamine. The assembly and disassembly processes can be monitored through the changes in the emission intensity of pyrene units as the emission of pyrene is quenched upon formation of vesicles but is recovered when the vesicles undergo disassembly in the presence of the competitive guests. It was also shown *in vitro* that these vesicles were effectively internalized by HeLa cells and that their toxicities could be regulated using an appropriate stimulus. The same group also demonstrated CB[8] ternary complexation–mediated formation of the polymeric peptide-amphiphile–based vesicle at the physiological temperature for the encapsulation and release of the basic fibroblast growth factor (bFGF).[32b]

A systematic study of the ternary complex formation process for aromatic amino acids using CB[8] and a viologen amphiphile shows that the affinity of the amino acid needs to be higher or in a comparable range to that of CB[8] for the amphiphile to form the ternary complex.[33] By taking these results into account, a supramolecular peptide amphiphile was prepared containing an azobenzene group at the N-terminus of the peptide to serve as a second guest for CB[8]. The vesicles formed by the self-assembly of this peptide amphiphile exhibit stimuli-responsive behavior toward a number of external stimuli such as light and competitive guests. Azobenzene groups can respond to the light through *cis–trans* isomerism; *trans*-isomer fits well in the cavity of CB[8], but *cis* isomer does not. Thus, irradiation of vesicles with UV-light at 365 nm causes their disassembly through *cis* isomer formation of azobenzene. Assembly and disassembly of the vesicles can be controlled by using an appropriate wavelength as well as the addition of the guest 1-adamantylamine or 2,6-dihydroxynaphthalene, both of which are known to have a high affinity toward CB[8].

It was also demonstrated by Jonkheijm *et al.* that CB[8]-based supramolecular amphiphiles-based vesicles can be employed for the encapsulation of proteins and their delivery into cells.[34] Vesicles about 200 nm in diameter were formed by the self-assembly of ternary complexes of CB[8], an alkylated paraquat derivative and a tetraethylene glycol–functionalized azobenzene (Figure 7.4). Their outer surfaces were functionalized with cell-targeting ligands, and these vesicles were utilized as supramolecular nanocarriers.

Liu *et al.* very recently reported highly stable giant supramolecular vesicles constructed by hierarchical self-assembly of CB[8]-based supra-amphiphiles for photoresponsive and targeted intracellular drug delivery.[35] Again, first a supra-amphiphile was constructed through the ternary complex formation of CB[8] with hydrophilic and hydrophobic blocks containing the guests' methyl viologen and photoresponsive azo moiety (Figure 7.5). This amphiphile simultaneously forms vesicles in water through self-assembly. The size and morphologies of these vesicles are determined by light microscopy and

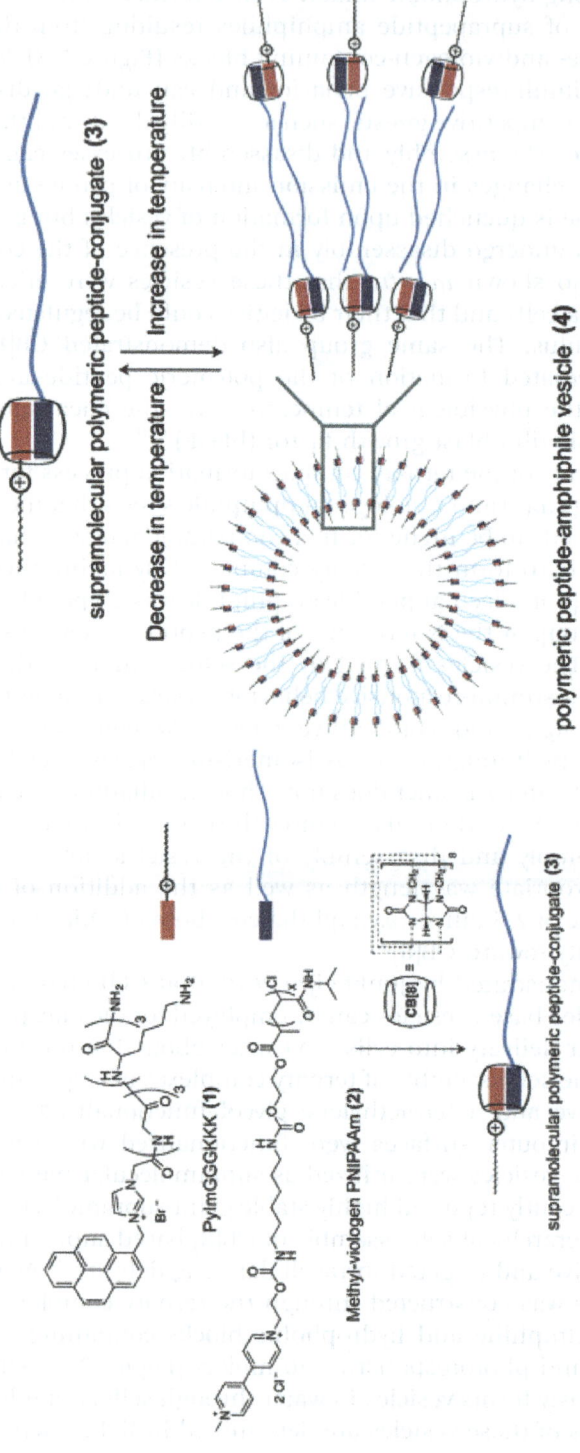

**Figure 7.3** The chemical structures of pyrene-imidazolium–labeled peptide and viologen-functionalized PNIPAAm; their ternary complex with CB[8] and the subsequent temperature-induced formation of a supramolecular polymeric peptide vesicle. Reproduced from ref. 32a with permission from John Wiley & Sons, Copyright © 2012 Wiley-VCH Verlag GmbH & Co. KGaA, Weinheim.

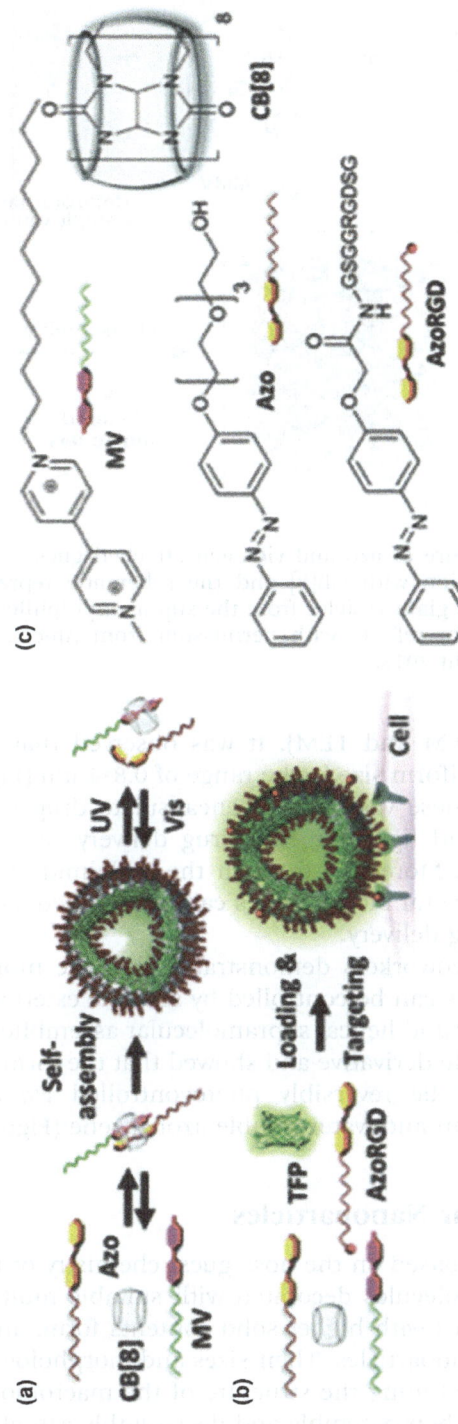

**Figure 7.4** (a) Formation of a supramolecular amphiphile through a ternary complex of CB[8] with methyl viologen linked to a hydrophobic alkyl chain (MV) and azobenzene linked to a hydrophilic oligo(ethylene glycol) chain (azo); (b) vesicles loaded with teal, yellow and red fluorescent proteins (TFP, YFP, TagRFP) as cargo and decorated with azoRGD peptide ligands for targeting; (c) molecular structures of MV, azo, azoRGD and CB[8]. Reproduced from ref. 34 with permission from the Royal Society of Chemistry.

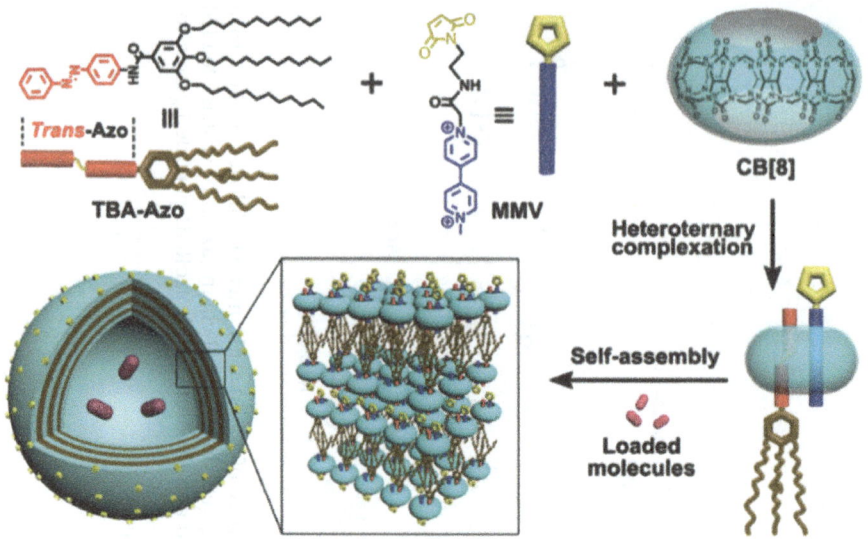

**Figure 7.5**   Chemical structure of azo and viologen attached guests, their ternary complex formation with CB[8] and the schematic representation of photoresponsive giant vesicles from the supra-amphiphiles.
Reproduced from ref. 35 with permission from American Chemical Society, Copyright 2018.

electron microscopies (SEM and TEM). It was observed that they had a spherical shape with a uniform size in the range of 0.8–1 µm (Figure 7.6). It was demonstrated that these vesicles can encapsulate drug molecules in high loading capacity, and light-triggered drug delivery can be achieved using these nanocarriers. Moreover, through the maleimide units on the surface of the vesicles, several biomolecules can be attached as a targeting group for the control drug delivery.

Recently, Yu Liu and coworkers demonstrated that the morphology of supramolecular aggregates can be controlled by suitable external stimuli.[36] They constructed lamellar and helical supramolecular assemblies using CBs and a naphthalene diimide derivative and showed that the formation of the lamellar assembly could be reversibly photocontrolled *via* competitive binding with α-cyclodextrin and water-soluble azobenzene (Figure 7.7).

## 7.2.2   Supramolecular Nanoparticles

If supra-amphiphiles are based on the host–guest chemistry of macrocyclic hosts with large macromolecules decorated with suitable multiple guests, less well defined structures with higher solid contents form, and these aggregates can be called nanoparticles. Their sizes and morphologies can also be controlled by carefully tuning the structure of the macromolecules and the reaction conditions. Their assembly and disassembly can also be easily

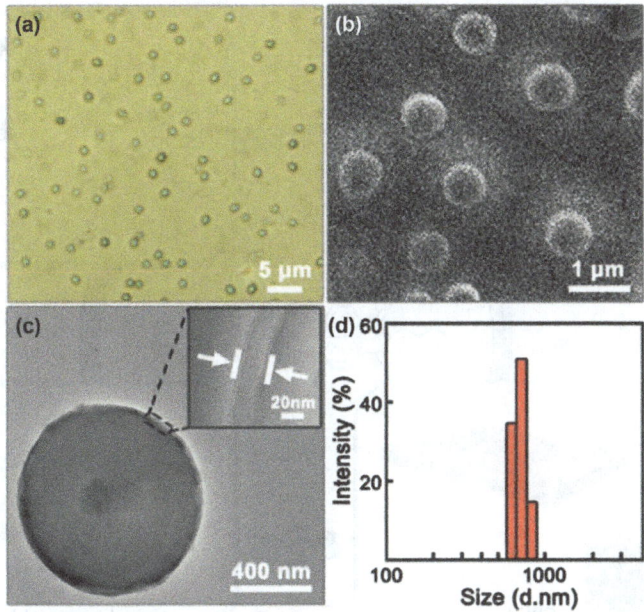

**Figure 7.6** Microscope images of giant photoresponsive vesicles (a) optical micro-scope; (b) SEM; (c) TEM; and (d) DLS histogram.
Reproduced from ref. 35 with permission from American Chemical Society, Copyright 2018.

controlled by triggering with suitable stimuli if they possess stimuli-responsive features. These stimuli-responsive nanoparticles can be utilized as biomedical delivery vehicles.

To this end, the host–guest chemistry of CB[8] was also successfully utilized in the preparation of single-chain nanoparticles by installing both guests on the same polymer chain. This approach proved to be very convenient for the synthesis of nanoparticles with well-defined shape, size and composition.[37] For the preparation of these nanoparticles, poly($N$-hydroxyethyl acrylamide) polymers were prepared by atom transfer radical polymerization (ATRP) and were functionalized using an isocyanate conjugation with guest moieties ($MV^{2+}$ and Np) for complexation with CB[8]. By tuning the concentration of polymers and CB[8], the size and dispersity can be controlled. When CB[7] was used as a host instead of CB[8], no nanoparticle formation was observed.

Limited examples have been found in literature where the strategy of controlled and reversible host–guest supramolecular chemistry was adapted to prepare core–shell polymeric microspheres. In the following example, CB[8]-based, core–shell polymeric microspheres with a cleavable shell is prepared in water, where CB[8] is used as supramolecular "handcuff" to lock 2-naphthol–functionalized linear acrylate polymers (shell) onto a methyl viologen–functionalized polymeric microsphere (core).[38] The polymeric shell

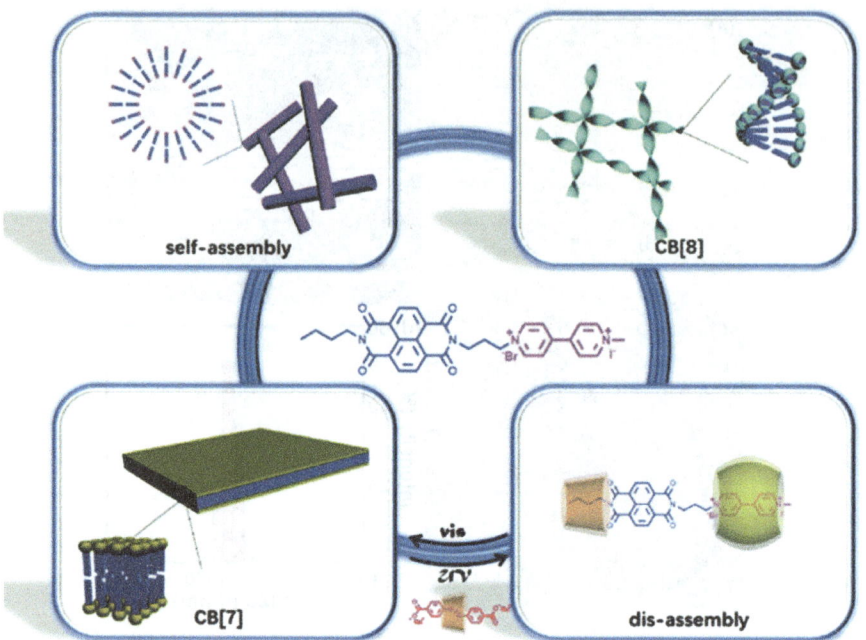

**Figure 7.7**    Formation of lamellar and helical supramolecular assemblies using
              CB[7], CB[8] and naphthalene diimide derivatives and the reversibility
              of the lamellar assembly *via* competitive binding with α-cyclodextrin and
              water-soluble azobenzene.
              Reproduced from ref. 36 with permission from the Royal Society of
              Chemistry.

and the microsphere core are linked *via* CB[8]-mediated ternary complex
with the residues of Np and $MV^{2+}$ (Figure 7.8). The strategy of switching the
cytotoxicity of the forming microspheres has extended the range of potential
applications in cancer therapy.

Huskens *et al.* developed a multicomponent supramolecular nanoparticle
fabrication method based on heteroternary CB[8] complexes.[21] In this
method, CB[8] was utilized as a cross-linker to link viologen and naphthol-
containing dendrimer and polymers, namely, methyl viologen-poly(ethylene
imine) (MV-PEI), naphthol-poly(ethylene glycol) (Np-PEG) and naphthol$_8$-
poly(amidoamine) (Np$_8$-PAMAM). The formation of nanoparticles was
thermodynamically controlled, and time and temperature were noted to
affect their formation rate.[39] Their sizes can also be controlled by varying the
concentrations of multivalent core and monovalent shell-forming stopper
molecules while keeping the ratio of MV/Np/CB[8] at 1 : 1 : 1. Disassembly of
these nanoparticles was shown to be achieved by using a reducing agent that
decreases the dicationic MV species to $MV^+$ radical cations, which undergo
stable homoternary complex formation in one CB[8] by releasing the Np
guests; this, in turn, causes the disassembly of the particles. As a follow-up of
this work, the same group substituted Np guest units with azo groups to add

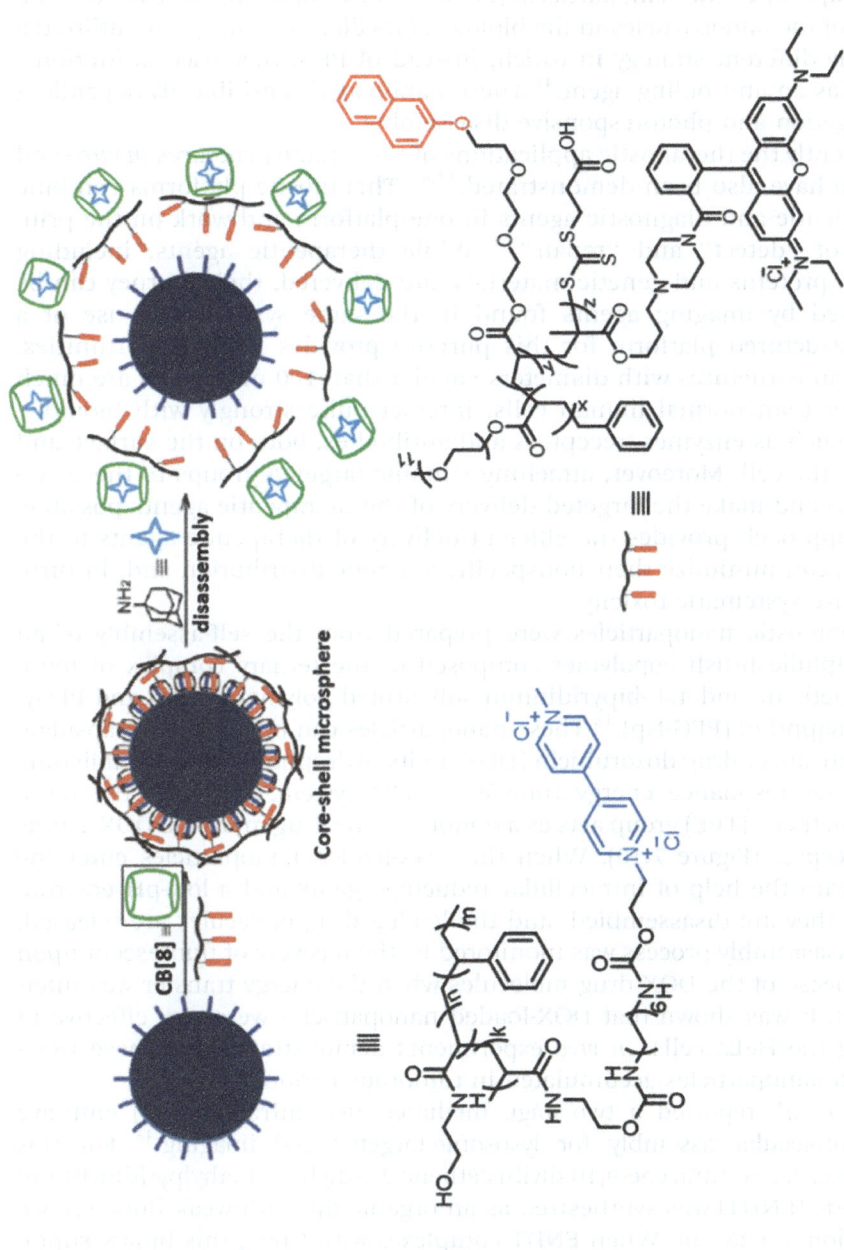

**Figure 7.8** Reversible preparation of core-shell polymeric microspheres *via* the formation and dissociation of CB[8] ternary complex. Reproduced from ref. 38 with permission from the Royal Society of Chemistry.

to their system photo responsiveness.[40] Light-triggered *cis–trans* isomerism of azo groups caused self-assembly and disassembly of the nanoparticles. Under UV light (350 nm), *trans–azo* isomer turned into *cis*-isomer by causing a disruption of the nanoparticles (Figure 7.9). In order to increase the stability of the nanoparticles in the biological media, the same group utilized a slightly different strategy in which, instead of PEG, they used zwitterionic motif as an antifouling agent.[41] These nanoparticles exhibit pH-dependent aggregation and photoresponsive disassembly.

Recently the theranostic applications of these nanostructures *in vitro* and *in vivo* have also been demonstrated.[42,43] Theranostic platforms combine therapeutic and diagnostic agents in one platform and work on the principle of "detect" and "repair".[51] While therapeutic agents, including drugs, proteins and genetic materials, are delivered, their journey can be followed by imaging agents found in the same system. The use of a nanostructured platform for this purpose provides many opportunities. The nanostructures with diameters smaller than 100 nm, which are much smaller than normal human cells, interact quite strongly with biomolecules such as enzymes, receptors and antibodies, both on the surface and inside the cell. Moreover, attaching suitable targeted groups to these systems would make the targeted delivery of the therapeutic agents possible. This approach provides the efficient delivery of therapeutic agents to the target, can minimize their nonspecific systemic distribution and, in turn, decrease systematic toxicity.

Theranostic nanoparticles were prepared from the self-assembly of an amphiphilic brush copolymer composed of the ternary complex of tetraphenylethene and 4,4'-bipyridinium substituted polymer (PTPE) and PEGylated naphthol (PEG-Np).[42] These nanoparticles can be used to encapsulate the anticancer drug doxorubicin (DOX) in its hydrophobic core, establishing a Förster resonance energy transfer (FRET) system, in which the tetraphenylethene (TPE) group acts as a donor and the drug molecule DOX acts as an acceptor (Figure 7.10). When the DOX-loaded nanoparticles enter the cells with the help of intracellular reducing agents and a low-pH environment, they are disassembled, and the loaded drug molecules are released. The disassembly process was monitored by the recovery of fluorescent upon the release of the DOX drug molecules when the energy transfer was interrupted. It was shown that DOX-loaded nanoparticles were very effective in killing the HeLa cells. *In vivo* experiments demonstrated that these DOX-loaded nanoparticles accumulated in tumorous regions.

Liu *et al.* reported a two-stage mediated near-infrared (NIR) emissive supramolecular assembly for lysosome-targeted cell imaging.[43] For this purpose, 4,4'-anthracene-9,10-diylbis(ethene-2,1-diyl)bis(1-ethylpyridin-1-ium) bromide (ENDT) was synthesized as an organic dye with weak fluorescence emission at 625 nm. When ENDT complexes with CB[8], this binary supramolecular complex assembles into nanorods with a near-infrared fluorescence emission (655 nm) and fluorescence enhancement as the first stage (Figure 7.11). Such supramolecular complexes interact with lower-rim

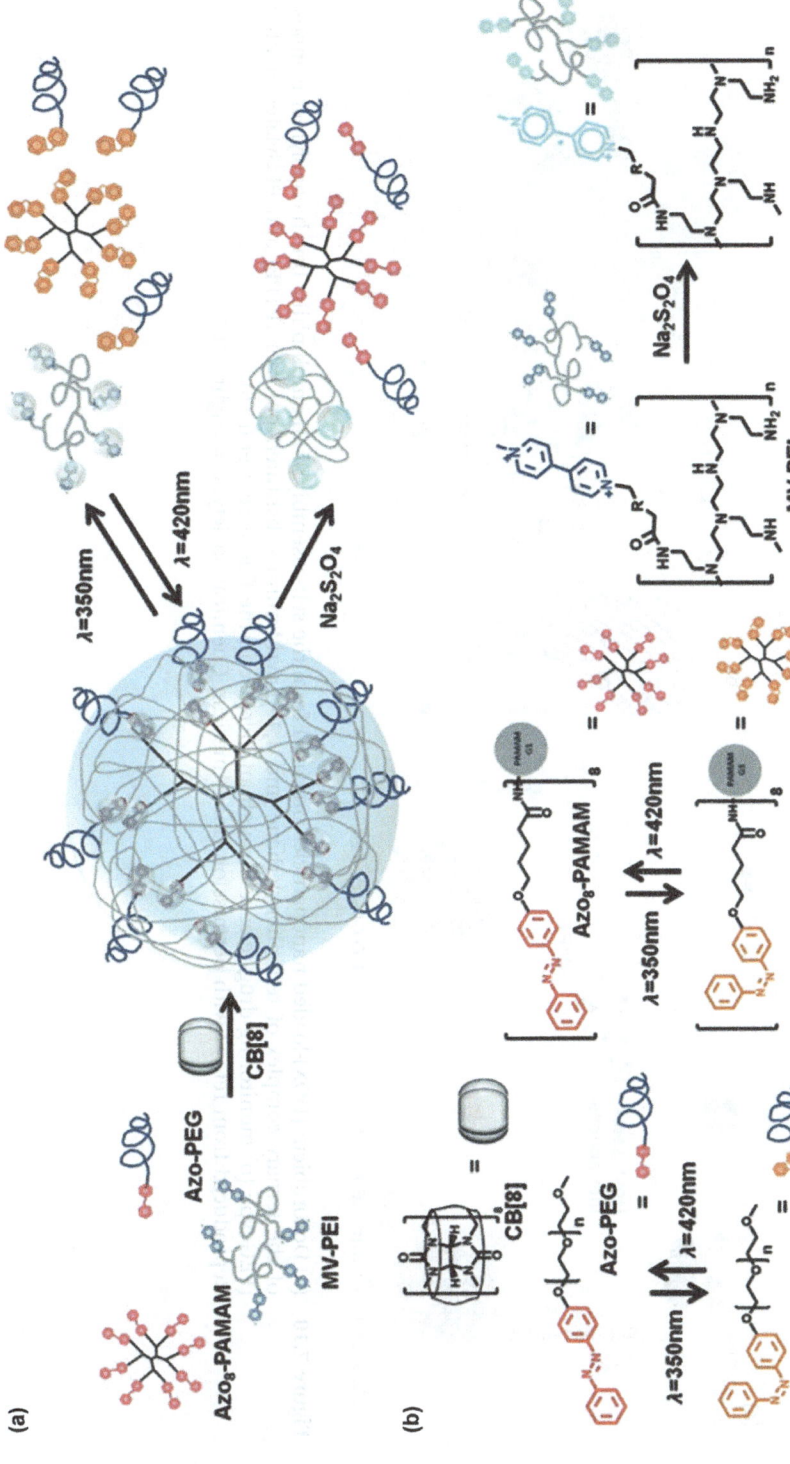

**Figure 7.9** (a) Schematic presentation of the supramolecular nanoparticle formation through ternary complex formation and disassembly process of the nanoparticles *via* light triggered *cis–trans* isomerism of azo groups and reduction of viologen with a reducing agent; (b) chemical structures of CB[8], azobenzene-functionalized poly(ethylene glycol) (azo-PEG), methyl viologen functionalized poly(ethylene imine) (MV-PEI), and azo8-poly(amidoamine). Reproduced from ref. 40 with permission from John Wiley & Sons, Copyright © 2017 Wiley-VCH Verlag GmbH & Co. KGaA, Weinheim.

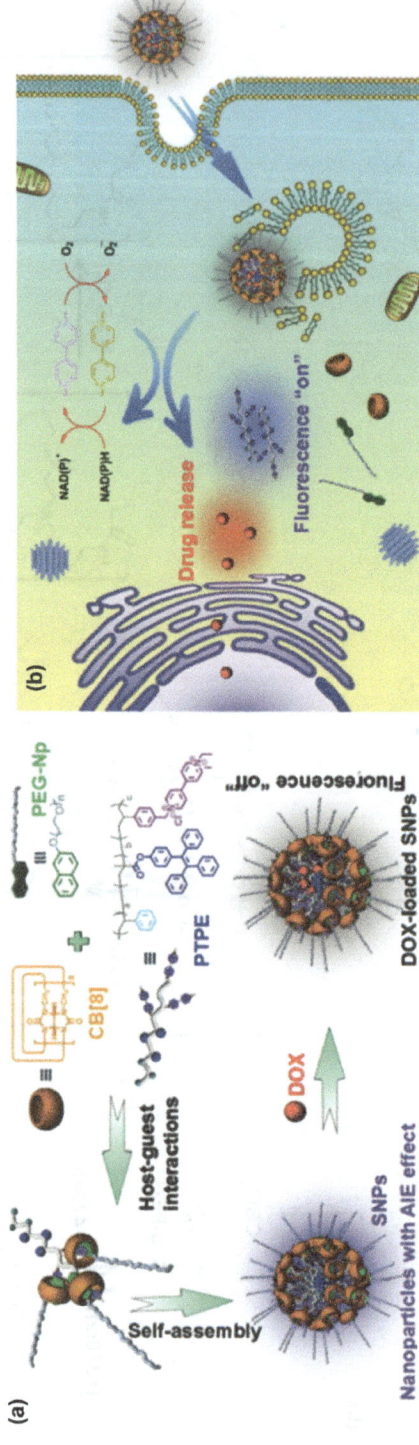

**Figure 7.10** (a) Doxorubicin (DOX)-loaded nanoparticles prepared from the self-assembly of an amphiphilic brush copolymer composed of the ternary complex of tetraphenylethene and 4,4′-bipyridinium substituted polymer (PTPE) and PEGylated naphthol (PEG-Np); (b) monitoring drug release through an increase in the fluorescent emission. Reproduced from ref. 42 with permission from American Chemical Society, Copyright 2017.

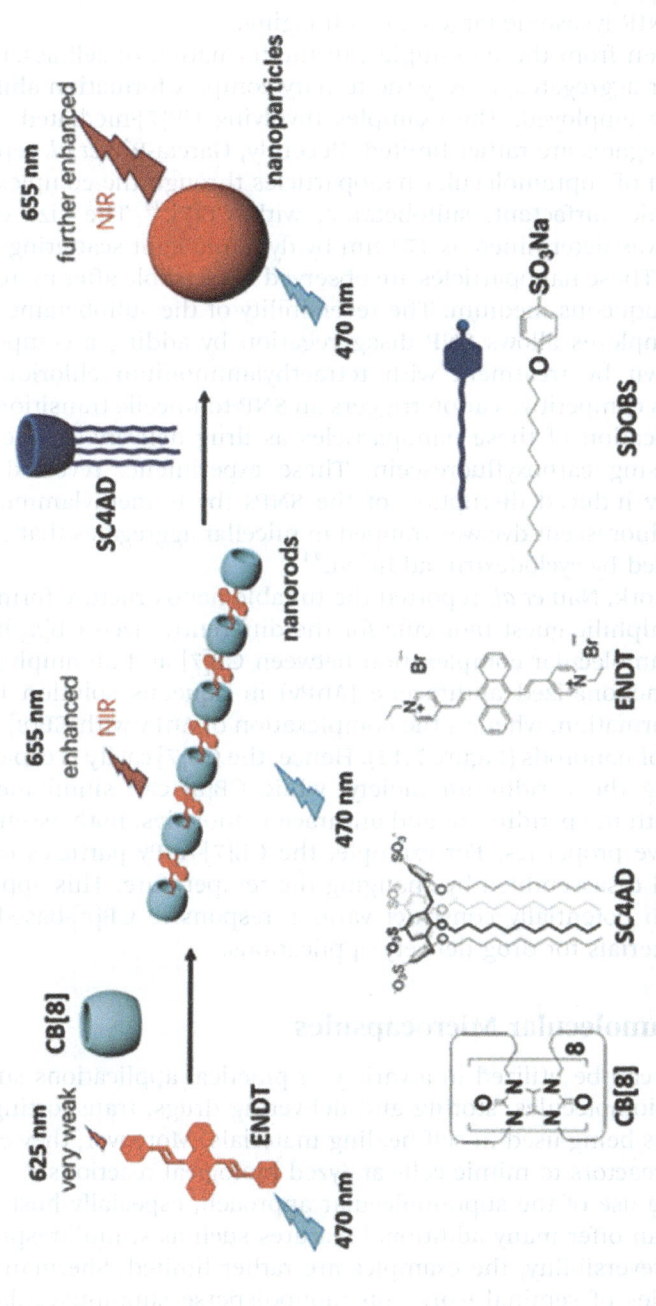

**Figure 7.11** Schematic presentation of a two-stage mediated near-infrared (NIR) emissive supramolecular assembly for lysosome-targeted cell imaging.
Reproduced from ref. 43 with permission from John Wiley & Sons, Copyright © 2018 Wiley-VCH Verlag GmbH & Co. KGaA, Weinheim.

dodecyl-modified sulfonatocalix-[4]arene (SC4AD) to form nanoparticles for further fluorescence enhancement as the second stage. Furthermore, based on a costaining experiment with LysoTracker™ Blue, such nanoparticles can be applied in NIR lysosome-targeted cell imaging.

As can be seen from these examples, in the formation of self-assembled supramolecular aggregates, mostly the ternary complex formation ability of CB[8] has been employed. The examples involving CB[7]-mediated supramolecular aggregates are rather limited. Recently, Garcia-Rio *et al.* reported the preparation of supramolecular nanoparticles through the complexation of a zwitterionic surfactant, sulfobetaine, with CB[7].[44] The size of the nanoparticles was determined as 172 nm by dynamic light scattering (DLS) and cryo-TEM. These nanoparticles are observed to be stable after more than 2 weeks in an aqueous medium. The reversibility of the sulfobetaine/CB[7] host–guest complexes allows SNP disaggregation by adding a competitive guest, as shown by treatment with tetraethylammonium chloride. The addition of this competitive cation triggers an SNP-to-micelle transition. The potential application of these nanoparticles as drug delivery vehicles was investigated using carboxyfluorescein. These experiments revealed that, upon externally induced disruption of the SNPs (by tetraethylammonium chloride), the fluorescent dye was trapped in micellar aggregates that can be further disrupted by cyclodextrin addition.[44]

In another work, Nau *et al.* reported the tunable nanostructure formation using an amphiphilic guest molecule for the differently sized CB[n] homologues.[45] Supramolecular complexation between CB[7] and an amphiphilic pyridinium–functionalized anthracene (AnPy) in aqueous solution led to nanoparticle formation, whereas the complexation of AnPy with CB[8] led to the formation of nanorods (Figure 7.12). Hence, the CB[7] cavity is capable of accommodating the pyridinium moiety, while CB[8] can simultaneously encapsulate both the pyridinium and anthracene moieties. Both assemblies show responsive properties. For example, the CB[7]-AnPy particles can be assembled and disassembled by changing the temperature. This approach can be used to potentially construct various responsive CB[n]-based self-assembled materials for drug delivery applications.

## 7.2.3    Supramolecular Microcapsules

Microcapsules can be utilized in a variety of practical applications such as sequestering biomolecules, storing and delivering drugs, transporting vaccines, as well as being used in self-healing materials. Moreover, they can be used as microreactors to mimic cell-catalyzed biological reactions.[52]

Although the use of the supramolecular approach, especially host–guest interactions, can offer many additional features such as stimuli-responsive behavior and reversibility, the examples are rather limited. Sherman *et al.* reported a series of seminal works on monodisperse supramolecular microcapsules, fabricated through the integration of traditional microfluidic techniques and interfacial host–guest chemistry, specifically CB-mediated

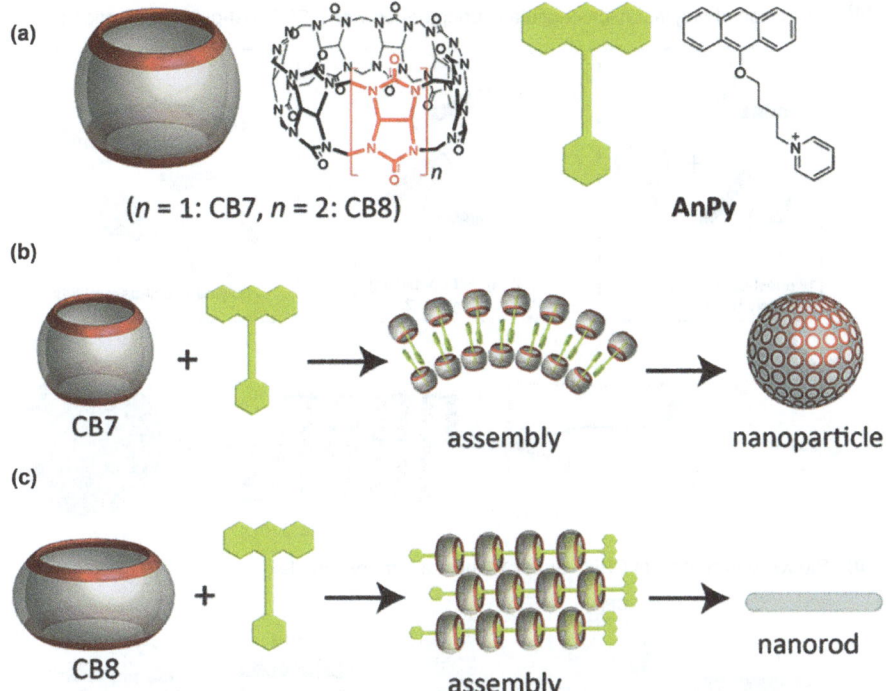

**Figure 7.12** The tunable nanostructure formation using an amphiphilic guest molecule for the differently sized CB[*n*] homologues.
Reproduced from ref. 45 with permission from the Royal Society of Chemistry.

host–guest interactions (Figure 7.13).[46,53–59] They employed three methods for the microcapsule fabrication: colloidal particle–driven assembly, interfacial condensation–driven assembly and electrostatic interaction–driven assembly. They also studied systematically the design criteria required for structural complexity with the desirable functionality and demonstrated their proof-of-principle applications in cargo delivery. On account of its dynamic nature, the CB-mediated host–guest complexation has demonstrated efficient response toward various external stimuli such as UV light, pH change, redox chemistry and competitive guests. It has also demonstrated different microcapsule modalities, which are engineered with a CB host–guest chemistry and also can be disrupted with the aid of external stimuli, for a triggered release of payloads.

## 7.3 CB Containing Functional Nanostructures: Through Functionalized-cucurbituril Derivatives

Although functionalized CB-based nanostructures with many interesting properties and features could be suitable in a variety of applications,

**(a)** Formation of polymer microcapsules strengthened with CB[*n*] host-guest interactions

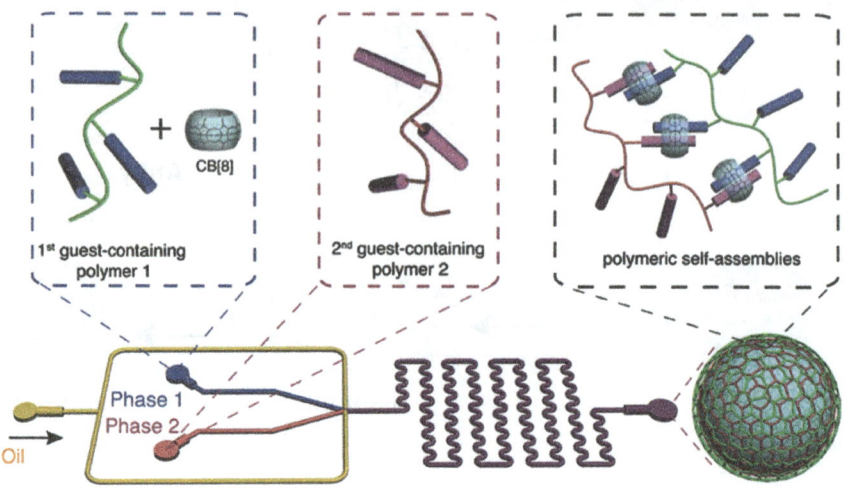

**(b)** Stepwise formation of CB[8]•Guest 1•Guest 2 ternary complex

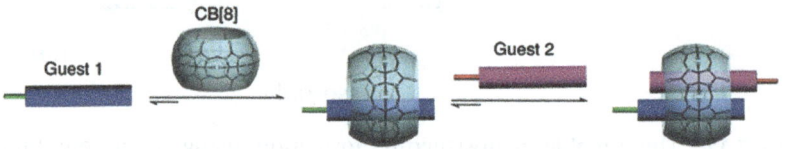

**Figure 7.13** (a) Schematic illustration of supramolecular polymer microcapsules assembled at the interface of microfluidic droplets. By using a microfluidic flow–focusing device, an aqueous phase carrying CB[8] and first guest–containing polymer 1 intersects with another phase consisting of second guest–containing polymer 2, at flow-focusing microchannel junctions to form a periodic flow of oil-in-water microdroplets. (b) Stepwise formation of a supramolecular heteroternary complexation of CB[8] and guest 1 (electron deficient, such as methyl viologen) and then guest 2 (electron rich, such as naphthol, azobenzene, benzyl, phenylalanine *etc.*).
Reproduced from ref. 46 with permission from American Chemical Society, Copyright 2017.

including theragnostic, photonics and catalysis, examples in the literature are rather limited. In this section, we are going to discuss the research efforts on functionalized CB-based nanoparticles, and in Chapter 10, Kim and coworkers discuss the CB[6]-based nanocapsules.

The first example of functionalized CB-based nanoparticles was reported by Kim and coworkers after they developed a convenient method for the synthesis of perhydroxylated CBs that opened the possibility of practical applications of CB[6]-based nanomaterials in a variety of applications, such as nanomedicine and catalysis.[47] Nanoparticles around 200 nm in diameter were prepared from functionalized CB[6]-derivative (3-(6-hydroxyhexanethio)-propan-1-oxy)$_{12}$CB[6],

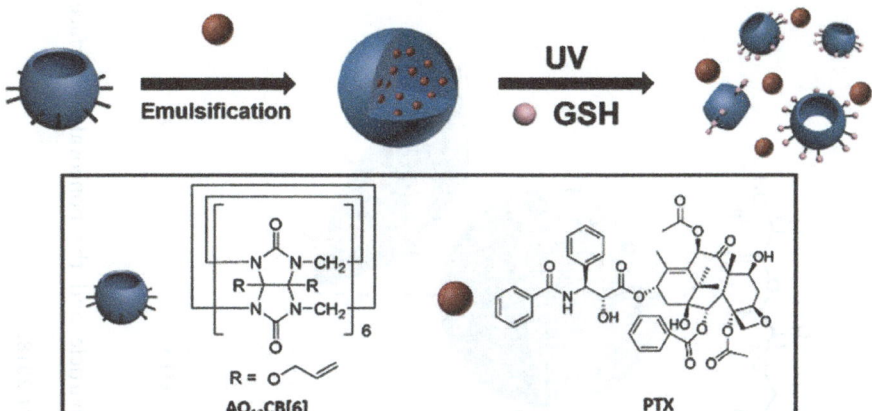

**Figure 7.14** Schematic presentation of UVA- (sunlight)-triggered, GSH-responsive PTX-loaded nanoparticle prepared from $AO_{12}CB[6]$ and their chemical structures.
Reproduced from ref. 48 with permission from Royal Society of Chemistry.

which was synthesized by photoreaction between (allyloxy)$_{12}$CB[6] ($AO_{12}CB[6]$) and 6-mercaptohexanol in methanol. They demonstrated that functionalized CB-based nanoparticles could be utilized as an efficient carrier for the delivery of hydrophobic drugs.

Wang *et al.* prepared nanoparticles from perallyoxy-CB[6] ($AO_{12}CB[6]$) using a mini emulsion method and loaded these nanoparticles with the anticancer drug paclitaxel (PTX) (Figure 7.14). These nanoparticles were observed to be light irradiated, glutathione (GSH) responsive. Under light and in the presence of GSH, allyl groups of CB[6] react with the thiol groups of GSH *via* thiol-ene click reaction, and the resulting functionalized CB[6] becomes more hydrophilic. This, in turn, causes the disassembly of the nanoparticles by the release drug molecules. The authors observed that drug-loaded nanoparticles not only exhibited efficient cellular uptake but also significantly increased the cytotoxicity and apoptosis rate of cancer cells, with remarkably reduced cytotoxicity against noncancerous cells under UVA light irradiation.

Following up on their previous work, Wang *et al.* developed a new strategy for the preparation of biocompatible nanoparticles. For this purpose, functionalized CB[7] was decorated with poly(lactic acid) (PLA)/poly(lactic-*co*-glycolic acid) (PLGA) and subsequently converted into nanoparticles (Figure 7.15).[49] The surface of these nanoparticles could be further functionalized due to the available cavity of CB[7] as a host. A variety of guests, including amantadine-conjugated folate, polyethylene glycol, and fluorescein isothiocyanate and drug molecules, were used to demonstrate the application of these nanoparticles as a theranostic platform.

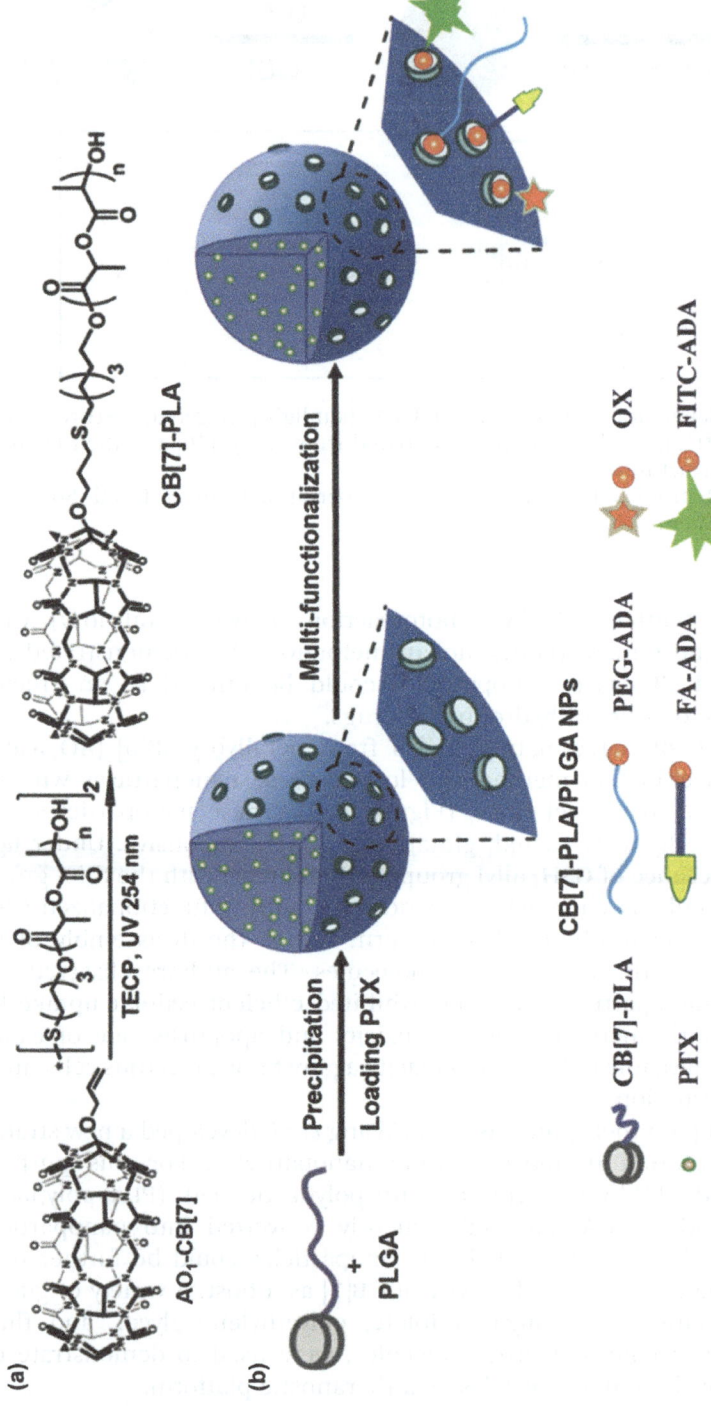

**Figure 7.15** (a) Synthesis of CB[7]-PLA; (b) schematic presentation of CB[7]-PLA-based nanoparticle and the noncovalent surface modification of nanoparticles with a variety of functionalities. Reproduced from ref. 49 with permission from American Chemical Society, Copyright 2018.

# 7.4 Conclusion

In this chapter, we discussed CB-based supramolecular nanostructures. In the first section, nanostructures prepared by taking advantage of the rich-host chemistry of CB homologues were reviewed. Particularly, the ternary complex formation of CB[8] with electron-rich and electron-deficient guests through charge transfer was extensively applied in the preparation of a variety of nanostructures, including micelles, vesicles and nanoparticles. In their preparation, first a supra-amphiphile, formed through the ternary complex formation of CB[8] with two other guests and subsequent self-assembly, turned into micelles, vesicles and nanoparticles in water. Depending on the structures of the supra-amphiphile, a number of different properties are added to these nanostructures. For instance, they can be photo or pH sensitive, and their assembly and disassembly processes can be controlled by triggering light, pH, redox potential and competitive guests. Applications of these nanostructures were also demonstrated in drug delivery and in general theranostics. Microcapsules were also prepared through the ternary complex formation ability of CB[8] and making use of the advantages of microfluidics. Resulting microcapsules were reversible and stimuli responsive.

In the second section of the chapter, nanostructures prepared from functionalized CB derivatives were discussed. Although this approach offers many unprecedented opportunities in the area of nanostructured materials, it is not well explored. Currently there are only a handful of examples in the literature, but we think there will be more to come in the near future.

## Acknowledgements

We gratefully acknowledge the financial support of the Scientific and Technological Research Council of Turkey (TUBITAK) grant number 215Z035.

## References

1. (a) J.-M. Lehn, *Supramolecular Chemistry: Concepts and Perspectives*, VCH, Weinheim, 1995; (b) J. L. Atwood, J. E. D. Davies, D. D. MacNicol, F. Vogtle and J.-M. Lehn, *Comprehensive Supramolecular Chemistry*, Pergamon Press, Oxford, 1996; (c) J. Steed and J. L. Atwood, *Supramol. Chem.*, Wiley, New York, 2000.
2. (a) J.-M. Lehn, *Angew. Chem., Int. Ed.*, 1990, **29**, 1304; (b) J.-M. Lehn, *Chem. Soc. Rev.*, 2007, **36**, 151.
3. X. Ma and H. Tian, *Acc. Chem. Res.*, 2014, **47**, 1971.
4. P. Wei, X. Yan and F. Huang, *Chem. Soc. Rev.*, 2015, **44**, 815.
5. J. Boekhoven and S. I. Stupp, *Adv. Mater.*, 2014, **26**, 1642.
6. X. Yan, F. Wang, B. Zheng and F. Huang, *Chem. Soc. Rev.*, 2012, **41**, 6042.
7. G. Yu, K. Jie and F. Huang, *Chem. Rev.*, 2015, **115**, 7240.

8. Q.-D. Hu, G.-P. Tang and P. K. Chu, *Acc. Chem. Res.*, 2014, **47**, 2017.
9. A. Harada, Y. Takashima and M. Nakahata, *Acc. Chem. Res.*, 2014, **47**, 2128.
10. Y. Kang, K. Guo, B. J. Li and S. Zhang, *Chem. Commun.*, 2014, **50**, 11083.
11. J. Murray, K. Kim, T. Ogoshi, W. Yao and B. C. Gibb, *Chem. Soc. Rev.*, 2017, **46**, 2479.
12. D.-S. Guo and Y. Liu, *Acc. Chem. Res.*, 2014, **47**, 1925.
13. (a) R. Behrend, E. Meyer and F. Rusche, *Justus Liebigs Ann. Chem.*, 1905, **339**, 1; (b) W. L. Mock, in *Comprehensive Supramolecular Chemistry*, ed. F. Vogtle, Pergamon Press, Oxford, 1996, vol. 2, p. 477;; (c) W. A. Freeman, W. L. Mock and N.-Y. Shih, *J. Am. Chem. Soc.*, 1981, **103**, 7367; (d) Y. M. Jeon, H. Kim, D. Whang and K. Kim, *J. Am. Chem. Soc.*, 1996, **118**, 9790; (e) D. Whang, Y. M. Jeon, J. Heo and K. Kim, *J. Am. Chem. Soc.*, 1996, **118**, 11333; (f) I.-S. Kim, J. Jung, S.-Y. Kim, E. Lee, J.-K. Kang, S. Sakamoto, K. Yamaguchi and K. Kim, *J. Am. Chem. Soc.*, 2000, **122**, 540.
14. D. Shetty, J. K. Khedkar, K. M. Park and K. Kim, *Chem. Soc. Rev.*, 2015, **44**, 8747.
15. S. J. Barrow, S. Kasera, M. J. Rowland, J. del Barrio and O. A. Scherman, *Chem. Rev.*, 2015, **115**, 12320.
16. L. Isaacs, *Acc. Chem. Res.*, 2014, **47**, 2052.
17. K. I. Assaf and W. M. Nau, *Chem. Soc. Rev.*, 2015, **44**, 394.
18. K. Kim, N. Selvapalam, Y. H. Ko, K. M. Park, D. Kim and J. Kim, *Chem. Soc. Rev.*, 2007, **36**, 267.
19. S. Gürbüz, M. Idris and D. Tuncel, *Org. Biomol. Chem.*, 2015, **13**, 330.
20. H. Zou, J. Liu, Y. Li, X. Li and X. Wang, *Small*, 2018, **14**, 1802234.
21. C. Stoffelen and J. Huskens, *Small*, 2016, **12**, 96.
22. A. Koc and D. Tuncel, *Isr. J. Chem.*, 2018, **58**, 334.
23. Y. J. Jeon, P. K. Bharadwaj, S. W. Choi, J.-W. Lee and K. Kim, *Angew. Chem., Int. Ed.*, 2002, **41**, 4474.
24. M. E. Bush, N. D. Bouley and A. R. Urbach, *J. Am. Chem. Soc.*, 2005, **127**, 14511.
25. Y. Ling, W. Wang and A. E. Kaifer, *Chem. Commun.*, 2007, 610.
26. J. Zhao, Y.-M. Zhang, H.-L. Sun, X.-Y. Chang and Y. Liu, *Chem. – Eur. J.*, 2014, **20**, 15108.
27. D.-D. Li, X.-C. Chen, K.-F. Ren and J. Ji, *Chem. Commun.*, 2015, **51**, 1576.
28. Y. Lan, Y. Wu, A. Karas and O. A. Scherman, *Angew. Chem., Int. Ed.*, 2014, **53**, 2166.
29. S. D. Choudhury, N. Barooah, V. K. Aswal, H. Pal, A. C. Bhasikuttana and J. Mohanty, *Soft Matter*, 2014, **10**, 3485.
30. X. J. Loh, J. del Barrio, P. P. C. Toh, T.-C. Lee, D. Jiao, U. Rauwald, E. A. Appel and O. A. Scherman, *Biomacromolecules*, 2012, **13**, 84.
31. Y. Wang, D. Li, H. Wang, Y. Chen, H. Han, Q. Jin and J. Ji, *Chem. Commun.*, 2014, **50**, 9390.
32. (a) D. Jiao, J. Geng, X. J. Loh, D. Das, T.-C. Lee and O. A. Scherman, *Angew. Chem., Int. Ed.*, 2012, **51**, 9633; (b) X. J. Loh, J. del Barrio, T.-C. Lee and O. A. Scherman, *Chem. Commun.*, 2014, **50**, 3033.

33. J. H. Mondal, S. Ahmed, T. Ghosh and D. Das, *Soft Matter*, 2015, **11**, 4912.
34. E. Cavatorta, J. Voskuhl, D. Wasserberg, J. Brinkmann, J. Huskens and P. Jonkheijm, *RSC Adv.*, 2017, 7, 54341.
35. C. Hu, N. Ma, F. Li, Y. Fang, Y. Liu, L. Zhao, S. Qiao, X. Li, X. Jiang, T. Li, F. Shen, Y. Huang, Q. Luo and J. Liu, *ACS Appl. Mater. Interfaces*, 2018, **10**, 4603.
36. C.-C. Zhang, Y.-M. Zhang and Y. Liu, *Chem. Commun.*, 2018, **54**, 13591.
37. E. A. Appel, J. Dyson, J. del Barrio, Z. Walsh and O. A. Scherman, *Angew. Chem., Int. Ed.*, 2012, **51**, 4185.
38. Y. Lan, X. J. Loh, J. Geng, Z. Walsh and O. A. Scherman, *Chem. Commun.*, 2012, **48**, 8757.
39. C. Stoffelen and J. Huskens, *Chem. Commun.*, 2013, **49**, 6740.
40. C. Stoffelen, J. Voskuhl, P. Jonkheijm and J. Huskens, *Angew. Chem., Int. Ed.*, 2014, **53**, 3400.
41. C. Stoffelen and J. Huskens, *Nanoscale*, 2015, 7, 7915.
42. D. Wu, Y. Li, J. Yang, J. Shen, J. Zhou, Q. Hu, G. Yu, G. Tang and X. Chen, *ACS Appl. Mater. Interfaces*, 2017, **9**, 44392.
43. X.-M. Chen, Y. Chen, Q. Yu, B.-H. Gu and Y. Liu, *Angew. Chem., Int. Ed.*, 2018, **57**, 12519.
44. J. Fernańdez-Rosas, M. Pessego, A. Acuña, C. Vazquez-Vazquez, J. Montenegro, M. Parajó, P. Rodríguez-Dafonte, F. Nome, and L. Garcia-Rio, *Langmuir*, 2018, **34**, 3485.
45. K. I. Assaf, M. A. Alnajjar and W. M. Nau, *Chem. Commun.*, 2018, **54**, 1734.
46. J. Liu, Y. Lan, Z. Yu, C. S. Y. Tan, R. M. Parker, C. Abell and O. A. Scherman, *Acc. Chem. Res.*, 2017, **50**, 208.
47. K. M. Park, K. Suh, H. Jung, D.-W. Lee, Y. Ahn, J. Kim, K. Baeka and K. Kim, *Chem. Commun.*, 2009, 71.
48. Q. Cheng, S. Li, C. Sun, L. Yue and R. Wang, *Mater. Chem. Front.*, 2019, **3**, 199.
49. C. Sun, H. Zhang, S. Li, X. Zhang, Q. Cheng, Y. Ding, L.-H. Wang and R. Wang, *ACS Appl. Mater. Interfaces*, 2018, **10**, 25090.
50. (a) M. S. Nikolic, C. Olsson, A. Salcher, A. Kornowski, A. Rank, R. Schubert, A. Fromsdorf, H. Weller and S. Förster, *Angew. Chem.*, 2009, **121**, 2790; (b) J. Richter, R. Seidel, R. Kirsch, M. Mertig, W. Pompe, J. Plaschke and H. K. Schackert, *Adv. Mater.*, 2000, **12**, 507; (c) Y. Wang, Z. Tang, S. Tan and N. A. Kotov, *Nano Lett.*, 2005, **5**, 243; (d) Q. Dai, J. G. Worden, J. Trullinger and Q. Huo, *J. Am. Chem. Soc.*, 2005, **127**, 8008.
51. E.-K. Lim, T. Kim, S. Paik, S. Haam, Y.-M. Huh and K. Lee, *Chem. Rev.*, 2015, **115**, 327.
52. (a) T. Bollhorst, K. Rezwan and M. Maas, *Chem. Soc. Rev.*, 2017, **46**, 2091; (b) R. Gref, Y. Minamitake, M. T. Peracchia, V. Trubetskoy, V. Torchilin and R. Langer, *Science*, 1994, **263**, 1600; (c) S. Gouin, *Trends Food Sci. Technol.*, 2004, **15**, 330; (d) C. S. Peyratout and L. Dähne, *Angew. Chem.*,

*Int. Ed.*, 2004, **43**, 3762; (e) R. Akiyama and S. Kobayashi, *Chem. Rev.*, 2009, **109**, 594.

53. Y. Zheng, Z. Yu, R. M. Parker, Y. Wu, C. Abell and O. A. Scherman, *Nat. Commun.*, 2014, **5**, 5772.
54. G. Stephenson, R. M. Parker, Y. Lan, Z. Yu, O. A. Scherman and C. Abell, *Chem. Commun.*, 2014, **50**, 7048.
55. Z. Yu, Y. Lan, R. M. Parker, W. Zhang, X. Deng, O. A. Scherman and C. Abell, *Polym. Chem.*, 2016, 7, 5996.
56. Z. Yu, J. Zhang, R. J. Coulston, R. M. Parker, F. Biedermann, X. Liu, O. A. Scherman and C. Abell, *Chem. Sci.*, 2015, **6**, 4929.
57. R. M. Parker, J. Zhang, Y. Zheng, R. J. Coulston, C. A. Smith, A. R. Salmon, Z. Yu, O. A. Scherman and C. Abell, *Adv. Funct. Mater.*, 2015, **25**, 4091.
58. Z. Yu, Y. Zheng, R. M. Parker, Y. Lan, Y. Wu, R. J. Coulston, J. Zhang, O. A. Scherman and C. Abell, *ACS Appl. Mater. Interfaces*, 2016, **8**, 8811.
59. Y. Lan, X. J. Loh, J. Geng, Z. Walsh and O. A. Scherman, *Chem. Commun.*, 2012, **48**, 8757.

CHAPTER 8

# Cucurbit[8]uril-based 2D and 3D Regular Porous Frameworks

HUI WANG,[a] DAN-WEI ZHANG,[a] XIN ZHAO*[b] AND
ZHAN-TING LI*[a]

[a] Department of Chemistry, Fudan University, Xinjiangwan Campus,
2205 Songhu Road, Shanghai 200438, China; [b] Shanghai Institute of
Organic Chemistry, Chinese Academy of Sciences, 345 Lingling Road,
Shanghai 200032, China
*Email: ztli@fudan.edu.cn; xzhao@sioc.ac.cn

## 8.1 Introduction

Porous structures have found extensive industrial applications in, for example, catalysis, separation, adsorption and pollutant cleanup and removal.[1,2] Early studies have focused on inorganic materials like zeolites and molecular sieves. Among others, the last several decades have witnessed the development of porous polymers,[3,4] metal-organic frameworks (MOFs)[5] or covalent-organic frameworks (COFs).[6-8] In particular, MOFs and COFs have received great attention in the construction of periodic crystalline porous architectures. However, due to their insolubility, investigations for their properties or functions have been limited to the solid state or interface.

Macrocyclic molecules have been widely studied for molecular recognition and self-assembly.[9] In principle, expanding the cavity of a macrocycle in the two- or three-dimensional (2D or 3D) space can lead to the formation of 2D or 3D porous architectures. However, this approach would not be easily realized for a single molecule due to the forbidding synthetic challenge. Self-assembly provides a feasible approach for this aim. The challenge comes

Smart Materials No. 36
Cucurbituril-based Functional Materials
Edited by Dönüs Tuncel
Published by the Royal Society of Chemistry, www.rsc.org

from how noncovalent interactions can be used to hold molecular components in regular supramolecular arrays. In this category, cucurbit[8]uril (CB[8])–enhanced dimerization of hydrophobic aromatic (CEDA) segments has been demonstrated to be an ideal binding motif to generate well-ordered 2D and 3D self-assembled porous structures, which we termed "supramolecular organic frameworks (SOFs)."[10,11] SOFs can be regarded as a family of structurally specific supramolecular polymers because the monomers are held together by noncovalent force and they can be reversibly formed or decomposed.[12–15] However, differently from conventionally main-chain or cross-linked supramolecular polymers, SOFs possess high structural regularity or periodicity that for the first time endows well-defined 2D or 3D porosity for supramolecular polymers.

## 8.2   CB[8] Encapsulation of Aromatic Dimers: Binding Motif for Regular Pores

CB[8] has a hydrophobic cavity with a 0.69-nm diameter,[16] which can host two identical or different aromatic segments in water through hydrophobicity.[17–24] This 1:2 binding stoichiometry was first reported for the dimer of viologen radical cation by Kim and coworkers and more recently has been extended for many other aromatic molecules.[25–29] In particular, cationic aromatic segments work well due to their electrostatic attraction to the carbonyl O atoms on the two rims of CB[8]. All the binding motifs exhibit very high stability in water, and some of them have been applied for the construction of supramolecular polymers in water. We found that the encapsulation of CB[8] for two 4-aryl-pyridinium units is very useful as a binding motif in water for the generation of periodic supramolecular porous structures.[30] The general strategy is to attach hydrophobic 4-aryl-pyridinium units to multitopic rigid molecular frameworks.[10,11] The multicationic molecules prepared in this way all exhibited good solubility in water. Although CB[8] has a very limited water-solubility (<0.01 mM),[16] its interaction with the multitopic molecules could remarkably increase its solubility in water. The resulting self-assembled frameworks, which reflected the rigidity and regularity of the structures as established for MOFs and COFs, also displayed water solubility on the millimolar scale.

## 8.3   2D Regular Porous Frameworks

The most important aim of conducting the CEDA-based self-assembly of supramolecular polymers was to build water-soluble homogeneous regular porous materials. For this aim, we prepared a rigid conjugated triangular molecule 1 (Figure 8.1),[30] which bears three 4-(4-pyridyl)pyridinium (PP) units. The molecule was highly soluble in water due to the existence of three cationic segments and the introduction of three hydrophilic glycol chains. Molecular modeling indicated that the introduction of the glycol chains also could force the three neighboring benzene rings to twist nearly

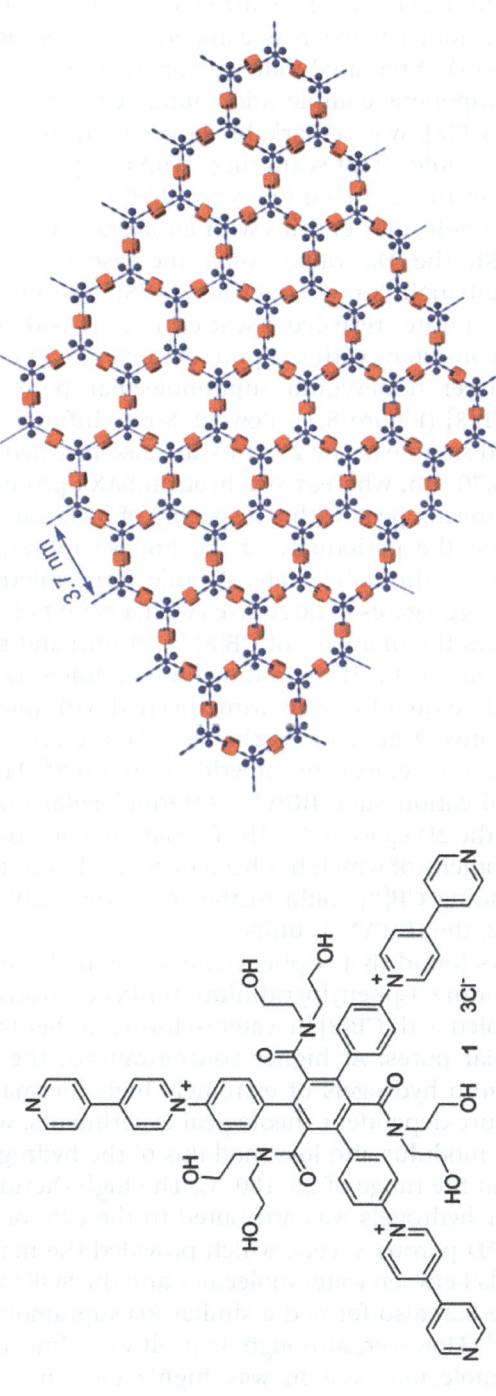

**Figure 8.1** Schematic representation of 2D honeycomb monolayer supramolecular organic framework formed by compound 1 and CB[8] (2:3). Adapted from ref. 30 with permission from American Chemical Society, Copyright 2013.

perpendicularly from the central benzene ring and thus inhibit the face-to-face stacking of the backbone. $^1$H NMR experiments in $D_2O$ supported the 2:3 binding stoichiometry, whereas competition $^1$H NMR experiments in $CD_3CO_2Na$-buffered $D_2O$ revealed that the apparent association constant ($K_a$) for the three-component complexation motif CB[8]⊂(PP·PP), formed by compound 1 and CB[8], was remarkably larger than that of mono- and ditopic controls. Dynamic light-scattering (DLS) experiments revealed that, at [1] = 1.0 mM in the 2:3 solution with CB[8] in water, the two compounds formed supramolecular entities with an average $D_H$ of 69 nm, and that, at [1] = 2.1 mM, the $D_H$ value could increase to about 200 nm. A solution-phase small-angle X-ray scattering (SAXS) experiment of 2:3 solution ([1] = 1.0 mM) in water revealed a scattering peak with a *d*-spacing of 3.61 nm, which was consistent with the pore diameter (3.70 nm) calculated for a regular monolayer honeycomb supramolecular pore formed from compounds 1 and CB[8] (Figure 8.1). Powder X-ray diffraction (PXRD) experiment for the dried sample of the 2:3 mixture also revealed a broad peak with a *d*-spacing of 3.70 nm, whereas synchrotron SAXS profile of the solid sample showed a scattering peak with a *d*-spacing of 3.78 nm. These results further supported that the periodicity of the honeycomb supramolecular pore was maintained in the solid state. Atomic force microscopy (AFM) image showed planar aggregates of large sizes with a height of 1.72 nm. This value perfectly matches the diameter of CB[8] (1.75 nm) and thus supports the monolayer character of the 2D periodic porous framework.

Compound 1 could be quantitatively trimethylated with methyl iodide to afford viologen derivative 2 after ion exchanges (Figure 8.2).[31] Upon treatment with sodium dithionate, its three bipyridinium ($BIPY^{2+}$) units could be reduced to its radical cation state $BIPY^{\bullet+}$. Intermolecular dimerization of this radical cation in the 2D space led to the formation of another monolayer regular SOF, the periodicity of which has been evidenced by a solution-phase SAXS experiment. Adding CB[8] could further raise the stability of the 2D pore by encapsulating the $(BIPY^{\bullet+})_2$ units.

Zhao and coworkers found that triphenylamine could also be used as the core.[32] The corresponding 4-phenylpyridinium (PhPy) derivatives 3a and 3b (Figure 8.2) coassembled with CB[8] in water to form another two monolayer regular supramolecular pores. At higher concentrations, the porous SOFs were found to turn into hydrogels of extremely high thermal stability, as revealed by temperature-dependent rheological experiments, which showed that both the storage modulus and loss modulus of the hydrogels remained nearly constant within the range of 30–180 °C. The high thermal stability of these supramolecular hydrogels was attributed to the extraordinarily large surface areas of the 2D porous sheets, which provided the maximum number of hydrogen bonds between water molecules and the SOF layers. Another compound 3c (Figure 8.2) also formed a similar 2D supramolecular porous structure with CB[8].[33] However, although 3c itself was almost nonemissive in water, the supramolecular system was highly emissive. This turn-on fluorescence has been attributed to the inhibition of charge-transfer

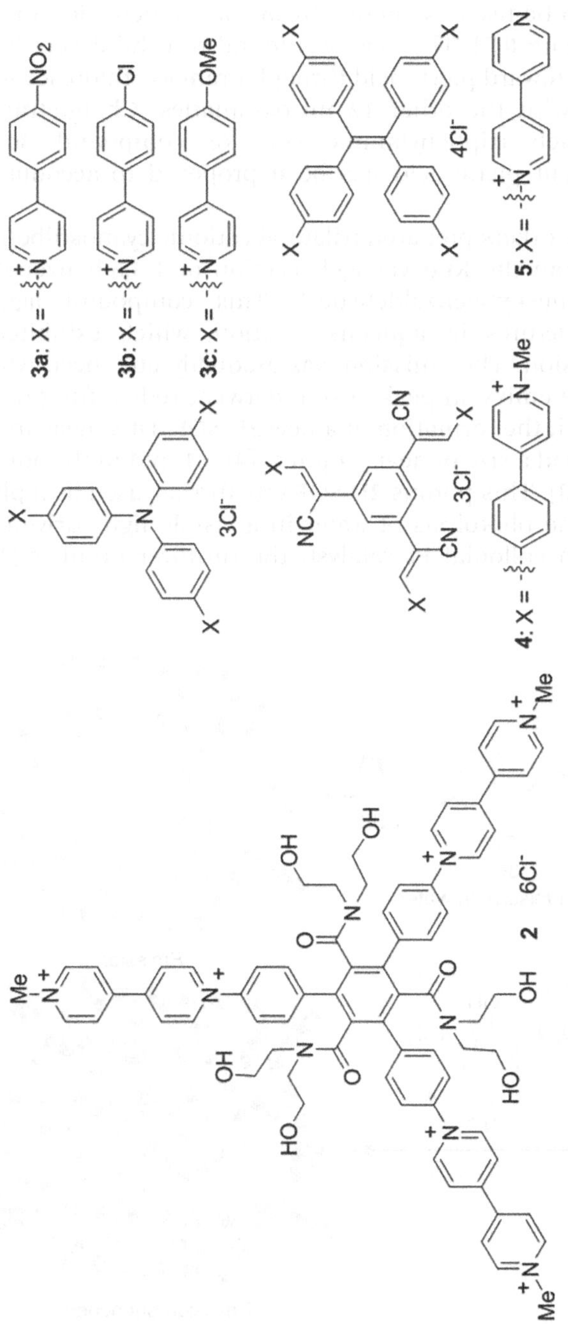

**Figure 8.2** The structure of compounds **2**, **3a–c**, **4** and **5**.

through the encapsulation of CB[8] for the MeOPhPy units, which possibly led to the aggregation-induced emission (AIE) effect. This fluorescent 2D SOF could also be used as chemosensor for the detection of nitroaromatic explosives (Figure 8.3). The SOF was found to exhibit very high selectivity and sensitivity toward picric acid through dramatic fluorescence quenching, as compared with the other 12 nitroaromatics. Charge-transfer between the electron-rich triphenylamine core of compound **3c** and highly electron-deficient picric acid has been proposed to account for the high selectivity.

Park and coworkers prepared trilateral cationic cyanostilbene monomer **4** (Figure 8.2) from the Knoevenagel reaction of 1,3,5-benzenetriacetonitrile with 4-(pyridine-4-yl)benzaldehyde.[34] This compound aggregated into nanosized structures in aqueous solution, which exhibited weak fluorescence emission. The emission was gradually enhanced when CB[8] was added, and the emission peak also underwent red-shifting from 520 nm to 585 nm through the formation of a new 2D SOF. DLS measurement showed that this SOF had a size of nearly 1 μm at [**4**] = 1 mM in the aqueous solution with CB[8] (2 : 3). This porous 2D SOF was further used as a photosensitizer template for the photolysis of water in a visible light–driven $H_2$ evolution system. With a colloidal Pt catalyst, the turnover number (TON) reached 300 in 6 h.

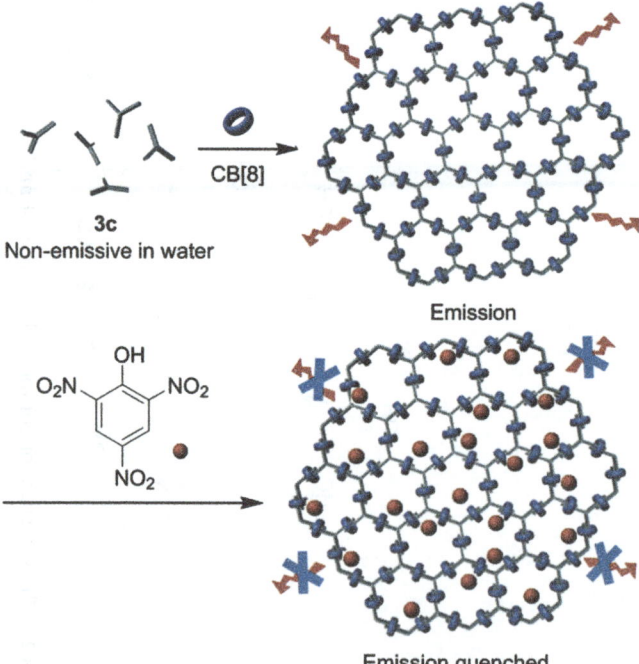

CB[8]

**3c**
Non-emissive in water

Emission

Emission quenched

**Figure 8.3**   Cartoon representation for the 2D SOF assembled from compound 3c and CB[8] and its sensing for picric acid.

Zhao and coworkers also reported that tetraphenylethene (TPE) derivative **5** (Figure 8.2) coassembled with CB[8] (1:2) in water to give rise to a monolayer 2D parallelogram porous SOF.[35] TPE has been widely used for studying the AIE effect.[36] It was found that compound **5** itself had no fluorescence emission in aqueous solution. However, the 2D porous SOF formed by **5** and CB[8] exhibited strong fluorescence due to the restriction of the free rotation of the TPE moiety. Moreover, the addition of organic solvents, such as THF, acetone, 1,4-dioxane and acetonitrile, further enhanced the fluorescence by promoting the aggregation of the single-layer MOFs through solvophobicity, which could inhibit the rotation of TPE units in the stacked layers. Liu and coworkers found that, at higher concentrations,[37] the same 2D parallelogram porous SOF further stacked to form hydrogel. The hydrogel was highly thermostable and could be maintained even at 200 °C. This supramolecular hydrogel was also responsive to acid–base, competitive guests as well as mechanical force. Notably, it exhibited large loading and removal capability for anionic dyes.

The encapsulation of CB[8] for the donor–acceptor interaction-driven heterodimers of electro-rich and deficient aromatic moieties has also been used to assemble 2D SOF pores. For example, Feng and coworkers prepared $C_3$-symmetric rigid tris(methoxynaphthyl)-substituted truxene **6** (Figure 8.4) as the vertex of the honeycomb framework and viologen-attached naphthalene diimide **7** as the edge.[38] Because **6** was insoluble in water, in order to realize its coassembly with **7** (Figure 8.4) and CB[8], a solution of **6** in toluene was placed over an aqueous solution of **7** and CB[8]. Under this condition, the three building blocks interacted at the liquid–liquid interface, which was driven by CB[8]-encapsulation-enhanced donor–acceptor interaction. After draining the solution from the bottom of the trough, SOF monolayers were isolated, which could be as large as 0.25 cm².

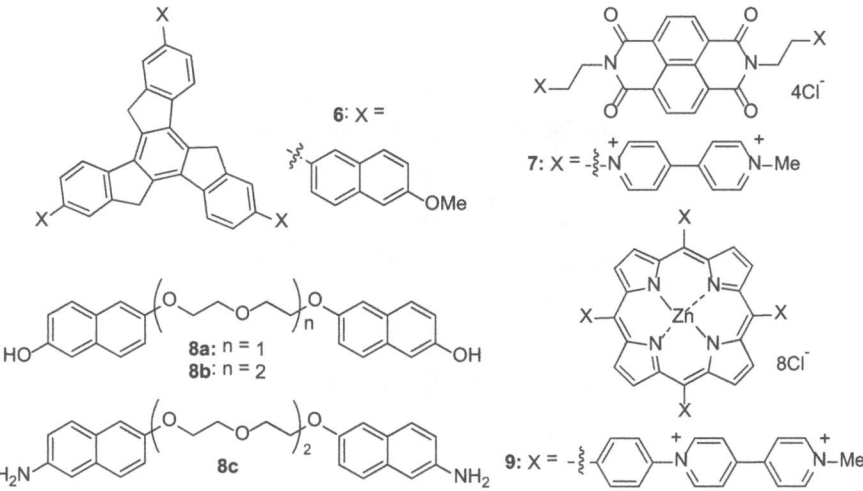

**Figure 8.4** The structure of compounds **6**, **7**, **8a–c** and **9**.

Rigid porphyrin derivative **9** (Figure 8.4) bears four viologen units and thus was highly water soluble. Zhao and coworkers found that its $1:4:4$ mixture with **8a** or **8b** (Figure 8.4) and CB[8] in water could lead to the formation of a new family of square-styled 2D SOF pores,[39] which was driven by CB[8]-encapsulation-enhanced donor–acceptor interaction between the naphthalene units of **8a** or **8b** and the BIPY$^{2+}$ units of **9**. Differently from these 2D SOF pores, these two monolayer systems exhibited a flexible feature but still exhibited large defined cavity (*d*-spacing: 5 or 6 nm, Figure 8.5) and periodicity. AFM measurement showed that the heights of the as-prepared SOFs were only 1.32 and 1.24 nm, respectively. The values were lower than the outer diameter of CB[8], suggesting a collapse of the films on the solid supporter, which had been attributed to the flexibility of the two linear building blocks.

Using a similar approach, Li and coworkers further utilized monomer **2** to generate 2D honeycomb monolayer SOF pores through coassembly with CB[8] and **8b** or **8c** (Figure 8.4).[40] The pore size was as large as 7 nm. The 2D SOF formed from **8c** was also found to be pH-responsive. That is, adding hydrochloric acid decomposed the supramolecular pores by protonating the two amino groups of **8c**, which decreased its electron richness as well as its hydrophobicity, whereas adding sodium hydroxide to neutralize the acid caused the regeneration of the SOF. This result also supported the reversibility of the coassembly of the three components. Both monolayer SOF pores also displayed antibacterial activities, which were evidenced by agar diffusion assays of methicillin-resistant *Staphylococcus aureus* as a test strain. None of the molecular components exhibited any detectable activity. Thus, the activity of the two SOFs was ascribed to the ion-enriching effect produced by the periodic framework.

Cao and coworkers further extended the utility of TPE-derived molecules for the formation of advanced supramolecular systems with CB[8].[41]

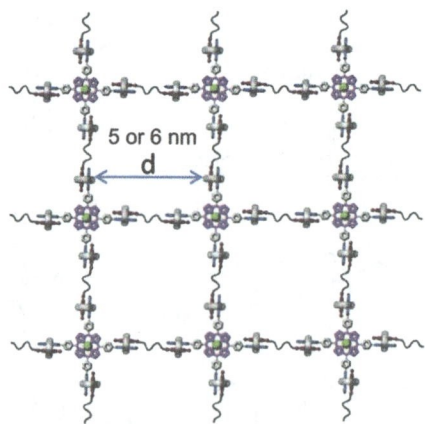

**Figure 8.5**  Schematic representation of the 2D supramolecular polymeric pattern formed by **8a** or **8b**, **9** and CB[8].

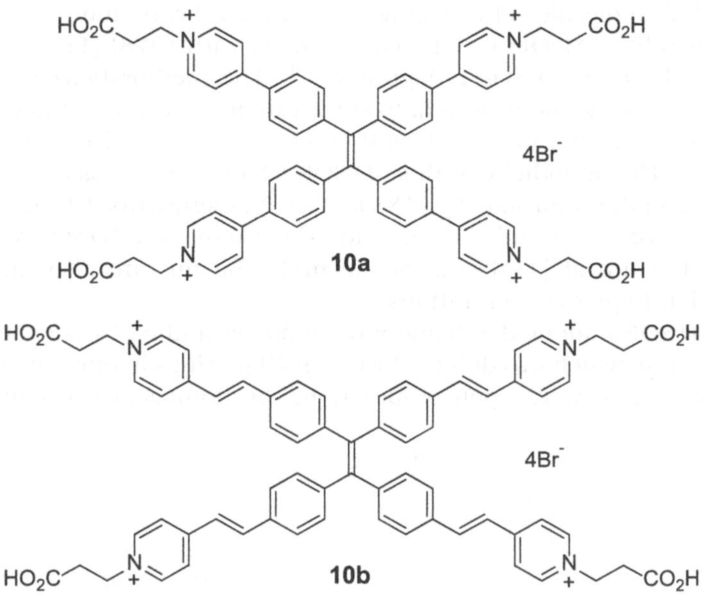

**Figure 8.6** The structure of compounds **10a** and **10b**.

The group prepared compounds **10a** and **10b** (Figure 8.6) and found that **10a** and CB[8] (1 : 2) coassembled into 2D SOF, which further stacked to afford supramolecular cuboids. In contrast, 2D SOF formed from **10b** and CB[8] aggregated to generate spheroids. This difference has been attributed to the increased flexibility of **10b** and the rotatability of its pyridylethylene moieties. Different from the TPE monomers, both SOFs exhibited large redshifting of their maximum emission, which could be tuned by changing pH. The turnoff and turnon fluorescence was further applied in cellular imaging. Adding competitive guests, such as 3,5-dimethyl-1-adamantylamine, to the fluorescence-quenched system could cause the fluorescence of the TPE monomers to recover in HeLa cells by replacing the monomer from the cavity of CB[8].

## 8.4 Tetraphenylmethane-cored 3D Regular Diamondoid Porous Frameworks

The generation of 2D regular SOFs from both rigid and flexible multitopic monomers demonstrated the robustness of the CEBA strategy. In principle, the ordered stacking of 2D monolayer frameworks could lead to the formation of regular 3D frameworks that possess defined deep pores. This strategy has been extensively applied for the construction of 2D MOFs or COFs, which formally should be considered 3D architectures in the solid state. However, currently it is still impossible to convert 2D SOFs into 3D porous structures in solution. In 2014, Li and coworkers reported that the tetrahedral molecule **11a** (Figure 8.7), which bears four 4-(4-methoxy-phenyl)pyridine (MeOPP) units, coassembled with CB[8] (1:2) in water to generate the first 3D SOF-1 (Figure 8.8).[42] The methoxybenzene unit was electron donating and expected, to some extent, to facilitate intermolecular head-to-tail stacking of two MeOPP units, as established by Zhao and co-workers.[43] The periodicity of this 3D SOF-1 had been characterized in both solution and the solid state by SAXS and XRD experiments. DLS experiments conducted for the 1:2 mixtures in water showed that the 3D SOF had a $D_H$ of 91 and 100 nm at [**11a**] = 1.5 or 2.0 mM. The mixture also turned into hydrogel at higher concentrations.

The 3D SOF-1 adopted a diamondoid topology and had a pore aperture of about 2.1 nm, which was defined by the six CB[8] rings in one self-assembled chair-like macrocycle. Upon evaporation, the homogeneous solution first

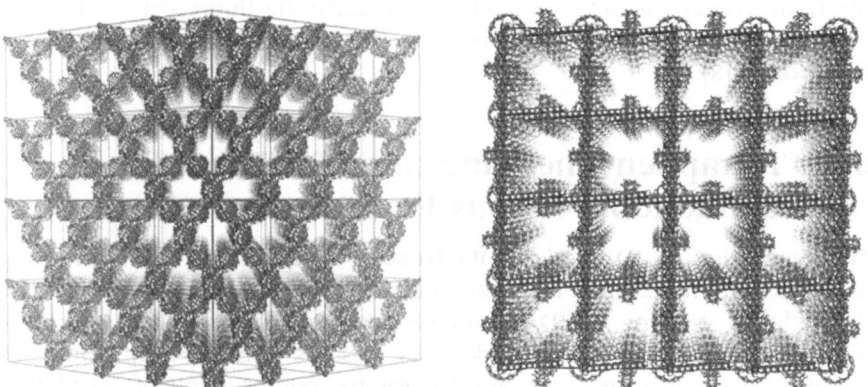

**11a:** R = OMe
**11b:** R = SMe
**11c:** R = NMe$_2$
**11d:** R = CN
**11e:** R = CH$_2$OH
**11f:** R = CHO

**12** 2Cl$^-$

**Figure 8.7**   The structure of compounds **11a–f** and **12**.

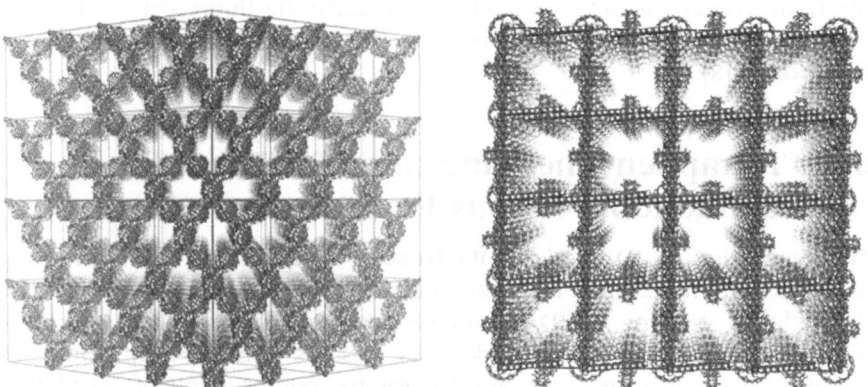

**Figure 8.8**   Modeled structures of (left) the 3D diamondoid SOF formed by tetra-hedral **11a** and CB[8] (1:2) and (right) the 3D cubic metal ion-cored SOF formed by octahedral **16** and CB[8] (1:3).

turned into hydrogel and then further solidified into microcrystals, which could dissolve only in boiling water. The microcrystals also exhibited periodicity that was similar to that in solution, and the periodic porosity could be observed from TEM images. The void volume of the 3D SOF-1 was *ca.* 77%. Fluorescent studies showed that this 3D SOF-1 efficiently adsorbed many different anionic organic guests, including dyes 1–d, drugs 1–3, peptides 1–4, sDNA-1–4, DNA-1 and DNA-2, and dendrimers 1–3 (Figure 8.9), through ion exchange. The monomer and the guests were soft acid and bases, whereas their counterions were hard bases or acids. In water, the guest soft bases preferred to form ion pairs with the soft acid part of the framework, leading their adsorption. The SOF microcrystals were insoluble in water at ambient temperature and could adsorb the preceding guests from their aqueous solution. XRD experiments showed that the periodicity of SOF-1 was maintained after the guest adsorption.

Li and coworkers further prepared compounds **11b–e** (Figure 8.7) and found that all these molecules could coassemble with CB[8] to produce similar 3D SOFs.[44] Dialysis experiments revealed that all the SOFs efficiently adsorbed anticancer drugs pemetrexed disodium and doxorubicin,[44,45] which was driven by electrostatic attraction and/or hydrophobicity. The *in situ*–prepared drug@SOFs were highly stable and could avoid important release of the drug during plasm circulation. Notably, drug@SOFs overcame the multidrug resistance of human breast MCF-7/Adr cancer cells to enter the cancer cells, whereas the acidic microenvironment of the cancer cells promoted the intracellular release of the drugs. *In vitro* and *in vivo* studies revealed considerably improved treatment efficacy of both drugs and led to a 6- to 18-fold reduction of the $IC_{50}$ values. This 3D SOF–based drug delivery strategy omits the complicated but indispensable loading process of liposome or of many nanoparticle-based delivery systems and in principle enables the simultaneous evaluation of different drugs against different tumors.

The aldehyde groups of compound **11f** in the related 3D SOF reacted with the $[Ru(bipy)_3]^{2+}$ derivative **12** to realize postmodification through the formation of hydrazone bonds (Figure 8.7).[46] The resultant $[Ru(bipy)_3]^{2+}$-appended 3D SOF worked as an efficient recyclable heterogeneous catalyst for visible light–induced reduction of aromatic azides to anilines in dichloromethane and *n*-hexane (1 : 1) in the presence of *N,N*-diisopropylethylamine and formic acid and with Hantzsch ester as reductive reagent. The catalyst was recycled for 11 runs and still exhibited considerable activity. For several azide derivatives, the catalysis efficiency was comparable with the control **12**, which catalyzed the reaction homogeneously. The XRD experiment indicated that after 11 runs, the SOF still maintained its periodicity, even though the appended $[Ru(bipy)_3]^{2+}$ moiety suffered partial loss probably due to the hydrolysis of the hydrazone bonds.

The fully rigid tetrahedral compound **13** (Figure 8.10) could also interact with CB[8] to afford a 3D SOF.[47] The modeled structure of this porous system, including the chloride anions, had approximately 67% of void volume,

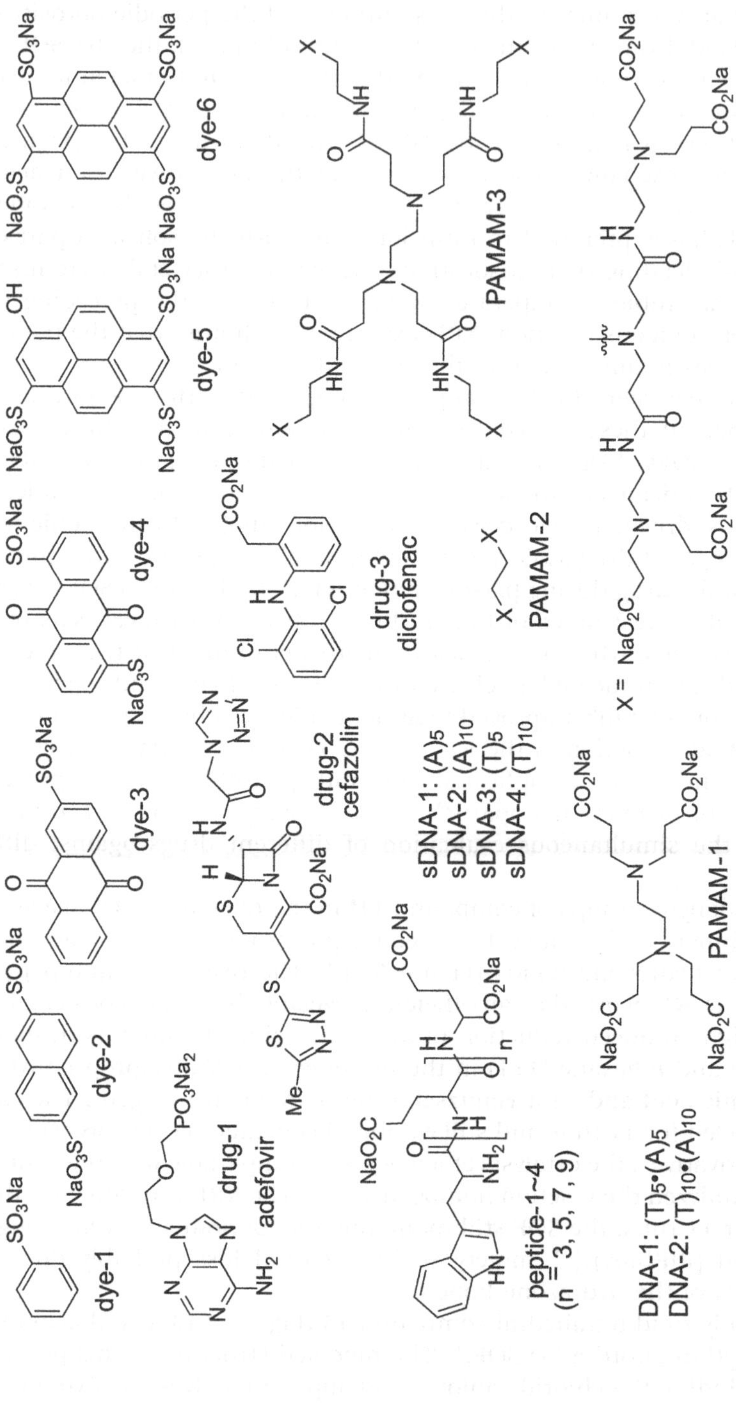

**Figure 8.9** The structure of the guests adsorbed by 3D SOF-1.

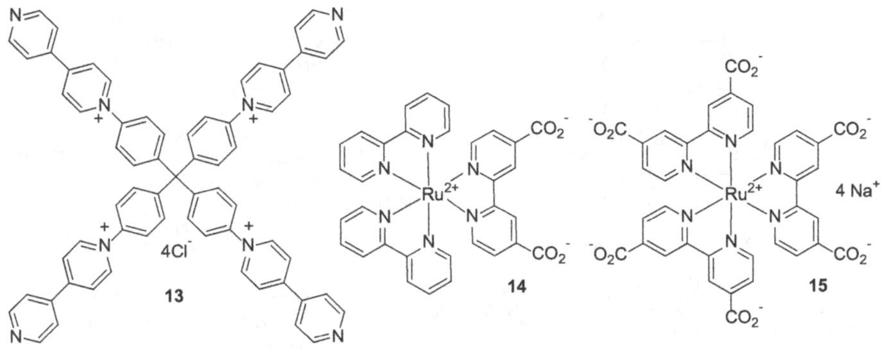

**Figure 8.10** The structure of compounds **13–15**.

and the chair-styled pore defined by six cyclically arranged CB[8] rings had a minimum aperture of approximately 2.3 nm. This very stable 3D SOF could simultaneously enrich both $[Ru(bpy)_3]^{2+}$-based photosensitizers (**14** or **15**) (Figure 8.10) and various polyoxometalate (POM) catalysts, such as Wells–Dawson-type $[P_2W_{18}O_{62}]^{6-}$ ($K^+$ salt) and Keggin-type $PW_{12}O_{40}{}^{3-}$, $SiW_{12}O_{40}{}^{3-}$ and $PMo_{12}O_{40}{}^{3-}$ ($Na^+$ salt), at very low loading. The enrichment of hydrophobic zwitterion **14** was proposed to be driven by hydrophobicity, as observed for doxorubicin by the 3D SOF formed by **11b–e**. The enrichment substantially increased the apparent concentration of both photosensitizer and catalyst in the interior of the framework, and the resulting system could realize recyclable, homogeneous, visible light–driven photocatalysis of the reduction of proton to hydrogen gas. Compared with the control, this new integrated catalysis system could exhibit 110-fold increase of TON.

## 8.5 $[Ru(bipy)_3]^{2+}$-cored 3D Regular Cubic Porous Frameworks

MOFs have been typically constructed with metal ions as vertexes connected by organic ligands through electrostatic or coordination interaction.[48] The crystalline frameworks are highly regular and possess well-defined pores but generally lack solubility. The formation of water-soluble diamondoid 3D SOFs from tetraphenylmethane-cored molecules strongly suggested the extension of the CEBA strategy to transition metal-derived precursors.[49,50] $[Ru(bipy)_3]^{2+}$ complexes not only possess high stability but also have been widely used as visible light sensitizers. Incorporation of $[Ru(bipy)_3]^{2+}$ moiety as a vertex to assemble porous frameworks through the CEBA strategy might produce new kinds of photoactive homogeneous regular pores. To address this possibility, Li and coworkers prepared $[Ru(bipy)_3]^{2+}$ derivative **16** (Figure 8.11).[51] This octacationic complex had a modest solubility in water. The six *N*-methyl-4-phenylpyridinium units were connected to the $[Ru(bipy)_3]^{2+}$ core through amide to ensure the rigidity of the whole molecule. Its 1 : 3 mixture with CB[8] led to the generation of the first

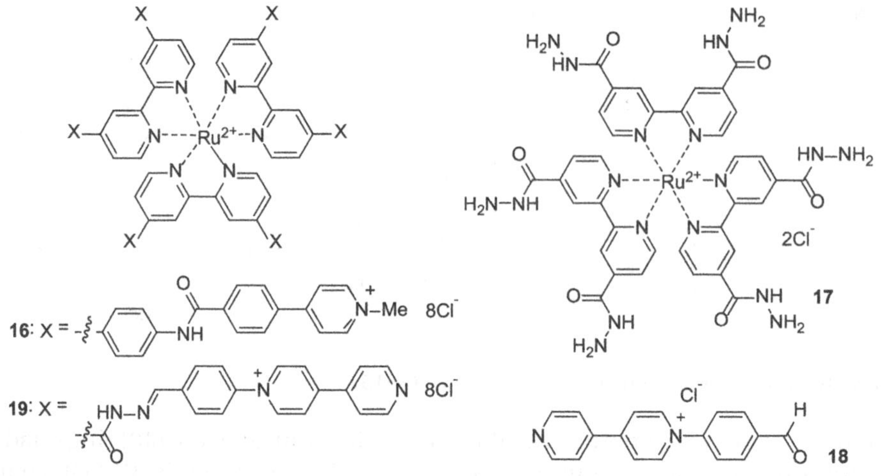

**Figure 8.11**  The structure of compounds **16–19**.

supramolecular metal-organic framework (SMOF) that was more soluble in water (Figure 8.11). This porous framework may be regarded as a hybrid of MOF and SOF. The periodicity of this SMOF in water was confirmed by synchrotron SAXS and XRD experiments. Upon evaporation, the solution of this SMOF also slowly solidified to afford microcrystals, which maintained the periodicity, as revealed by both the SAXS and the XRD experiments. The void volume of this cubic SMOF was estimated to be 80%, whereas the pore aperture, which was defined by the four CB[8] units in one self-assembled macrocycle that adopted a square-like conformation, was calculated to be approximately 1.5 nm.

As a specific supramolecular cationic polyelectrolyte, this SMOF also accommodated Wells–Dawson–type polyoxoanion $[P_2W_{18}O_{62}]^{6-}$ (WD-POM), which has a width of about 1.1 nm. Moreover, this encapsulation occurred in one-cage/one-POM manner. As a result, every Ru(bpy)$_3^{2+}$ or encapsulated WD-POM cluster was actually mutually surrounded by 8 counterparts at the vertices of a cubic cage. Such an arrangement was ideal for visible light–initiated electron transfer from the excited Ru(bpy)$_3^{2+}$ to the redox-active WD-POM. Thus, visible light (500 nm)–initiated sensitization of $[Ru(bpy)_3]^{2+}$ for the catalytic reduction of protons to produce hydrogen gas by WD-POM was realized in acidic aqueous solution (pH = 1.8), using methanol as the sacrificial electron donor. At $[Ru^{2+}] = 0.3$ mM and $[POM] = 2$ μM, the TON, defined as $n(0.5H_2)/n$POM, for $H_2$ production reached 392, which corresponded to a $H_2$ evolving rate, *i.e.*, a turnover frequency (TOF), of 3553 μmol g$^{-1}$ h$^{-1}$ (based on WD-POM). This level of $H_2$ production was about 4 times higher than that of an earlier reported heterogeneous WD-POM@$[Ru(bpy)_3]^{2+}$-MOF system that contains identical amounts of WD-POM and $[Ru(bpy)_3]^{2+}$ units.[52] The high efficiency of the new WD-POM@SMOF system has been attributed to its unique one-cage/one-POM arrangement and homogeneity, which should allow for quicker diffusion and close contact of hydronium and methanol molecules and thus facilitate the electron transfer from excited $[Ru(bpy)_3]^{2+}$ to WD-POM.

Li and coworkers further prepared compounds **17** and **18**.[53] Their reaction, with a molar ratio of 1:6, in water afforded complex **19** through the formation of six hydrazine bonds. However, the yield was as low as 4.2%. In the presence of CB[8] (3 equiv., relative to compound **17**), the reaction took place quantitatively due to the formation of another cubic SOF by **19** and CB[8]. This cubic SOF combined the feature of MOF and COF in that it possessed metal ion vertexes as well as dynamic hydrazine bonds. The periodicity of this hybrid porous structure in water and in the solid state was confirmed by synchrotron SAXS and XRD experiments. Moreover, it could enrich discrete anionic POMs and maintain periodicity in an acidic medium. The enrichment remarkably facilitated visible light–induced electron transfer from its excited $[Ru(bpy)_3]^{2+}$ units to the enriched POMs, which led to enhanced catalysis of the POMs for the reduction of proton to $H_2$ in both aqueous (homogeneous) and organic (heterogeneous) media. Probably because the introduction of the hydrazine bonding decreased the stability, the catalysis efficiency was notably lower than that of the SOF formed by compound **16**.

## 8.6  Conclusion

In the past decade, CB[8] has been demonstrated to be a very useful rigid macrocyclic receptor for molecular recognition. Its encapsulation for two identical or discrete hydrophobic aromatic subunits in water provides ideal binding patterns for the generation of advanced porous materials in water. Preliminary studies for the 3D diamondoid SOFs have established them as useful carriers for drug delivery. Such carriers are highly stable and homogeneous in water, and the synthesis of the molecular components, which are not available for the clinically used liposomes, are generally straightforward. One possible extension of this function is the delivery of nucleic acids. Efficient postmodification may also allow for the development of stimuli-responsive drug delivery and controlled release. For $[Ru(bpy)_3]^{2+}$-based SOFs, more applications in visible light–induced (asymmetric) catalysis of organic reactions can be expected. Other transition metal ions, particularly $Fe^{2+}$, $Fe^{3+}$ and $Co^{2+}$, may also be used as the vertexes. For efficient recyclable heterogeneous catalysis, large apertures should be required, which need the design and synthesis of multitopic monomers with longer arms.

## Acknowledgements

We thank the National Natural Science Foundation of China (Nos. 21432004 and 21529201) for financial support, the Shanghai Synchrotron Radiation Facility for providing BL16B1 and BL14B1 beamlines for collecting the synchrotron X-ray scattering and diffraction data, and the SIBYLS Beamline

12.3.1 of the Advanced Light Source (ALS), Lawrence Berkeley National Laboratory, for collecting solution-phase synchrotron small-angle X-ray scattering data.

# References

1. *Handbook of Porous Media*, ed. K. Vafai, CRC Press, Boca Raton, 3rd edn, 2015, pp. 939.
2. *Porous Materials: Processing and Applications*, ed. P. Liu and G. F. Chen, Butterworth-Heinemann, Oxford, 2014, pp. 576.
3. S. Qiu and T. Ben, *Porous Polymers: Design, Synthesis and Applications*, Royal Society of Chemistry, Cambridge, 2016, pp. 311.
4. *Porous Polymers*, ed. M. S. Silverstein, N. R. Cameron and M. A. Hillmyer, John Wiley & Sons, Hoboken, 2011, pp. 454.
5. *Metal-Organic Framework Materials*, ed. L. R. MacGillivray and C. M. Lukehart, John Wiley & Sons, Chichester, 2014, pp. 563.
6. X. Feng, X. Ding and D. Jiang, *Chem. Soc. Rev.*, 2012, **41**, 6010.
7. S. Y. Ding and W. Wang, *Chem. Soc. Rev.*, 2013, **42**, 548.
8. P. J. Waller, F. Gandara and O. M. Yaghi, *Acc. Chem. Res.*, 2015, **48**, 3053.
9. B. Dietrich, P. Viout and J. M. Lehn, *Macrocyclic Chemistry: Aspects of Organic and Inorganic Supra Molecular Chemistry*, VCH, Weinheim, 1993, pp. 384.
10. J. Tian, L. Chen, D. W. Zhang, Y. Liu and Z. T. Li, *Chem. Commun.*, 2016, **52**, 6351.
11. J. Tian, H. Wang, D. W. Zhang, Y. Liu and Z. T. Li, *Natl. Sci. Rev.*, 2017, **4**, 426.
12. J. M. Lehn, *Polym. Int.*, 2002, **51**, 825.
13. L. Brunsveld, B. J. B. Folmer, E. W. Meijer and R. P. Sijbesma, *Chem. Rev.*, 2001, **101**, 4071.
14. J. M. Lehn, *Prog. Polym. Sci.*, 2005, **30**, 814.
15. L. Yang, X. Tan, Z. Wang and X. Zhang, *Chem. Rev.*, 2015, **115**, 7196.
16. J. Lagona, P. Mukhopadhyay, S. Chakrabarti and L. Isaacs, *Angew. Chem., Int. Ed.*, 2005, **44**, 4844.
17. Y. H. Ko, E. Kim, I. Hwang and K. Kim, *Chem. Commun.*, 2007, 1305.
18. F. Biedermann, W. M. Nau and H. J. Schneider, *Angew. Chem., Int. Ed.*, 2014, **53**, 11158.
19. S. J. Barrow, S. Kasera, M. J. Rowland, J. del Barrio and O. A. Scherman, *Chem. Rev.*, 2015, **115**, 12320.
20. H. Yang, B. Yuan, X. Zhang and O. A. Scherman, *Acc. Chem. Res.*, 2014, **47**, 2106.
21. L. Isaacs, *Acc. Chem. Res.*, 2014, **47**, 2052.
22. X. Yang, F. Liu, Z. Zhao, F. Liang, H. Zhang and S. Liu, *Chin. Chem. Lett.*, 2018, **29**, 1560.
23. S. Wu, J. Li, H. Liang, L. Wang, X. Chen, G. Jin, X. Xu and H. H. Yang, *Sci. China: Chem.*, 2017, **60**, 628.

24. T. T. Li, L. L. Wen, H. L. Ji, F. Y. Liu and S. G. Sun, *Chin. Chem. Lett.*, 2017, **28**, 463.
25. W. S. Jeon, H. J. Kim, C. Lee and K. Kim, *Chem. Commun.*, 2002, 1828.
26. S. Y. Kim, Y. H. Ko, J. W. Lee, S. Sakamoto, K. Yamaguchi and K. Kim, *Chem. - Asian J.*, 2007, **2**, 747.
27. Z. J. Zhang, Y. M. Zhang and Y. Liu, *J. Org. Chem.*, 2011, **76**, 4682.
28. Y. Liu, H. Yang, Z. Wang and X. Zhang, *Chem. - Asian J.*, 2013, **8**, 1626.
29. Z. J. Yin, Z. Q. Wu, F. Lin, Q. Y. Qi, X. N. Xu and X. Zhao, *Chin. Chem. Lett.*, 2017, **28**, 1167.
30. K. D. Zhang, J. Tian, D. Hanifi, Y. Zhang, A. C. H. Sue, T. Y. Zhou, L. Zhang, X. Zhao, Y. Liu and Z. T. Li, *J. Am. Chem. Soc.*, 2013, **135**, 17913.
31. L. Zhang, T. Y. Zhou, J. Tian, H. Wang, D. W. Zhang, X. Zhao, Y. Liu and Z. T. Li, *Polym. Chem.*, 2014, **5**, 4715.
32. T.-Y. Zhou, Q.-Y. Qi, Q.-L. Zhao, J. Fu, Y. Liu, Z. Ma and X. Zhao, *Polym. Chem.*, 2015, **6**, 3018.
33. Y. Zhang, T. G. Zhan, T. Y. Zhou, Q. Y. Qi, X. N. Xu and X. Zhao, *Chem. Commun.*, 2016, **52**, 7588.
34. H. J. Lee, H. J. Kim, E. C. Lee, J. Kim and S. Y. Park, *Chem. - Asian J.*, 2018, **13**, 390.
35. S. Q. Xu, X. Zhang, C. B. Nie, Z. F. Pang, X. N. Xu and X. Zhao, *Chem. Commun.*, 2015, **51**, 16417.
36. J. Mei, N. L. C. Leung, R. T. K. Kwok, J. W. Y. Lam and B. Z. Tang, *Chem. Rev.*, 2015, **115**, 11718.
37. X. M. Chen, Y. M. Zhang and Y. Liu, *Supramol. Chem.*, 2016, **28**, 817.
38. M. Pfeffermann, R. Dong, R. Graf, W. Zajaczkowski, T. Gorelik, W. Pisula, A. Narita, K. Müllen and X. Feng, *J. Am. Chem. Soc.*, 2015, **137**, 14525.
39. X. Zhang, C. B. Nie, T. Y. Zhou, Q. Y. Qi, J. Fu, X. Z. Wang, L. Dai, Y. Chen and X. Zhao, *Polym. Chem.*, 2015, **6**, 1923.
40. L. Zhang, Y. Jia, H. Wang, D.-W. Zhang, Q. Zhang, Y. Liu and Z. T. Li, *Polym. Chem.*, 2016, **7**, 1861.
41. Y. Li, Y. Dong, X. Miao, Y. Ren, B. Zhang, P. Wang, Y. Yu, B. Li, L. Isaacs and L. Cao, *Angew. Chem., Int. Ed.*, 2018, **57**, 729.
42. J. Tian, T. Y. Zhou, S. C. Zhang, S. Aloni, M. V. Altoe, S. H. Xie, H. Wang, D. W. Zhang, X. Zhao, Y. Liu and Z. T. Li, *Nat. Commun.*, 2014, **5**, 5574.
43. Y. Zhang, T. Y. Zhou, K. D. Zhang, J. L. Dai, Y. Y. Zhu and X. Zhao, *Chem. - Asian J.*, 2014, **9**, 1530.
44. J. Tian, C. Yao, W. L. Yang, L. Zhang, D. W. Zhang, H. Wang, F. Zhang, Y. Liu and Z. T. Li, *Chin. Chem. Lett.*, 2017, **28**, 798.
45. C. Yao, J. Tian, H. Wang, D. W. Zhang, Y. Liu, F. Zhang and Z. T. Li, *Chin. Chem. Lett.*, 2017, **28**, 893.
46. Y. P. Wu, B. Yang, J. Tian, S. B. Yu, H. Wang, D. W. Zhang, Y. Liu and Z. T. Li, *Chem. Commun.*, 2017, **53**, 13367.
47. S. B. Yu, Q. Qi, B. Yang, H. Wang, D. W. Zhang, Y. Liu and Z. T. Li, *Small*, 2018, **14**, 1801037.

48. M. Eddaoudi, D. B. Moler, H. Li, B. Chen, T. M. Reineke, M. O'Keeffe and O. M. Yaghi, *Acc. Chem. Res.*, 2001, **34**, 319.
49. H. Wang, D. W. Zhang and Z. T. Li, *Acta Polym. Sin.*, 2017, 19.
50. Y. Chen, F. Huang, Z. T. Li and Y. Liu, *Sci. China: Chem.*, 2018, **61**, 979.
51. J. Tian, Z. Y. Xu, D. W. Zhang, H. Wang, S. H. Xie, D. W. Xu, Y. H. Ren, H. Wang, Y. Liu and Z. T. Li, *Nat. Commun.*, 2016, 7, 11580.
52. Z. M. Zhang, T. Zhang, C. Wang, Z. Lin, L. S. Long and W. Lin, *J. Am. Chem. Soc.*, 2015, **137**, 3197.
53. X. F. Li, S. B. Yu, B. Yang, J. Tian, H. Wang, D. W. Zhang, Y. Liu and Z. T. Li, *Sci. China: Chem.*, 2018, **61**, 830.

CHAPTER 9

# Supramolecular Interactions of Cucurbit[n]uril Homologues and Derivatives with Biomolecules and Drugs

QIAN CHENG, HANG YIN, IAN W. WYMAN AND
RUIBING WANG*

State Key Laboratory of Quality Research in Chinese Medicine, and
Institute of Chinese Medical Sciences, University of Macau, Taipa,
Macau SAR, China
*Email: rwang@umac.mo

## 9.1 Introduction

As an important family of macrocyclic compounds, cucurbit[$n$]urils (CB[$n$]s) have drawn much attention because of their unique binding properties. CB[$n$]s ($n = 5$–8, 10, 13–15) consist of $n$ glycoluril units that are connected by $2n$ methylene groups to form a pumpkin-shape with two oxygen-laced hydrophilic portals and a hydrophobic cavity.[1] The members of the CB[$n$] family have different portal and cavity sizes, which give them various recognition properties. For instance, on account of its relatively small cavity, CB[5] has very few biomedical applications. In contrast, CB[7] and its derivatives can host relatively large molecules that are too big to be encapsulated by CB[6] and have thus attracted significant attention for their potential applications in pharmaceutical sciences and drug delivery.[2]

Smart Materials No. 36
Cucurbituril-based Functional Materials
Edited by Dönüs Tuncel
© The Royal Society of Chemistry 2020
Published by the Royal Society of Chemistry, www.rsc.org

Meanwhile, CB[6] and its derivatives are known to be able to host alkyl ammonium ions, which have been rarely studied directly as drug carriers, partly due to the small cavity size of CB[6] that cannot accommodate many common drug molecules.[3] However, CB[6] is often the dominant product obtained *via* CB[n] synthesis, and its functionalization is often more readily achievable.[4] With a relatively larger cavity, CB[8] is even able to host two molecules simultaneously by forming a 1 : 2 host–guest complex that has been widely used in polymer/material sciences[5] and biomedical sciences.[6] As a member of the CB[n] family with the largest cavity, CB[10] shows its unique molecular recognition properties that are rather distinct from other CB[n]s; however, its extremely poor water solubility has limited its applications.[7] Meanwhile, CB[14] and CB[15] have unusual circular cavities showing a 180° twist and structural flexibility, giving them two small cavities instead of a large one, resulting in unusual binding behaviors with several guest molecules.

To overcome some of the limitations of CB[n]s, such as poor water solubility, scientists have made great efforts to introduce functional groups into these macrocycles. A variety of potential applications of functionalized CB[n]s have been reported, such as CB[n]-based ion channels,[8] supramolecular vesicles,[9] cross-linked polymers,[10] drug delivery platforms[11] and bioimaging systems.[12] In this review, we will summarize the supramolecular interactions of CB homologues and their derivatives with biomolecules and small organic drug molecules based on our recent studies and related studies by peers in this research area. The works regarding the applications of CB[7] and CB[8], as well as their derivatives in the pharmaceutical and biomolecular sciences, that has been summarized by our previous review[13] will not be repeated in this chapter.

## 9.2 Interactions of CB[6] and Its Derivatives with Biomolecules and Drugs

### 9.2.1 Interactions of CB[6] with Biomolecules and Drugs

As the first structurally identified homologue, CB[6] is the most abundant member of the CB[n]s family. With a relatively narrow portal diameter, CB[6] is commonly known to selectively encapsulate alkyl ammonium ions with relatively high binding affinity that is attributed to the ion–dipole interactions between the positively charged protonated amine and the electronegative carbonyl portals, as well as the hydrophobic effect involving the alkyl chains and the hydrophobic cavity.[14] As a new strategy to increase the bioavailability and to decrease the side effects of chemotherapeutic drugs, CB[6] can serve as vehicles to transport drugs directly in their cavities. Due to the poor aqueous solubility of CB[6], extensive work has been carried out in saline solutions, which is compatible with physiological systems. For instance, Sevilla *et al.* found that the water solubility of the otherwise poorly soluble antitumor drug emodin and nonsteroidal anti-inflammatory drug

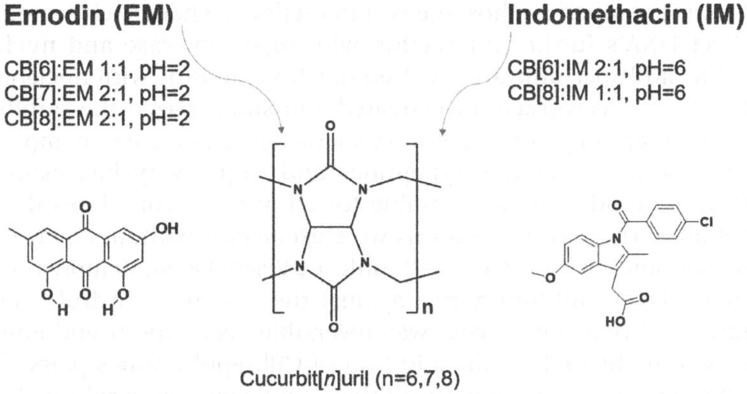

**Emodin (EM)**　　　　　　　　　　　　**Indomethacin (IM)**

CB[6]:EM 1:1, pH=2　　　　　　　　　　　CB[6]:IM 2:1, pH=6
CB[7]:EM 2:1, pH=2　　　　　　　　　　　CB[8]:IM 1:1, pH=6
CB[8]:EM 2:1, pH=2

Cucurbit[*n*]uril (n=6,7,8)

**Figure 9.1**　Cucurbit[*n*]urils ($n = 6$–8) used as host molecules for supramolecular complexes formation with two different drugs: emodin and indomethacin.

indomethacin exhibited much improved water solubility upon their complexations by CB[6], and their stability was also improved significantly as a consequence of the supramolecular interactions (Figure 9.1).[15]

Their study revealed that emodin could form complexes with CB[6] with a 1:1 (host:guest) stoichiometry only in acidic environments at pH = 2. In the case of indomethacin, complexes were formed with CB[6] at weakly acid conditions (pH = 6), with a stoichiometry of 2:1. Similarly, Sharma *et al.* also studied the noncovalent complexation behaviors of sodium ascorbate (SA) with CB[6] at neutral pH in an aqueous $Na_2SO_4$ solution employing NMR, UV–Vis, fluorescence, TGA and DRS characterization techniques, and this study might provide a solution for vitamin C formulation.[16] In addition, the supramolecular inclusion of pyridine-2-aldoxime (a drug to treat alkylphosphate poisoning) by CB[6] in an aqueous saline medium was investigated by Roy and coworkers and may also provide a promising solution for supramolecular formulation of this drug molecule.[17]

Besides small organic molecule–based drugs, proteins and enzymes could also be complexed or partially encapsulated by CB[6]. For instance, Liu and coworkers reported a pH-responsive artificial selenoenzyme that was constructed by reversible binding between this organoselenium compound and CB[6] to form a pseudorotaxane-based molecular switch in response to pH stimuli.[18] The glutathione peroxidase (GPx) activity of the artificial selenoenzyme can be switched on/off in a mild and body-suitable environment between pH = 6 and 7. Kim and coworkers demonstrated that CB[6] could control fibril assemblies by a phase transfer of amyloid proteins from soluble state to insoluble state when the amount of CB[6] exceeded its solubility limit in solution, as the phase transfer of proteins postponed the amyloid nucleation kinetics, and the nuclei formed during the early stage of this process were uniformly assembled into fibrils.[19] In addition, Nakamura *et al.* also utilized CB[6] to regulate the interactions of spermidine and spermine with DNA *via* competitive host–guest interactions, where such a modulation may affect DNA's further interaction with topoisomerase and nuclease as spermidine and spermine are involved in DNA's binding with nucleobases.[20] Schollmeyer and coworkers investigated and summarized the stability constants and thermodynamic parameters of supramolecular complexations between a series of short polypeptides and CB[6]. Very interestingly, entropies were found to be comparable for all of these complexes due to the fact that all of these complexations were exclusions without engagement of the hydrophobic cavity of CB[6].[21] Similarly, Pischel *et al.* demonstrated that CB[6] may play an inhibitory role against the enzymatic hydrolysis of DNA substrates and that the process was reversible, as a type II endonuclease's activity was modulated *via* the addition of CB[*n*]-polyamines pairs.[22] Meanwhile, protein modification with synthetic molecules allowed for the development of various functions that cannot be achieved by the individual components. Francis and coworkers synthesized a series of protein conjugates *via* a new type of CB[6]-catalyzed azide–alkyne click chemistry, including protein–peptide, protein–DNA, protein–polymer, and protein–drug conjugates, which provided a promising new approach for the development of protein conjugates (Figure 9.2).[23]

Nanomaterials have received much attention in recent years because of their various applications in many areas, including drug delivery and bioimaging. Therefore, CB[*n*]s have also been investigated for their roles in nanomaterials. Liu and coworkers constructed a biocompatible polysaccharide nanoparticle based on host–guest interactions of an adamantane-bis(diamine) moiety of the polysaccharide and CB[6]. Very interestingly, CB[6] significantly reduced the nonspecific cytotoxicity of the carrier and enhanced the cationic density through macrocycle-induced p$K_a$ up-shift, which facilitated the subsequent siRNA binding and cargo delivery into malignant human prostate PC-3 cells.[24] Furthermore, Liu's research group constructed nanocarriers *via* CB[6]-mediated self-assembly of an amphiphilic guest molecule, which possesses an alkyl head and two ammonium tails. These carriers could encapsulate hydrophobic payloads

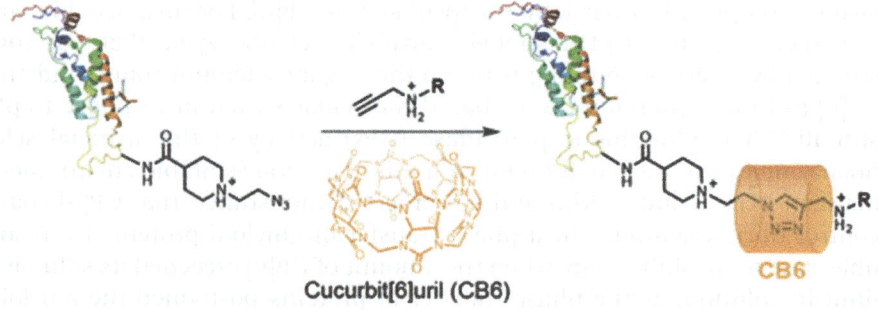

Cucurbit[6]uril (CB6)

**Figure 9.2**   A protein bioconjugation strategy using CB[6].
Reproduced from ref. 23 with permission from American Chemical Society, Copyright 2017.

including dyes and drugs such as doxorubicin (DOX). When α-cyclodextrin was used to include the hydrophobic regions, the nanocarriers were destroyed, and the cargo was released.[25] In terms of drug delivery systems, CB[n]-based supramolecular hydrogels have also attracted much attention recently due to their good biocompatibility and stimuli-responsiveness. For example, Tan and coworkers constructed a thermo-responsive supramolecular hydrogel consisting of CB[6] and butan-1-aminium 4-methylbenzenesulfonate (BAMB). When a small amount of CB[6] was added to the BAMB solution, the butylamine units of BAMB were included, forming a polypseudorotaxane, and, upon cooling, the material was turned into a hydrogel and turned into a sol state when the temperature was raised again to 25 °C or higher.[26]

## 9.2.2 Interactions of CB[6] Derivatives with Biomolecules and Drugs

To further extend the use of CB[n] and biomolecules, functionalization is often required, so that (1) the limited solubility of CB[n]s could be improved, (2) novel supramolecular structures could be constructed through the co-valent conjugation of CB[n]s with functional groups, and (3) CB[n]s would allow further modular functionalization through host–guest complexations. A variety of functionalized CB[6]s have been prepared, and their applications have been extensively investigated in areas such as biosensors,[4c,27] drug delivery[28] and tissue engineering,[29] which are summarized in the next section.

### 9.2.2.1   Derivatized CB[6]-based Biosensors

The first example of a CB[6]-based biosensor was achieved *via* anchoring of a CB[6] derivative, perallyloxyCB[6], onto a thiol-terminated glass surface through UV light–initiated "thiol–ene" click reaction, which could be employed for the noncovalent conjugation of fluorescent guests bearing spermidine. The fluorescein-derived FITC-spermine has been used as a tag to confirm the covalent linkage of CB6 to silica surfaces (Figure 9.3).[4c]

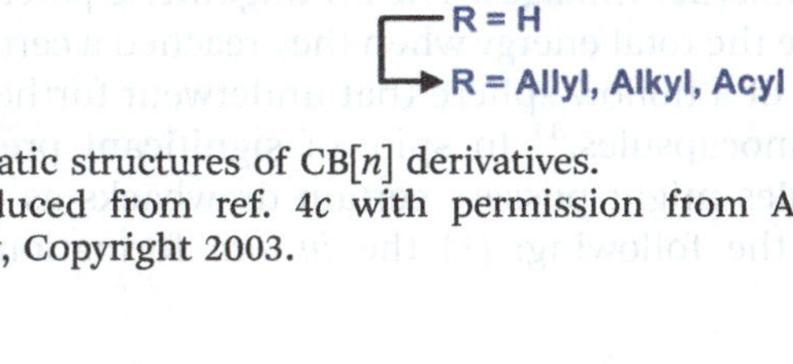

CB[n] (n = 5 - 8)          (RO)$_{2n}$CB[n]

R = H
R = Allyl, Alkyl, Acyl

**Figure 9.3**   Schematic structures of CB[n] derivatives.
Reproduced from ref. 4c with permission from American Chemical Society, Copyright 2003.

Furthermore, through spin-coating, a stable and homogeneous perallyl-oxyCB[6] layer was formed on a water-stable p-channel semiconductor film, which could selectively capture (thus detect) acetylcholine (Ach+), a neurotransmitter in the human central nervous system. This yielded a detection limit as low as $1 \times 10^{-12}$ M, which was up to six orders of magnitude lower than the previously reported value.[27]

### 9.2.2.2    Derivatized CB[6]-based Drug Delivery Systems (DDS)

Drug delivery systems based on functionalized CB[6] are often nanomaterials with unique advantages, as CB[6] on the surface of the nanoparticles allows for noncovalent modular modification with targeting or imaging tags through host–guest interactions between CB[6] and spermine-conjugated tags. Wei and coworkers discovered that the complexation of thiabendazole (TBZ), a chemical to prevent the growth of mold, with symmetrical tetra-methylCB[6]- and metahexamethyl-substituted CB[6] promoted the bioactivity of TBZ against the growth of *F. graminearum* mycelia. The improved bioactivity of TBZ could possibly be explained by the improved cellular uptake of TBZ caused by an enhancement of its water solubility in the presence of the host molecule.[30] Kim and coworkers reported the preparation of CB[6]-based nanoparticles through a precipitation method, and these were employed as highly efficient drug delivery carriers with high loading capacities for hydrophobic drugs.[28] Various "tag" units can be introduced on the surface of CB[6] nanoparticles, and the attachment of folate-spermidine remarkably facilitated the efficient delivery of hydrophobic drugs *via* folate receptor–mediated endocytosis. Vesicles formed by amphiphilic CB[6] derivatives could also serve as efficient drug delivery vehicles (Figure 9.4).[31] By following the same principle, various tag units with targeting and imaging functions were attached to the surface of vesicles simultaneously for imaging-guided, targeted payload delivery. In 2010, Kim and coworkers prepared disulfide bond-containing amphiphilic CB[6] derivatives, which were subsequently self-assembled into reduction-sensitive vesicles.[32] The loaded DOX could be delivered to cancer cells in a targeted manner by the surface-decorated targeting-ligand *via* the receptor-mediated endocytosis and subsequent reduction-triggered drug release into cytoplasm.

Another exciting achievement in developing CB[6]-based drug delivery systems is the design and preparation of hollow nanocapsules through the cross-linking of perallyloxyCB[6] by dithiols.[33] Upon UV irradiation, the disk-shaped allyloxyCB[6] were cross-linked to form dimers and trimers, which would subsequently grow into two-dimensional (2D) oligomeric patches that were linked by thioether linkages. The 2D oligomeric patches began to bend in order to reduce the total energy when they reached a certain size, resulting in the formation of a hollow sphere that underwent further cross-linking to afford robust nanocapsules.[34] In spite of significant preliminary success, these nanocapsules might possess certain drawbacks as drug delivery carriers, including the following: (1) the *in situ* formation of cargo-loaded

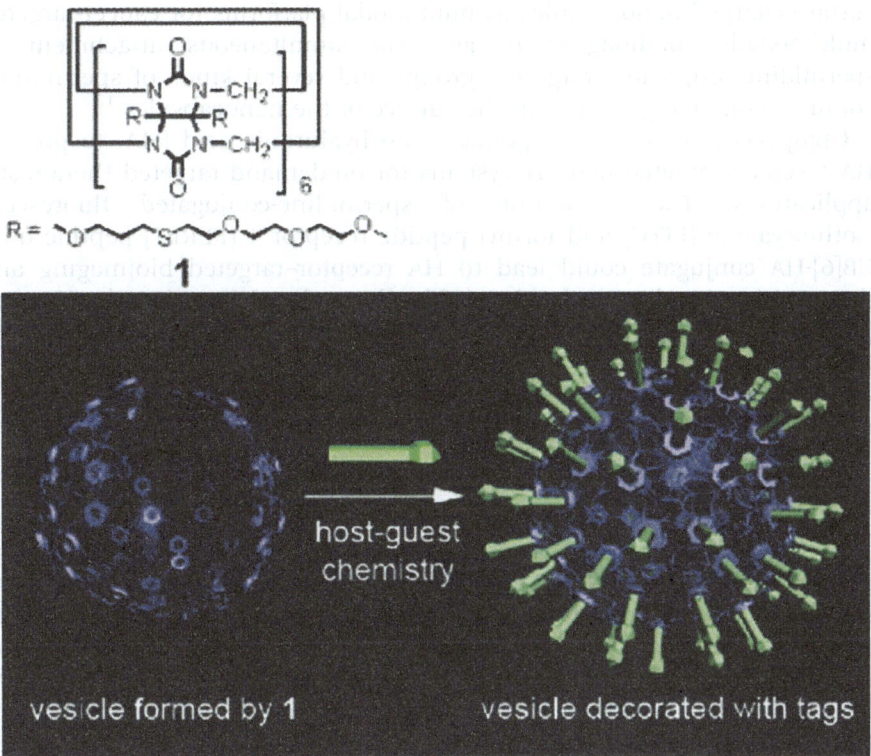

**Figure 9.4**  Pictorial illustration of the novel vesicle formed by an amphiphilic CB[6] derivative, the surface of which can be easily modified *via* host – guest interactions.

Reproduced from ref. 31 with permission from American Chemical Society, Copyright 2005.

nanocapsules required the cargo to be exposed to strong UV irradiation, which might decompose the cargo, particularly UV-sensitive cargo; (2) the entrapped cargo was not readily released due to the covalent cross-linking of the nanocapsules; (3) the nanocapsules were stable only in organic solvents and would aggregate in aqueous solution, which greatly limited its applicability in real-world scenarios. To address the first two problems, Kim *et al.* upgraded the nanocapsules by using an amino-terminated CB[6] derivative, which was cross-linked by *N*-hydroxysuccinimide (NHS)-activated dicarboxylic acid with disulfide bonds in the middle of the chain in order to afford reduction-labile polymer nanocapsules that can deliver specific payloads inside cancer cells due to the enriched glutathione species.[35] To improve the stability of the nanocapsules in aqueous solutions, methyl iodide ($CH_3I$) was added after the formation of nanocapsules to induce a positive charge on sulfur atoms, resulting in a zeta potential increase for the nanocapsules, which could ultimately increase the aqueous stability.[36] Very recently, Kim and coworkers successfully demonstrated the utilization of surface-charged nanocapsules as multimodal platforms for cancer-targeted multimodality bioimaging *in vivo via* simultaneous attachment of spermidine-conjugated targeting groups and several kinds of spermidine-conjugated imaging groups on the surface of the nanocapsules.[12]

CB[6] has also been incorporated into hyaluronic acid (HA) to prepare HA receptor–targeted delivery systems for on-demand targeted theranostic applications. The decoration of spermidine-conjugated fluorescein isothiocyanate (FITC) and formyl peptide receptor 1 (FPRL1) peptide on a CB[6]-HA conjugate could lead to HA receptor–targeted bioimaging and therapeutic signal transduction with elevated calcium ion ($Ca^{2+}$) and phosphorylated extracellular signal-regulated kinases (pERK) levels in FPRL1/MCF-7 cells.[37]

### 9.2.2.3 Derivatized CB[6]-based Hydrogels for Tissue Engineering

The application of CB[6]-based hydrogels for tissue engineering has also been demonstrated by Kim's research group.[29] CB[6]-HA and diaminohexane-HA (DAH-HA) could undergo supramolecular assembly to form a biocompatible hydrogel (CB[6]/DAH-HA) in the presence of cells with tunable mechanical properties through the strong and selective encapsulation of DAH by CB[6] (Figure 9.5).[29] The hydrogel could be modularly modified with various functional tags *via* noncovalent interactions between DAH and fluorescent dye–conjugated CB[6] and peptide-conjugated CB[6]. Cell proliferation experiments suggested that the treatment of CB[6]/DAH-HA hydrogels with c(RGDyK)-CB[6] (a cyclic derivative of RGD having a high affinity for αvβ3 integrin, RGD for arginylglycylaspartic acid) not only led to the capture of c(RGDyK) peptides but also reconstructed a stable RGD environment for cellular adhesion and proliferation with high efficiency. *In vivo* experiments suggested that the CB[6]/DAH-HA hydrogel formed *in situ* under the nude mice's skin within a few minutes postadministration,

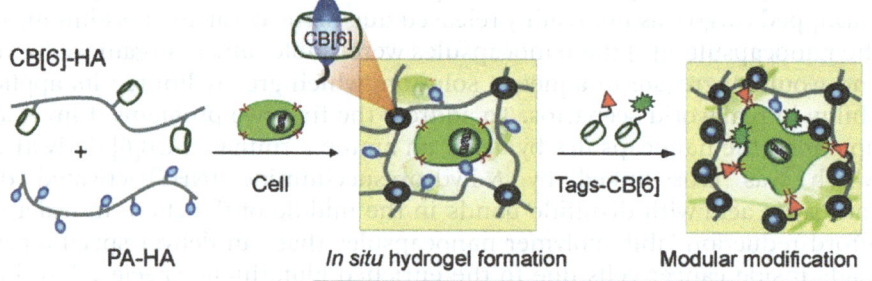

CB[6]-HA            CB[6]

+

PA-HA            Cell        *In situ* hydrogel formation        Tags-CB[6]        Modular modification

Strong host-guest interaction

**Figure 9.5**    Illustration of the *in situ* formation of a supramolecular hydrogel and its modular decoration.
Reproduced from ref. 29 with permission from American Chemical Society, Copyright 2012.

and its shape was retained for more than 2 weeks. After *in situ* modification of hydrogel with FITC-CB[6], the hydrogel residing in the mouse maintained fluorescence for 11 days, while the hydrogel modified with carboxy-fluorescein without CB[6] conjugation lost its fluorescence within 1 day. This strategy was subsequently employed to prepare CB[6]/DAH-HA hydrogels modified with a drug-CB[6] conjugate for the facial chondrogenesis control of human mesenchymal stem cells and long-term mesenchymal stem cell cancer therapy *in vitro* and *in vivo*.[38]

## 9.3 Interaction of CB[7] and Its Derivatives with Biomolecules and Drugs

### 9.3.1 Interaction of CB[7] with Biomolecules and Drugs

With a relatively high biocompatibility[39] and the capacity to encapsulate a wide range of different-sized guest molecules with higher affinity than that with β-CDs, CB[7] emerges as a potential pharmaceutical excipient in the biomedical research field.[13,40] The complexations of drug molecules by CB[7] often improved the stability and solubility of the guest drugs, up-shifted their $pK_a$ values and positively modulated their therapeutic efficacies with reduced side effects. For instance, one of the major biologically active compounds extracted from the Areca nut, arecoline hydrobromide, has shown profound links to hepatotoxicity.[41] CB[7] could form 1:1 host–guest inclusion complexes with arecoline hydrobromide, with a binding constant $K_a$ of $6.59(\pm 0.23) \times 10^4$ M$^{-1}$, and the hepatotoxicity of the drug was significantly alleviated (up to fourfold) upon its encapsulation by CB[7] against human liver cell line L02.[41] This supramolecular strategy was employed to alleviate the inherent hepatotoxicity of the antidepressant Trazodone *via* encapsulation with CB[7].[42] In addition to hepatotoxicity, cardiotoxicity is also one of the severe side effects of chemotherapy drugs. Bedaquiline is a newly approved antituberculosis drug for the treatment of multidrug-resistant tuberculosis, but it has poor aqueous solubility and poses a potential risk of cardiotoxicity.[43] Upon CB[7]-Bedaquiline complex formation, the solubility was increased in a linear relationship with the concentration of CB[7], and the complexation significantly reduced the cardiac malfunctions otherwise caused by Bedaquiline in *in vivo* zebrafish models, and the antimycobacterial efficacy of the drug was well maintained.[44] Likewise, the complexation of the bioactive halonium ion, diphenyleneiodonium, by CB[7] was demonstrated to alleviate the cardiotoxicity of the halonium species both *in vitro* and *in vivo* and meanwhile improved its inhibitory activities against ROS generation *in vitro*.[45] In addition, CB[7] also constrained the teratogenicity of pesticides,[46] improved the biological properties of tuftsin,[47] enhanced the photostability of the antibiotic drug ciprofloxacin[48] and modulated the fluorescence quantum yield, excited-state lifetime and photostability of fluoroquinolone drugs and enhanced the antibacterial activity of these drugs through host–guest encapsulation.[49]

Meanwhile, complexation of CB[7] with the polycationic drug hexadimethrine bromide (HB), was found to alleviate the serious blood coagulation side effects associated with this drug. HB was encapsulated by CB[7] at a 1:1 binding ratio between the repeating hexyl unit of HB and the host under neutral conditions with a strong complexation constant of $1.04(\pm 0.19) \times 10^7$ $M^{-1}$. Both *in vitro* hemagglutination assay and *in vivo* evaluations demonstrated that CB[7] may significantly alleviate polycation-induced blood coagulation effects as well as thrombus and hypoxic damage of the lungs.[50] In 2018, Cheng *et al.* established a gold nanostar (GNS) platform, where CB[7] was employed not only to improve the stability of GNS but also to encapsulate the anticancer drug camptothecin (CPT) *via* host–guest chemistry. A synergistic treatment can be achieved by the chemotherapeutic effect of CPT and the photothermal effect of GNS after triggering by NIR light irradiation.[51]

The competitive binding of CB[7] by spermine that is overexpressed in specific tumor cells leads to the controlled release of the cargoes from the CB[7] cavity. Based on this principle, Zhang and coworkers successfully demonstrated the controlled release of dimethyl viologen (MV) and ox-aliplatin from CB[7] in tumor cells with spermine overexpresed.[2b,52] *In vitro* studies suggested that the cytotoxicity of the cargoes to normal cell lines was significantly decreased, while the activity of the cargoes toward cancer cell lines was recovered through their displacement of the drug by the over-expressed spermine (Figure 9.6).

Research with CB[7] could also help provide valuable insight with regard to the recognition and selectivity for biological systems. Among them, DNA plays a key role in the field of life sciences. Z-DNA has been difficult to study probably due to its spontaneity of transition to B-DNA.[53] Positively charged spermine can decrease the electrostatic repulsion and induce the adoption

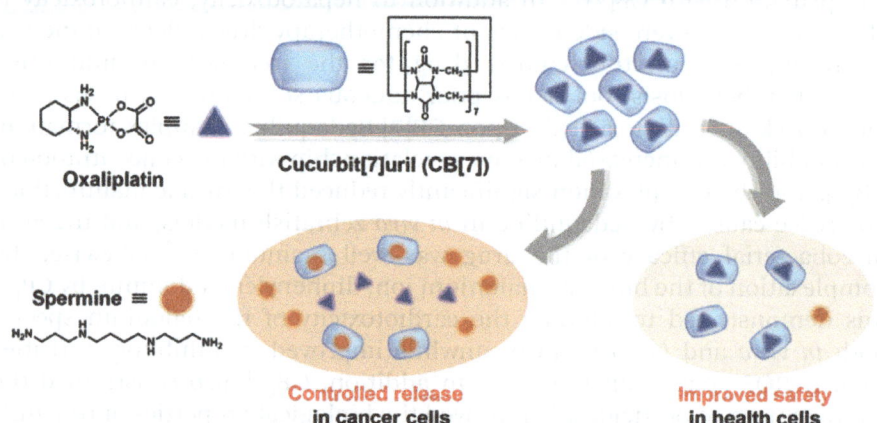

**Figure 9.6**    Illustration depicting the controlled release of CB[7]-included oxaliplatin toward spermine over expressed tumor cells.
Reproduced from ref. 2*b*, https://pubs.acs.org/doi/10.1021/acsami.7b01157, with permission from American Chemical Society, Copyright 2017.

of the Z-form, which can be encapsulated by CB[7] to cause a Z-B DNA transition, and the replacement of the 1-adamantanamine resulted in the reversible switching of DNA (Figure 9.7).[54] A CB[7]-indicator multicolor DNA recognition assay has been developed by Shao *et al.* using an indicator displacement assay. The indicators, alkaloids such as coptisine and palmatine, could form complexes with CB[7], and compete with the pyrimidines of DNAs for recognition of DNA *via* indicator displacement assay.[55]

Fibrillation of proteins/peptides, often involved in several disease progressions, is an important challenge for biotechnological drug development, which may be solved *via* supramolecular interactions between CB[7] and proteins/peptides to inhibit their fibrillation. In 2018, Li and coworkers utilized CB[7] to inhibit the fibrillation of human calcitonin (hCT), which would significantly reduce the bioavailability and therapeutic potency of hCT. The complex not only exhibited low cytotoxicity, as demonstrated by reduced blood calcium levels in rats and decreased immunogenicity of hCT, but also enhanced the osteogenic capacity and osteoblast proliferation.[56] Another harmful aggregation for humans is one of the causative agents for Alzheimer's disease, the aggregation of amyloid β (Aβ). Through ThT kinetics and transmission electron microscope experiments, Li *et al.* have found that CB[7] could inhibit the aggregation of both $Aβ_{4-40}$ and $Aβ_{1-40}$ species through host–guest interactions with fibril ends to prevent the elongation process that leads to Alzheimer's disease.[57]

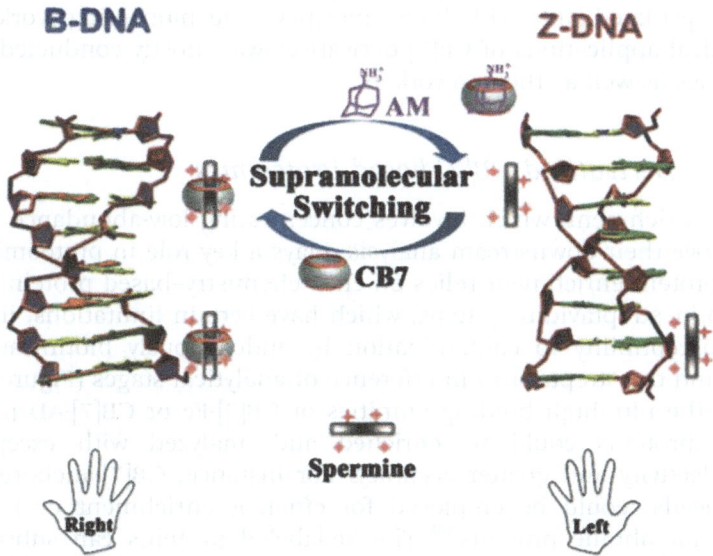

**Figure 9.7** The CB[7]-based supramolecular approach for reversible B/Z DNA transition.
Reproduced from ref. 54, https://doi.org/10.1002/advs.201800231, under the terms of the CC BY 4.0 licence, https://creativecommons.org/licenses/by/4.0/.

On the other hand, CB[7] often serves to obstruct certain intermolecular interactions, such as by preventing an enzyme from binding to its substrate. A reversible intervention tool was developed by Zhou *et al.* to control enzymatic recognition at epigenetic marker 5-formylcytosine (5fC) sites by modifying the 5fC with adamantane (5fC-AD), which could undergo complexation with CB[7]. This strategy may inhibit enzyme cleavage but may not significantly affect the hydrogen-bonding properties of duplex DNA natural nucleobases. The protein–nucleic acid interactions will be deblocked after the addition of an emulative guest to dissociate CB[7], thus providing a chemical approach for an epigenetic target through supramolecular tag–based label.[58]

## 9.3.2 Interaction of CB[7] Derivatives with Biomolecules and Drugs

CB[7] has exceptionally high binding affinities ($K_a \sim 10^{12}$–$10^{17}$ $M^{-1}$) with ammonium-containing derivatives of adamantane (AD) and ferrocene (Fc). This supramolecular latching system with ultrahigh specificity that could potentially replace the avidin-biotin system has great potential for various biomedical applications, such as protein enrichment for proteomics, biosensing, as well as drug delivery.[59] More importantly, the modification of CB[7] with functional groups such as targeting molecules, imaging molecules and PEG groups could endow CB[7] as a versatile carrier for delivering the encapsulated APIs with high efficiency. The pioneering work on the biomedical applications of CB[7] derivatives was mostly conducted by Kim and Isaacs as well as their coworkers.

### 9.3.2.1 *Derivatized CB[7]-based Proteomics*

Protein enrichment, which involves concentrating low-abundance proteins to improve their downstream analysis, plays a key role in proteomics. Normally, protein enrichment relies on click chemistry–based protein labeling and biotin–streptavidin systems, which have certain limitations, including their susceptibility to contamination by endogenously biotinylated molecules and the streptavidin interference of analytical stages (Figure 9.8).[60]

With the ultrahigh binding affinities of CB[7]-Fc or CB[7]-AD pairs, the labeled proteins could be enriched and analyzed with exceptionally high selectivity and greater accuracy. For instance, CB[7]-anchored beads (CB[7]-beads) could be employed for efficient enrichment of Fc-labeled plasma membrane proteins.[60] The Fc-labeled proteins can subsequently be released from the beads upon the addition of a competitive guest (bis(trimethylammoniomethyl)ferrocene)) with a higher binding affinity for CB[7]. Recently, CB[7] beads were further employed in a more complex, intracellular proteome for the efficient enrichment of AD-labeled histone deacetylases (HDACs).[61] Moreover, the anchoring of CB[7] onto an

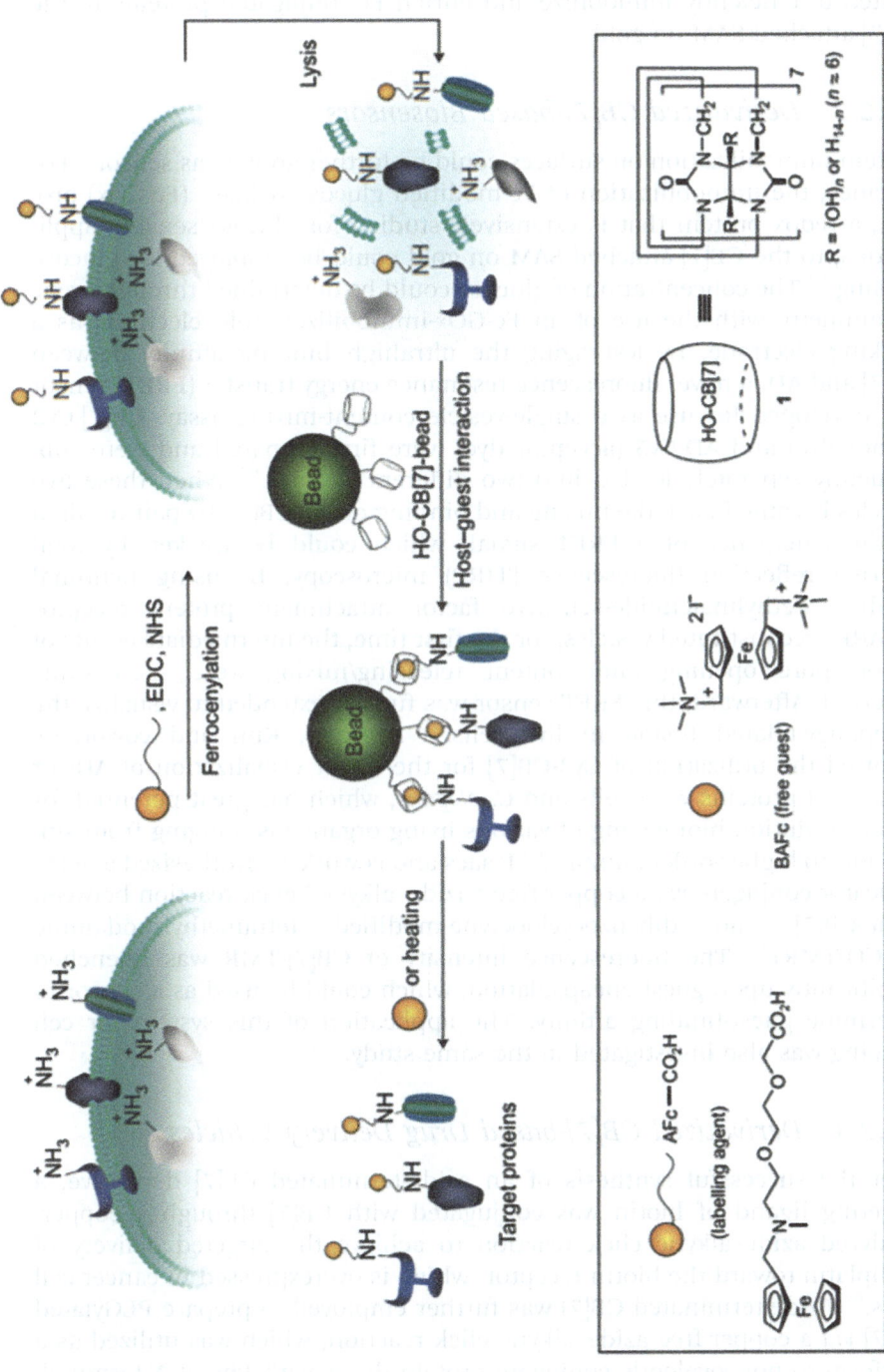

**Figure 9.8**   Strategy for isolation of plasma membrane proteins using an ultrastable synthetic binding pair. Reproduced from ref. 60 with permission from Springer Nature, Copyright 2010.

alkanethiolate self-assembled monolayer (SAM) on gold was also demon-strated to efficiently immobilize and enrich Fc-conjugated proteins to the CB[7]-attached SAM on gold.[62]

### 9.3.2.2  Derivatized CB[7]-based Biosensors

Protein immobilization on surfaces could be further applied as sensors. For instance, the immobilization of Fc-modified glucose oxidase (Fc-GOx) pro-tein, a redox protein that is extensively studied for glucose sensing appli-cations, to the CB[7]-attached SAM on gold could be employed for glucose sensing.[62] The concentration of glucose could be determined through cyclic voltammetry with the use of an Fc-GOx-immobilized gold electrode as a working electrode. By leveraging the ultrahigh binding affinity between CB[7] and AD, a novel fluorescence resonance energy transfer (FRET) sensor was developed for use as a single-vesicle content-mixing assay. CB[7]-Cy3 (donor dye) and AD-Cy5 (acceptor dye) were first prepared and were sub-sequently separately loaded into two different vesicles.[63] When these two vesicles became fused, the mixing and binding of the CB[7]-AD pair resulted in the emergence of a FRET signal, which could be tracked by total internal reflection fluorescence (TIRF) microscopy. By using neuronal soluble *N*-ethylmaleimide-sensitive factor attachment protein receptor (SNARE)–reconstituted vesicles, for the first time, the intermediate events of fusion pore opening and content releasing/mixing were successfully observed. Afterward, this FRET sensor was further extended to visualize the autophagy-related fusion in live cells.[64] Recently, Kim and coworkers reported the utilization of Cy3-CB[7] for the direct visualization of AD- or Fc-labeled proteins with cells and *C. elegans*, which has great potential for super-resolution bioimaging of various living organisms, ranging from sin-gle cells to higher-order animals.[65] Isaacs and coworkers synthesized a host–indicator conjugate *via* a copper-free "azide–alkyne" click reaction between azide-CB[7]   and   dibenzocyclooctyne-modified   tetramethylrhodamine (DBCO-TMR).[66] The fluorescence intensity of CB[7]-TMR was quenched significantly upon guest encapsulation, which could be used as a sensor to determine guest-binding affinity. The application of this system for cell imaging was also investigated in the same study.

### 9.3.2.3  Derivatized CB[7]-based Drug Delivery Vehicles

After the successful synthesis of an azide-terminated CB[7] derivative, a targeting ligand of biotin was conjugated with CB[7] through a copper-catalyzed azide–alkyne click reaction to achieve the targeted delivery of oxaliplatin toward the biotin receptor, which is overexpressed in cancer cell lines.[67] Azide-terminated CB[7] was further employed to prepare PEGylated CB[7] *via* a copper-free azide–alkyne click reaction, which was utilized as a platform to noncovalently conjugate protein drugs with Phe at *N*-terminal, such as insulin, and to facilitate protein delivery by extending their systemic circulation time and improving their stability.[68] Moreover, a bis(pyridine)-CB[7] derivative was synthesized from a clickable azide CB[7] precursor, which then underwent self-assembly with palladium nitrate (Pd(NO$_3$)$_2$) to afford a metal organic polyhedron (MOP).[69] The inner cavity of MOP was rendered hydrophobic *via* CB[7]'s encapsulation of a diaminoalkane or a photoresponsive diamine, which was subsequently employed to load a hydrophobic dye, Nile Red, or a hydrophobic drug, DOX. The cargos could be released upon external stimulation such as competitive guest addition, pH changes or photoirradiation.

Zhang and coworkers prepared a CB[7]-based main-chain polymeric PEG, poly-CB[7], *via* a click reaction between a bis-alkynyl-functionalized CB[7] and α,ω-diazide-PEG.[70] The encapsulation of OxPt by poly-CB[7] yielded a supramolecular polymeric drug with reduced cytotoxicity toward normal cells and increased cytotoxicity to spermine-overexpressed cancer cells.[52] Kim and coworkers reported the direct vesicle formation from monoallyloxyCB[7] ((AO)$_1$CB[7]), which may be disrupted by photoirradiation due to the "thiol–ene" reaction between (AO)$_1$CB[7] and GSH.[71] *In vitro* studies involving cancer cells (with intracellular GSH enriched) indicated that the prepared vesicles not only facilitated the cellular uptake of DOX but also exhibited a controlled cargo release upon light irradiation in cytosol. In fact, the relatively large size (~1 μm) and unstable morphology of the vesicles in aqueous media make them unsuitable for real-world drug delivery applications.

Very recently, we prepared the first CB[7]-conjugated poly(lactic acid) and demonstrated CB[7]-PLA/PLGA – based nanoparticles (NPs) as a novel and versatile drug delivery platform with a noncovalently tailorable surface.[72] CB[7] on the surface of NPs allowed facile, modular surface modification of various functional tags, including but not limited to folate-AD (a targeting tag), FITC-AD (a fluorescent conjugate), and PEG-AD (a hydrophilic polymer), and a guest drug molecule, offering a unique opportunity for synergistic therapy.

## 9.4 Interactions of CB[8] with Biomolecules and Drugs

On account of its modest water solubility and its relatively large cavity that can often simultaneously accommodate two guest molecules,[1] CB[8] has been widely used in the fields of material sciences[73] and biomedical sciences.[74] For instance, some specific peptides with Trp or Phe residues (particularly those on the *N*-terminals) could be dimerized by CB[8] *via* homoternary complexation.[6] Such peptide dimerization could be applied to dimerize proteins and manipulate their enzyme activity. For instance, proteins incorporating a phenyalanine-glycine-glycine (FGG) peptide motif can interact with CB[8] to form a redox-responsive protein self-assembly system with a reversible morphology. Thus, the FGG-modified protein FGIG was utilized as the building block to interact with CB[8], exhibiting morphological transformation from nanowires to nanorings.[75] Dankers and coworkers also immobilized fluorescent proteins *via* the ternary complex formation of FGG-modified and 2-ureido-4-pyrimidinone (UPy)–based materials with CB[8].[76] Very recently, a peptide-based cross-linked polymer that underwent homoternary complexation with CB[8] exhibited efficient cell adhesion and proliferation properties, as was demonstrated *via in vitro* experiments.[77] The backbone of the polymer consists of *N*-terminal FGG units that enable its complexation with CB[8]. Meanwhile, two cysteine residues may undergo cross-linking through disulfide bonds, and the C-terminal RGDS sequence is a well-known cell-adhesive unit.

Similarly, Scherman and coworkers successfully demonstrated the preparation of a tunable supramolecular hydrogel by CB[8]-mediated cross-linking of a Phe-functionalized polysaccharide, which may find applications in controlled drug delivery and tissue engineering.[78]

Recently, binding target proteins with high affinity and selectivity by using CB[8] has been achieved in chemical biology. Urbach *et al.* reported the molecular recognition of peptides containing an *N*-terminal methionine (Met) by the CB[8] with high binding affinity. As shown in Figure 9.9, peptides containing large and hydrophobic or cationic residues (*i.e.*, Leu, Lys, Arg, Tyr or Phe) that lack branching at the β carbon are ideally recognized for targeting Met-containing peptides by CB[8], and usually a relatively small third residue would minimize steric interactions.[79]

Scherman *et al.* utilized CB[8] to inhibit the toxicity of amyloid β (Aβ) through a supramolecular strategy. Due to the complexation of Phe residues in neighboring strands of αβ42 with CB[8] in a 2 : 1 ratio, the supramolecularly self-assembled aggregations, bypassing the toxic oligomeric state, reduced its toxicity in the neuronal cell line SH-SY5Y, providing a new approach to counteract Alzheimer's disease.[80]

After the discovery of CB[8]-mediated heteroternary complexation of MV and azobenzene (azo), photoresponsive drug/payload delivery vehicles were developed by incorporating the specific recognition pair into polymeric or amphiphilic systems (Figure 9.10).[81] We have recently reported a user-friendly herbicide formulation, derived from photoresponsive supramolecular vesicles

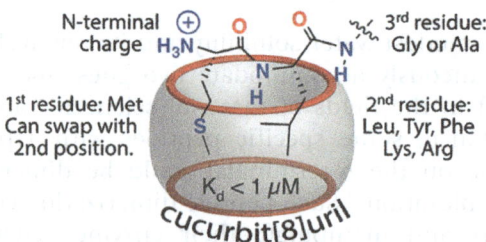

**Figure 9.9**     Summary schematic of the sequence determinants that lead to high-affinity binding to CB[8].
Reproduced from ref. 79 with permission from American Chemical Society, Copyright 2018.

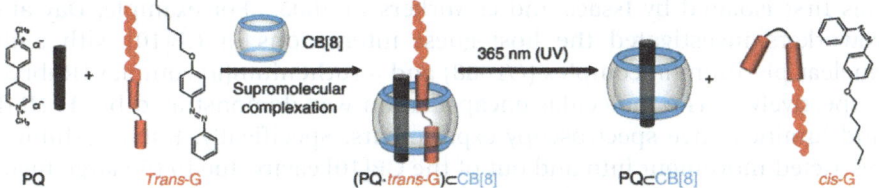

**Figure 9.10** Schematic illustration of CB[8]-mediated complexation with PQ as the first guest and *trans*-G as the second guest, as well as the photo-driven, reversible transition between complexation with the *trans*-G isomer and decomplexation with the *cis*-G isomer.
Reproduced from ref. 82, https://doi.org/10.1038/s41467-018-05437-5, under the terms of the CC BY 4.0 licence, http://creativecommons.org/licenses/by/4.0/.

formed by CB[8], paraquat and an azobenzene derivative. Upon UV (or sunlight) irradiation, the vesicle will release paraquat due to the dissociation of the ternary complex triggered by the *trans*-to-*cis* conformational change of azobenzene.[82]

Recently, an amphiphilic supramolecular brush copolymer was constructed by a CB[8], electron-rich PEG-Nap and 4,4′-bipyridinium derivative linked with PTPE (polytetraphenylethene, an aggregation–induced emission (AIE) agent).[83] The resulting copolymer could self-assemble into nanoparticles in aqueous solution, which could serve as a nanocarrier to encapsulate the anticancer drug doxorubicin. In addition, the anthracyl pyridinium derivative ENDT is a synthetic organic dye with weak fluorescence emission, which could self-assemble to nanorods through supramolecular interaction with CB[8], resulting in fluorescence enhancement and a near-infrared fluorescence emission. Such enhanced fluorescence may provide a possibility for NIR-light lysosome-targeted cellular imaging.[84] This supramolecular assembly would break apart through the addition of sulfonatocalix-[4]arene (SC4AD) as a competitive host, and the morphology would be converted back to nanoparticles.[84]

Very recently, the 2:1 binding between a halonium guest species, DPI, and CB[8] was reported for the first time.[45] CB[8] was shown to enhance the inhibitory activity of DPI against the generation of reactive oxygen species (ROS) in macrophage cell lines and to alleviate the toxicity of DPI on cardiac myoblast cell lines.

In addition, CB[8] has often been used as a biosensor or probe for the identification of natural amino acids,[85] detection of addictive drugs in urine,[86] monitoring of steroid depletion in bacterial cultures[87] and sensing of benzimidazole fungicides through host–guest interactions.[88]

## 9.5 Interactions of Other CB[n] (n = 10, 13–15) with Biomolecules and Drugs

Possessing the largest cavity in the CB[n]s family, CB[10] exhibits molecular recognition properties that are rather distinct from the rest of CB[n]s, since it was first isolated by Issacs and coworkers in 2005.[7] For example, Day and coworkers investigated the host–guest interactions of CB[10] with a dinuclear platinum(ɪɪ) complex (CT008) and a ruthenium(ɪɪ) complex (Rubb5), respectively.[89] The molecular encapsulation was demonstrated by [1]H NMR and luminescence spectroscopy experiments. Specifically, Rubb5 exhibited restricted movement into and out of the CB[10] cavity due to the large metal centers, resulting in binding kinetics that were rather slow on both the [1]H NMR and biological timescales, suggesting that CB[10] could be an ideal drug delivery vehicle for the controlled release of potential anticancer drugs.[89a] Day *et al.* also investigated the influence of CB[10]'s encapsulation on the antimicrobial activity and toxicity of Rubb12. It was found that the host–guest interaction of CB[10] and Rubb12 exhibited little influence on protein binding resistance and on the antimicrobial activity of the drug but resulted in a twofold decrease in its nonspecific toxicity.[89b,89c]

Larger CB[*n*] homologues, namely CB[13], CB[14] and CB[15], isolated by Tao *et al.*, have further broadened the potential applications of CB[*n*] (Figure 9.11).[90] For instance, Leonard F. Lindoy, Gang Wei and coworkers studied the interactions of 20 standard amino acids (AAs) with twisted CB[14].[91] AAs with either cationic or aromatic ring-containing side chains may bind with CB[14] more readily, whereas those AAs with hydrophobic or polar side chains exhibited much weaker binding with CB[14]. Furthermore, Huang *et al.* reported that the high-affinity supramolecular interaction of paraquat and CB[15] is mainly enthalpy driven. An amphiphilic guest containing 4,4′-bipyridinium moiety may be recognized by CB[15] and undergo self-assembly to form solid nanospheres in aqueous solution. With the addition of Ba[+], the nanospheres became converted into micelles, demonstrating a successful construction of barium ions–responsive supraamphiphilic complexes.[92]

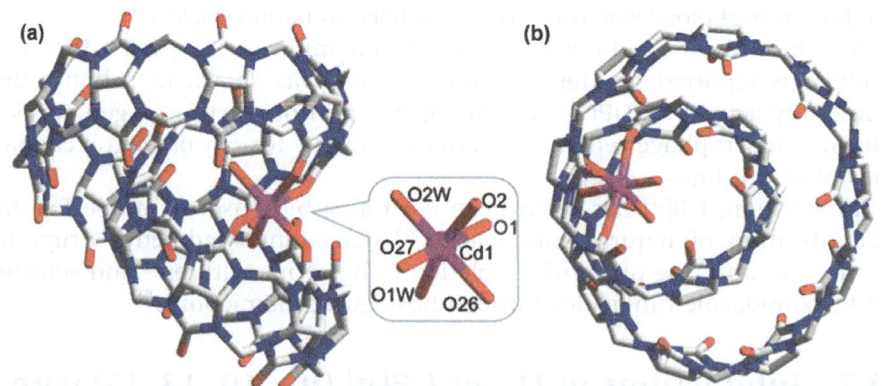

**Figure 9.11**   X-ray crystal structure of 2: (a) side view and (b) top view of the $Cd^{2+}/tCB[15]$ complex.
Reproduced from ref. 90*b* with permission from American Chemical Society, Copyright 2016.

## 9.6 Conclusion

In summary, CB[n]s and their derivatives, as well as their interactions with biomolecules and drug molecules, have been extensively investigated for their potential applications in the pharmaceutical and the biochemical and biomedical sciences. For example, CB[n]s can help minimize the side effects of drugs or inhibit the toxicity of certain compounds or reverse the effects of encapsulated toxins and molecules of biological interest. The CB[n]s can also encapsulate biomolecules such as DNA, genes, peptides, proteins and enzymes and modulate their various biological activities. For more precise and versatile applications in these areas, a variety of functionalized CB[n]s were designed and synthesized. Supramolecular interactions of CB[n] derivatives and selected guests may self-assemble into nanomaterials or hydrogels geared with excellent biocompatibility, targeting properties and stimuli-responsiveness for drug delivery and tissue engineering. In spite of these preliminary successes, the current studies have only scratched the surface about the applications of CB[n]s and their derivatives in the pharmaceutical, biochemical and biomedical sciences, and there is still a long way to go in order to realize some of these applications in real-world scenarios. Our research group will continue to work along this line of research, promoting the applications of CB[n]s and derivatives in pharmaceutical and biomedical sciences.

## Acknowledgements

The National Science Foundation of China (Grant No.: 21871301), Science and Technology Development Fund (FDCT), Macau SAR (Grant No.: 030/2017/A1 and 121/2018/A3) and Research Committee at University of Macau (Grant No.: MYRG2016-00165-ICMS-QRCM, MYRG2016-00008-ICMS-QRCM and MYRG2017-00010-ICMS) are gratefully acknowledged for providing financial support to this work.

## References

1. S. J. Barrow, S. Kasera, M. J. Rowland, J. del Barrio and O. A. Scherman, *Chem. Rev.*, 2015, **115**, 12320.
2. (a) D. Shetty, J. K. Khedkar, K. M. Park and K. Kim, *Chem. Soc. Rev.*, 2015, **44**, 8747; (b) Y. Chen, Z. Huang, H. Zhao, J.-F. Xu, Z. Sun and X. Zhang, *ACS Appl. Mater. Interfaces*, 2017, **9**, 8602; (c) X. Yang, S. Li, Q.-W. Zhang, Y. Zheng, D. Bardelang, L.-H. Wang and R. Wang, *Nanoscale*, 2017, **9**, 10606; (d) Q. Huang, K. I. Kuok, X. Zhang, L. Yue, S. M. Y. Lee, J. Zhang and R. Wang, *Nanoscale*, 2018, **10**, 10333; (e) H. Chen, Y. Chen, H. Wu, J.-F. Xu, Z. Sun and X. Zhang, *Biomaterials*, 2018, **178**, 697.
3. W. Mock and N. Shih, *J. Org. Chem.*, 1983, **48**, 3618.
4. (a) N. Zhao, G. O. Lloyd and O. A. Scherman, *Chem. Commun.*, 2012, **48**, 3070; (b) M. M. Ayhan, H. Karoui, M. Hardy, A. Rockenbauer,

L. Charles, R. Rosas, K. Udachin, P. Tordo, D. Bardelang and O. Ouari, *J. Am. Chem. Soc.*, 2015, **137**, 10238; (c) S. Y. Jon, N. Selvapalam, D. H. Oh, J.-K. Kang, S.-Y. Kim, Y. J. Jeon, J. W. Lee and K. Kim, *J. Am. Chem. Soc.*, 2003, **125**, 10186.

5. (a) J. Liu, C. S. Y. Tan, Z. Yu, Y. Lan, C. Abell and O. A. Scherman, *Adv. Mat.*, 2017, **29**, 1604941; (b) J. Liu, C. S. Y. Tan, Z. Yu, N. Li, C. Abell and O. A. Scherman, *Adv. Mat.*, 2017, **29**, 1605325.

6. L. C. Smith, D. G. Leach, B. E. Blaylock, O. A. Ali and A. R. Urbach, *J. Am. Chem. Soc.*, 2015, **137**, 3663.

7. S. Liu, P. Y. Zavalij and L. Isaacs, *J. Am. Chem. Soc.*, 2005, **127**, 16798.

8. Y. J. Jeon, H. Kim, S. Jon, N. Selvapalam, D. H. Oh, I. Seo, C.-S. Park, S. R. Jung, D.-S. Koh and K. Kim, *J. Am. Chem. Soc.*, 2004, **126**, 15944.

9. H.-K. Lee, K. M. Park, Y. J. Jeon, D. Kim, D. H. Oh, H. S. Kim, C. K. Park and K. Kim, *J. Am. Chem. Soc.*, 2005, **127**, 5006.

10. (a) Y. Xu, M. Guo, X. Li, A. Malkovskiy, C. Wesdemiotis and Y. Pang, *Chem. Commun.*, 2011, **47**, 8883; (b) Y. Liu, H. Yang, Z. Wang and X. Zhang, *Chem. Asian J.*, 2013, **8**, 1626; (c) S. Edmondson and W. T. S. Huck, *Adv. Mater.*, 2004, **16**, 1327.

11. Q. Cheng, S. Li, C. Sun, L. Yue and R. Wang, *Mater. Chem. Front.*, 2019, **3**, 199.

12. S. Kim, G. Yun, S. Khan, J. Kim, J. Murray, Y. M. Lee, W. J. Kim, G. Lee, S. Kim, D. Shetty, J. H. Kang, J. Y. Kim, K. M. Park and K. Kim, *Mater. Horiz.*, 2017, **4**, 450.

13. H. Yin and R. Wang, *Isr. J. Chem.*, 2018, **58**, 188.

14. W. L. Mock and N. Y. Shih, *J. Org. Chem.*, 1983, **48**, 3618.

15. E. Corda, M. Hernandez, S. Sanchez-Cortes and P. Sevilla, *Colloids Surf., A*, 2018, **557**, 66.

16. S. Pandey, V. K. Soni, G. Choudhary, P. R. Sharma and R. K. Sharma, *Supramol. Chem.*, 2017, **29**, 387.

17. K. Roy, S. Saha, B. Datta, L. Sarkar and M. N. Roy, *Z. Phys. Chem.*, 2018, **232**, 281.

18. J. Li, C. Si, H. Sun, J. Zhu, T. Pan, S. Liu, Z. Dong, J. Xu, Q. Luo and J. Liu, *Chem. Commun.*, 2015, **51**, 9987.

19. T. S. Choi, H. H. Lee, Y. H. Ko, K. S. Jeong, K. Kim and H. I. Kim, *Sci. Rep.*, 2017, **7**, 5710.

20. H. Isobe, S. Sato, J. W. Lee, H.-J. Kim, K. Kim and E. Nakamura, *Chem. Commun.*, 2005, 1549.

21. H.-J. Buschmann, L. Mutihac, R.-C. Mutihac and E. Schollmeyer, *Thermochim. Acta.*, 2005, **430**, 79.

22. C. P. Carvalho, A. Norouzy, V. Ribeiro, W. M. Nau and U. Pischel, *Org. Biomol. Chem.*, 2015, **13**, 2866.

23. J. A. Finbloom, K. Han, C. C. Slack, A. L. Furst and M. B. Francis, *J. Am. Chem. Soc.*, 2017, **139**, 9691.

24. Y.-M. Zhang, Y. Yang, Y.-H. Zhang and Y. Liu, *Sci. Rep.*, 2016, **6**, 28848.

25. X. Wu, Y.-M. Zhang and Y. Liu, *RSC Adv.*, 2016, **6**, 99729.

26. H. Yang, Y. Tan and Y. Wang, *Soft Matter*, 2009, **5**, 3511.

27. M. Jang, H. Kim, S. Lee, H. W. Kim, J. K. Khedkar, Y. M. Rhee, I. Hwang, K. Kim and J. H. Oh, *Adv. Funct. Mater.*, 2015, **25**, 4882.
28. K. M. Park, K. Suh, H. Jung, D.-W. Lee, Y. Ahn, J. Kim, K. Baek and K. Kim, *Chem. Commun.*, 2009, 71.
29. K. M. Park, J.-A. Yang, H. Jung, J. Yeom, J. S. Park, K.-H. Park, A. S. Hoffman, S. K. Hahn and K. Kim, *ACS Nano*, 2012, **6**, 2960.
30. Q. Liu, Q. Tang, Y.-Y. Xi, Y. Huang, X. Xiao, Z. Tao, S.-F. Xue, Q.-J. Zhu, J.-X. Zhang and G. Wei, *Supramol. Chem.*, 2015, **27**, 386.
31. H.-K. Lee, K. M. Park, Y. J. Jeon, D. Kim, D. H. Oh, H. S. Kim, C. K. Park and K. Kim, *J. Am. Chem. Soc.*, 2005, **127**, 5006.
32. K. M. Park, D. W. Lee, B. Sarkar, H. Jung, J. Kim, Y. H. Ko, K. E. Lee, H. Jeon and K. Kim, *Small*, 2010, **6**, 1430.
33. D. Kim, E. Kim, J. Kim, K. M. Park, K. Baek, M. Jung, Y. H. Ko, W. Sung, H. S. Kim and J. H. Suh, *Angew. Chem., Int. Ed.*, 2007, **119**, 3541.
34. D. Kim, E. Kim, J. Lee, S. Hong, W. Sung, N. Lim, C. G. Park and K. Kim, *J. Am. Chem. Soc.*, 2010, **132**, 9908.
35. E. Kim, D. Kim, H. Jung, J. Lee, S. Paul, N. Selvapalam, Y. Yang, N. Lim, C. G. Park and K. Kim, *Angew. Chem., Int. Ed.*, 2010, **49**, 4405.
36. G. Yun, Z. Hassan, J. Lee, J. Kim, N. S. Lee, N. H. Kim, K. Baek, I. Hwang, C. G. Park and K. Kim, *Angew. Chem., Int. Ed.*, 2014, **53**, 6414.
37. H. Jung, K. M. Park, J.-A. Yang, E. J. Oh, D.-W. Lee, K. Park, S. H. Ryu, S. K. Hahn and K. Kim, *Biomaterials*, 2011, **32**, 7687.
38. (a) J. Yeom, S. J. Kim, H. Jung, H. Namkoong, J. Yang, B. W. Hwang, K. Oh, K. Kim, Y. C. Sung and S. K. Hahn, *Adv. Healthcare Mater.*, 2015, **4**, 237; (b) H. Jung, J. S. Park, J. Yeom, N. Selvapalam, K. M. Park, K. Oh, J.-A. Yang, K. H. Park, S. K. Hahn and K. Kim, *Biomacromolecules*, 2014, **15**, 707.
39. X. Zhang, X. Xu, S. Li, L.-H. Wang, J. Zhang and R. Wang, *Sci. Rep.*, 2018, **8**, 8891.
40. K. I. Kuok, S. Li, I. W. Wyman and R. Wang, *Ann. N. Y. Acad. Sci.*, 2017, **1398**, 108.
41. S. Li, X. Yang, Y. Niu, G. L. Andrew, D. Bardelang, X. Chen and R. Wang, *ChemistrySelect*, 2017, **2**, 2219.
42. Q. Huang, S. Li, H. Yin, C. Wang, S. M. Lee and R. Wang, *Food Chem. Toxicol.*, 2018, **112**, 421.
43. K. Andries, P. Verhasselt, J. Guillemont, H. W. H. Göhlmann, J.-M. Neefs, H. Winkler, J. Van Gestel, P. Timmerman, M. Zhu, E. Lee, P. Williams, D. de Chaffoy, E. Huitric, S. Hoffner, E. Cambau, C. Truffot-Pernot, N. Lounis and V. Jarlier, *Science*, 2005, **307**, 223.
44. K. I. Kuok, P. C. I. Ng, X. Ji, C. Wang, W. W. Yew, D. P. Chan, J. Zheng, S. M. Lee and R. Wang, *Food Chem. Toxicol.*, 2018, **119**, 425.
45. H. Yin, Q. Huang, W. Zhao, D. Bardelang, D. Siri, X. Chen, S. M. Lee and R. Wang, *J. Org. Chem.*, 2018, **83**, 4882.
46. X. Yang, S. Li, Z. Wang, S. M. Lee, L. H. Wang and R. Wang, *Chem. - Asian J.*, 2018, **13**, 41.
47. E. A. Kovalenko, E. A. Pashkina, L. Y. Kanazhevskaya, A. N. Masliy and V. A. Kozlov, *Int. Immunopharmacol.*, 2017, **47**, 199.

48. D. R. Boraste, G. Chakraborty, A. K. Ray, G. S. Shankarling and H. Pal, *J. Photochem. Photobiol., A*, 2018, **358**, 26.
49. H. S. El-Sheshtawy, S. Chatterjee, K. I. Assaf, M. N. Shinde, W. M. Nau and J. Mohanty, *Sci. Rep.*, 2018, **8**, 13925.
50. Q. Huang, Q. Cheng, X. Zhang, H. Yin, L.-H. Wang and R. Wang, *ACS Appl. Bio Mater.*, 2018, **1**, 544.
51. P. Xu, Q. Feng, X. Yang, S. Liu, C. Xu, L. Huang, M. Chen, F. Liang and Y. Cheng, *Bioconjugate Chem.*, 2018, **29**, 2855.
52. Y. Chen, Z. Huang, J.-F. Xu, Z. Sun and X. Zhang, *ACS Appl. Mater. Interfaces*, 2016, **8**, 22780.
53. J. Lee, Y.-G. Kim, K. K. Kim and C. Seok, *J. Phys. Chem. B*, 2010, **114**, 9872.
54. S. R. Wang, J. Q. Wang, G. H. Xu, L. Wei, B. S. Fu, L. Y. Wu, Y. Y. Song, X. R. Yang, C. Li and S. M. Liu, *Adv. Sci.*, 2018, **5**, 1800231.
55. Y. Zhou, L. Gao, X. Tong, Q. Li, Y. Fei, Y. Yu, T. Ye, X.-S. Zhou and Y. Shao, *Anal. Chem.*, 2018, **90**, 13183.
56. H. Shang, A. Zhou, J. Jiang, Y. Liu, J. Xie, S. Li, Y. Chen, X. Zhu, H. Tan and J. Li, *Acta Biomater.*, 2018, **78**, 178.
57. G. Li, W.-Y. Yang, Y.-F. Zhao, Y.-X. Chen, L. Hong and Y.-M. Li, *Chem.: Eur. J.*, 2018, **24**, 13647.
58. S.-R. Wang, Y.-Y. Song, L. Wei, C.-X. Liu, B.-S. Fu, J.-Q. Wang, X.-R. Yang, Y.-N. Liu, S.-M. Liu, T. Tian and X. Zhou, *J. Am. Chem. Soc.*, 2017, **139**, 16903.
59. M. V. Rekharsky, T. Mori, C. Yang, Y. H. Ko, N. Selvapalam, H. Kim, D. Sobransingh, A. E. Kaifer, S. Liu and L. Isaacs, *Proc. Natl. Acad. Sci. U. S. A.*, 2007, **104**, 20737.
60. D.-W. Lee, K. M. Park, M. Banerjee, S. H. Ha, T. Lee, K. Suh, S. Paul, H. Jung, J. Kim and N. Selvapalam, *Nat. Chem.*, 2011, **3**, 154.
61. J. Murray, J. Sim, K. Oh, G. Sung, A. Lee, A. Shrinidhi, A. Thirunarayanan, D. Shetty and K. Kim, *Angew. Chem., Int. Ed.*, 2017, **129**, 2435.
62. I. Hwang, K. Baek, M. Jung, Y. Kim, K. M. Park, D.-W. Lee, N. Selvapalam and K. Kim, *J. Am. Chem. Soc.*, 2007, **129**, 4170.
63. B. Gong, B.-K. Choi, J.-Y. Kim, D. Shetty, Y. H. Ko, N. Selvapalam, N. K. Lee and K. Kim, *J. Am. Chem. Soc.*, 2015, **137**, 8908.
64. M. Li, A. Lee, K. L. Kim, J. Murray, A. Shrinidhi, G. Sung, K. M. Park and K. Kim, *Angew. Chem., Int. Ed.*, 2018, **130**, 2142.
65. K. L. Kim, G. Sung, J. Sim, J. Murray, M. Li, A. Lee, A. Shrinidhi, K. M. Park and K. Kim, *Nat. Commun.*, 2018, **9**, 1712.
66. A. T. Bockus, L. C. Smith, A. G. Grice, O. A. Ali, C. C. Young, W. Mobley, A. Leek, J. L. Roberts, B. Vinciguerra and L. Isaacs, *J. Am. Chem. Soc.*, 2016, **138**, 16549.
67. L. Cao, G. Hettiarachchi, V. Briken and L. Isaacs, *Angew. Chem., Int. Ed.*, 2013, **52**, 12033.
68. M. J. Webber, E. A. Appel, B. Vinciguerra, A. B. Cortinas, L. S. Thapa, S. Jhunjhunwala, L. Isaacs, R. Langer and D. G. Anderson, *Proc. Natl. Acad. Sci. U. S. A.*, 2016, **113**, 14189.

69. S. K. Samanta, J. Quigley, B. Vinciguerra, V. Briken and L. Isaacs, *J. Am. Chem. Soc.*, 2017, **139**, 9066.
70. H. Chen, Y. Chen, H. Wu, J.-F. Xu, Z. Sun and X. Zhang, *Biomaterials*, 2018, **178**, 697.
71. K. M. Park, K. Baek, Y. H. Ko, A. Shrinidhi, J. Murray, W. H. Jang, K. H. Kim, J. S. Lee, J. Yoo and S. Kim, *Angew. Chem., Int. Ed.*, 2018, **57**, 3132.
72. C. Sun, H. Zhang, S. Li, X. Zhang, Q. Cheng, Y. Ding, L.-H. Wang and R. Wang, *ACS Appl. Mater. Interfaces*, 2018, **10**, 25090.
73. (a) J. Liu, C. S. Y. Tan, Z. Yu, Y. Lan, C. Abell and O. A. Scherman, *Adv. Mat.*, 2017, **29**, 1604951; (b) J. Liu, C. S. Y. Tan, Z. Yu, N. Li, C. Abell and O. A. Scherman, *Adv. Mat.*, 2017, **29**, 1605325.
74. (a) L. C. Smith, D. G. Leach, B. E. Blaylock, O. A. Ali and A. R. Urbach, *J. Am. Chem. Soc.*, 2015, **137**, 3663; (b) S. Li, N. Jiang, W. Zhao, Y.-F. Ding, Y. Zheng, L.-H. Wang, J. Zheng and R. Wang, *Chem. Commun.*, 2017, **53**, 5870.
75. R. Wang, S. Qiao, L. Zhao, C. Hou, X. Li, Y. Liu, Q. Luo, J. Xu, H. Li and J. Liu, *Chem. Commun.*, 2017, **53**, 10532.
76. O. J. G. M. Goor, R. P. G. Bosmans, L. Brunsveld and P. Y. W. Dankers, *J. Polym. Sci. A Polym. Chem*, 2017, **55**, 3607.
77. P. Dowari, S. Saha, B. Pramanik, S. Ahmed, N. Singha, A. Ukil and D. Das, *Biomacromolecules*, 2018, **19**, 3994.
78. M. J. Rowland, C. C. Parkins, J. H. McAbee, A. K. Kolb, R. Hein, X. J. Loh, C. Watts and O. A. Scherman, *Biomaterials*, 2018, **179**, 199.
79. Z. Hirani, H. F. Taylor, E. F. Babcock, A. T. Bockus, C. D. Varnado Jr, C. W. Bielawski and A. R. Urbach, *J. Am. Chem. Soc.*, 2018, **140**, 12263.
80. S. Sonzini, H. F. Stanyon and O. A. Scherman, *Phys. Chem. Chem. Phys.*, 2017, **19**, 1458.
81. C. Hu, N. Ma, F. Li, Y. Fang, Y. Liu, L. Zhao, S. Qiao, X. Li, X. Jiang, T. Li, F. Shen, Y. Huang, Q. Luo and J. Liu, *ACS Appl. Mater. Interfaces*, 2018, **10**, 4603.
82. C. Gao, Q. Huang, Q. Lan, Y. Feng, F. Tang, M. P. M. Hoi, J. Zhang, S. M. Y. Lee and R. Wang, *Nat. Commun.*, 2018, **9**, 2967.
83. D. Wu, Y. Li, J. Yang, J. Shen, J. Zhou, Q. Hu, G. Yu, G. Tang and X. Chen, *ACS Appl. Mater. Interfaces*, 2017, **9**, 44392.
84. X.-M. Chen, Y. Chen, Q. Yu, B.-H. Gu and Y. Liu, *Angew. Chem., Int. Ed.*, 2018, **57**, 12519.
85. B. Wang, J. Han, N. M. Bojanowski, M. Bender, C. Ma, K. Seehafer, A. Herrmann and U. H. F. Bunz, *ACS Sens.*, 2018, **2018**, 1562.
86. G. H. Aryal, K. W. Hunter and L. Huang, *Org. Biomol. Chem.*, 2018, **16**, 7425.
87. A. Stahl, A. I. Lazar, V. N. Muchemu, W. M. Nau, M. S. Ullrich and A. Hennig, *Anal. Bioanal. Chem.*, 2017, **409**, 6485.
88. Q. Tang, J. Zhang, T. Sun, C.-H. Wang, Y. Huang, Q. Zhou and G. Wei, *Spectrochim. Acta A Mol. Biomol. Spectrosc.*, 2018, **191**, 372.

89. (a) M. J. Pisani, Y. Zhao, L. Wallace, C. E. Woodward, F. R. Keene, A. I. Day and J. G. Collins, *Dalton Trans.*, 2010, **39**, 2078; (b) F. Li, M. Feterl, J. M. Warner, A. I. Day, F. R. Keene and J. G. Collins, *Dalton Trans.*, 2013, **42**, 8868; (c) F. Li, A. K. Gorle, M. Ranson, K. L. Vine, R. Kinobe, M. Feterl, J. M. Warner, F. R. Keene, J. G. Collins and A. I. Day, *Org. Biomol. Chem.*, 2017, **15**, 4172.

90. (a) X.-J. Cheng, L.-L. Liang, K. Chen, N.-N. Ji, X. Xiao, J.-X. Zhang, Y.-Q. Zhang, S.-F. Xue, Q.-J. Zhu, X.-L. Ni and Z. Tao, *Angew. Chem., Int. Ed.*, 2013, **125**, 7393; (b) Q. Li, S.-C. Qiu, J. Zhang, K. Chen, Y. Huang, X. Xiao, Y. Zhang, F. Li, Y.-Q. Zhang, S.-F. Xue, Q.-J. Zhu, Z. Tao, L. F. Lindoy and G. Wei, *Org. Lett.*, 2016, **18**, 4020.

91. J. Zhang, Y.-Y. Xi, Q. Li, Q. Tang, R. Wang, Y. Huang, Z. Tao, S.-F. Xue, L. F. Lindoy and G. Wei, *Chem. Asian J.*, 2016, **11**, 2250.

92. Q. Li, J. Sun, J. Zhou, B. Hua, L. Shao and F. Huang, *Org. Chem. Front.*, 2018, **5**, 1940.

CHAPTER 10

# Cucurbit[6]uril-based Polymer Nanocapsules

JAMES MURRAY,[a] SUNGWAN KIM[a,b] AND KIMOON KIM*[a,b]

[a] Center for Self-assembly and Complexity (CSC), Institute of Basic Science (IBS), Pohang, Republic of Korea; [b] Department of Chemistry, Pohang University of Science and Technology (POSTECH), Pohang, Republic of Korea
*Email: kkim@postech.ac.kr

## 10.1 Introduction

Nature is adept at creating nanometer-sized capsules that are hollow and able to encapsulate cargos such as biomolecules and biochemicals. Such capsules can store, transport and deliver important biomolecules that are vital for a cell's function and metabolism. For example, ferritin is a 5 nm-sized capsule that is nearly ubiquitous in living organisms and is used to transport, store and control the release of iron to maintain normal iron metabolism.[1] Ferritin capsules are composed of 24 identical ferritin protein subunits. Virus capsids are another biological capsule and are among the most well-known, in part, because they are functionally complex, despite being (like ferritin) composed of many identical protein subunits, which self-assemble into a capsule.[2] A capsid's functions include recognizing receptors on a perspective host cell's surface and then delivering viral genetic material to the host cell. In this regard, virus capsids are protective shells for the targeted delivery of genetic material. Nature has formed such nanocapsules, by self-assembly from biomolecules, to perform highly specific functions. Taking inspiration from nature, the synthesis of artificial

Smart Materials No. 36
Cucurbituril-based Functional Materials
Edited by Dönüs Tuncel
© The Royal Society of Chemistry 2020
Published by the Royal Society of Chemistry, www.rsc.org

nanocapsules that mimic the structure and function of biological capsules may be a promising new avenue for novel biomedical applications.

A number of methods to create synthetic nanocapsules mimic the form and function of biological capsules, including template synthesis,[3–6] self-assembly[7–11] and emulsion polymerization.[12–14] For example, Möhwald and coworkers reported a method for constructing hollow polyelectrolyte shells by colloid-templated consecutive polyelectrolyte adsorption (layer-by-layer (LBL) technique), followed by the decomposition of the templating core.[3] Another approach is based on the formation of micelle by self-assembly; Wooley and coworkers prepared a spherical micelle with a core–shell morphology using diblock copolymers. Then a network structure was formed throughout the hydrophilic shell layer by covalent cross-linking; the core of the shell was chemically degraded and extracted to leave a hollow nanocapsule with a membrane-like exterior.[7] Each of these methods has its own merits, but they all need either a preorganized structure or template to shape a hollow shell structure. Furthermore, they require time-consuming and laborious multistep processes including the removal of the core or templates, repeated centrifugation or filtration, cross-linking of specially designed vesicular species or separation of large quantities of surfactants. The complexity of these methods illustrates the challenges of synthesizing nanocapsules; thus it is desirable to prepare nanocapsules in a facile manner with easy and fast purification.

Over the last 10 years, we have developed a new approach that allows one-pot, direct synthesis of hollow nanocapsules without the use of a template, emulsifier or preorganized structure.[15–22] This method directly produces polymer nanocapsules with a highly stable structure and a relatively narrow distribution and seems to be applicable to any monomers with a flat core and multiple polymerizable groups at the periphery. Various building blocks, including cucurbit[$n$]uril (CB[$n$], $n = 5$ or 6),[23–29] triphenylene[22] and phthalocyanine derivatives,[21] were polymerized with ditopic or tritopic linkers through thiol–ene "click" reaction, acetal formation, amide formation and $N$-alkylation reaction or without linkers through olefin metathesis. These blocks can successfully grow into flat oligomeric patches through irreversible covalent bond formation without the aid of preorganization or templates.[30] Further growth of the patches with curvature generation drives the system to the formation of polymer nanocapsules. This approach can also be extended to synthesize other nanostructured materials including two-dimensional (2D) polymer films[31] and toroidal nanotubular microrings.[32]

In particular, we extensively studied the one-pot, direct synthesis of hollow polymer nanocapsules (**PNCs**) made of CB[6] derivatives and their applications.[15,16] The most interesting feature of the **PNCs** is that the polymer shell, which comprises a host molecule, allows facile tailoring of its surface properties in a noncovalent and modular manner by virtue of the unique recognition properties of the accessible molecular cavities exposed on the surface. In this chapter, we discuss how such nanocapsules are synthesized and modified, their mechanism of formation, their properties and applications.

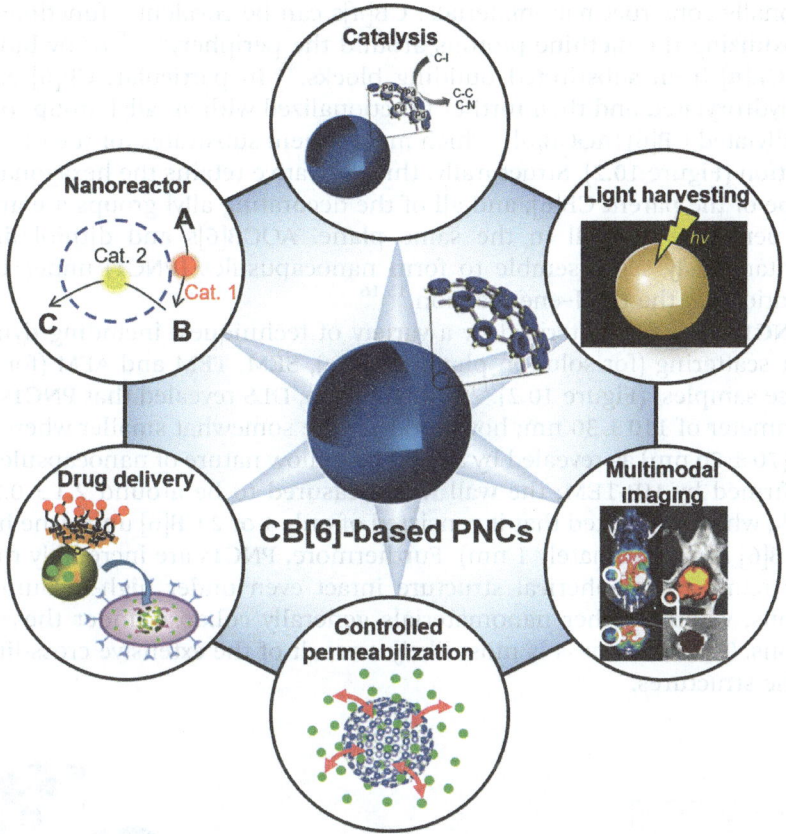

**Figure 10.1** Applications of CB[6]-based **PNCs**.
The image in the catalysis section is reproduced from ref. 20 with permission form John Wiley & Sons, Copyright 2014 WILEY-VCH Verlag GmbH & Co. KGaA, Weinheim. The image in the drug delivery section is reproduced from ref. 17 with permission from John Wiley & Sons, Copyright © 2010 WILEY-VCH Verlag GmbH & Co. KGaA, Weinheim. The image in multimodal imaging section is reproduced from ref. 19 with permission from the Royal Society of Chemistry.

Most notably, we highlight their use and potential as a biomedical material and as a catalytic platform. In the final section, we also suggest some new avenues in which we believe CB[6]-based **PNCs** may be useful (Figure 10.1).

## 10.2 Self-assembly of CB[6]-based Polymer Nanocapsules Through Covalent Bond Formation

### 10.2.1 Synthesis of CB[6]-based Polymer Nanocapsules

Since CB[*n*]s have a robust structure and well-defined shape, giving them a functionalizable periphery is an avenue that can be pursued in order to rationally construct nanomaterials. CB[*n*]s can be covalently functionalized by oxidizing the methine protons around the periphery[33–35] or by building the CB[*n*] from substituted building blocks.[36] In particular, CB[6] can be perhydroxylated and then further functionalized with an allyl group to yield perallylated CB[6] (AOCB[6]), which makes them substrates for the thiol–ene reaction (Figure 10.2). Structurally, this derivative retains the hexagonal disc shape of the parent CB[6], and all of the decorating allyl groups are around the periphery and all in the same plane. AOCB[6]s and dithiol linkers spontaneously self-assemble to form nanocapusules (**PNC1**) under UV irradiation *via* the thiol–ene reaction.[15,16]

**PNC1**s were characterized by a variety of techniques, including dynamic light scattering (for solution phase studies), SEM, TEM and AFM (for solid phase samples) (Figure 10.2).[15,16] In solution, DLS revealed that **PNC1**s have a diameter of $110 \pm 30$ nm; however, they are somewhat smaller when dried out ($70 \pm 20$ nm) as revealed by SEM. The hollow nature of nanocapsules was confirmed by HR-TEM; the wall was measured to be around $2.1 \pm 0.3$ nm thick, which suggested that it consisted of only 1 or 2 CB[6] units (the height of CB[6] is approximately 1 nm). Furthermore, **PNC1**s are incredibly robust; they retain their spherical structure intact even under high-vacuum conditions, whereas other nanomaterials generally collapse under these conditions. The robustness is most likely a result of the extensive cross-linking in the structures.

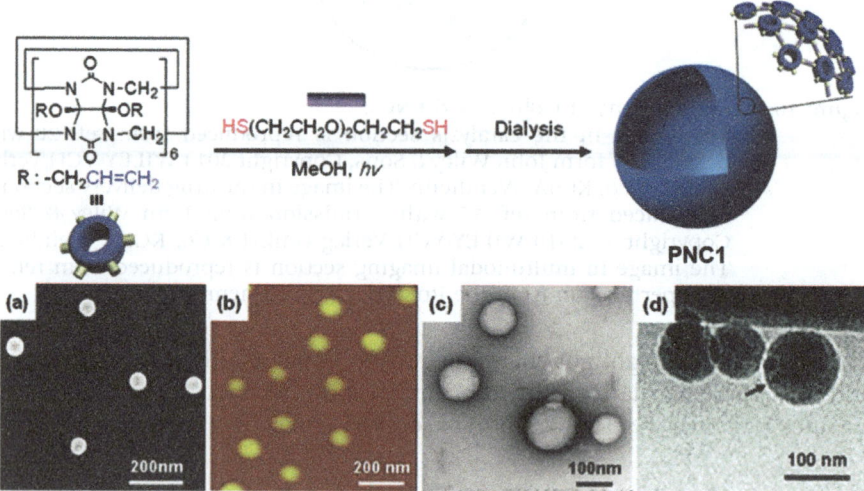

**Figure 10.2**   Synthesis and characterization of **PNC1**. (a) SEM image, (b) AFM image, (c) HR-TEM image and (d) cryo-TEM image.
The images in panels (a)–(c) are reproduced with permission from ref. 15 with permission from John Wiley & Sons, Copyright 2007 WILEY-VCH Verlag GmbH & Co. KGaA, Weinheim. The image in panel (d) is reproduced from ref. 16 with permission from American Chemical Society, Copyright 2010.

The diameter of the nanocapsules can be tailored by changing the length of the dithiol linker or by changing the reaction solvent. To illustrate, using ethanedithiol as the linker results in nanocapsules with a diameter of $50 \pm 10$ nm, whereas performing the reaction in chloroform yields nanocapsules with an average diameter of over 500 nm. Generally, better solvents and longer linkers result in larger nanocapsules. In a poor solvent such as methanol, an energy cost is associated with a solvent-exposed surface area, so small spheres form. In better solvents, the penalty is lower, so larger spheres form. The mechanism of formation is discussed in more detail in the next section.

The linking chemistry is not limited to thiol–ene reactions; peramine-decorated CB[6] and disuccinimide linker also form nanocapsules without the aid of a template, *via* amide bonds (**PNC3** and **PNC4**).[17] Furthermore, dynamic covalent bonds such as disulfides can also be used to construct nanocapsules (Table 10.1).[37] An interesting feature of such nanocapsules is that they can shift morphology in response to solvent. In methanol, a capsule (**PNC5**) is formed, but when the solvent is changed to dichloromethane, the nanocapsules become a 2D polymer sheet. Other polymerization methods, such as olefin metathesis, acetal formation, *N*-alkylation, also yield nanomaterials using different building blocks.[30]

## 10.2.2 Experimental and Theoretical Insights into the Mechanism of Formation

Mechanistic studies, using SEM and DLS, of the formation of these nanocapsules suggested that the key intermediates are slightly curved oligomeric patches, which then assemble into spherical nanocapsules (Figure 10.3).[15,16] During the early stages of assembly, the CB[6] "disks" form dimers and trimers, which then grow in lateral dimensions into 2D patches. These patches continue to grow and then begin to curve; covalent cross-linking of these patches generates the final spherical nanocapsule. This mechanism is quite similar to the way that vesicles form, in that vesicles also form a linear film first before curving and forming a spherical, cell-like structure.[38]

Theoretical studies of the energies of these systems give some insight into the shape and solvent preferences of the nanomaterial's structure. The energy in the system can be described by eqn (10.1):

$$E = -3\varepsilon \left[ \frac{8R^2}{d^2} (1 - \cos \theta) - \frac{2\pi R}{d} \sin \theta \right] + 4\pi\kappa(1 - \cos \theta) = 3\varepsilon f(R, \theta) \quad (10.1)$$

where $R$ is the radius of the curvature, $d$ is the distance between two adjacent monomer units, $\theta$ is the angle of the curvature, $\kappa$ is the bending modulus and $\varepsilon$ is the bond energy. The equation describes how the energy of the system is determined by three factors: (1) the energy of the surface, (2) the cost of exposing the rim to solvent, and (3) the bending energy. In the case of a 2D oligomeric patch, when $R$ is small, the total energy increases with $\theta$; thus

**Table 10.1** Various building blocks and linkers for the preparation of PNCs.

| PNCs | PNC1 and PNC2 | PNC3 and PNC4 | PNC5 |
|---|---|---|---|
| Building block | | | |
| Linker | |  PNC4 = X = S  PNC5 = X = CH$_2$ | No linker (Thiol is deprotected.) |
| Polymerization methods | Thiol–ene reaction | Amide bond formation | Disulfide bond formation |

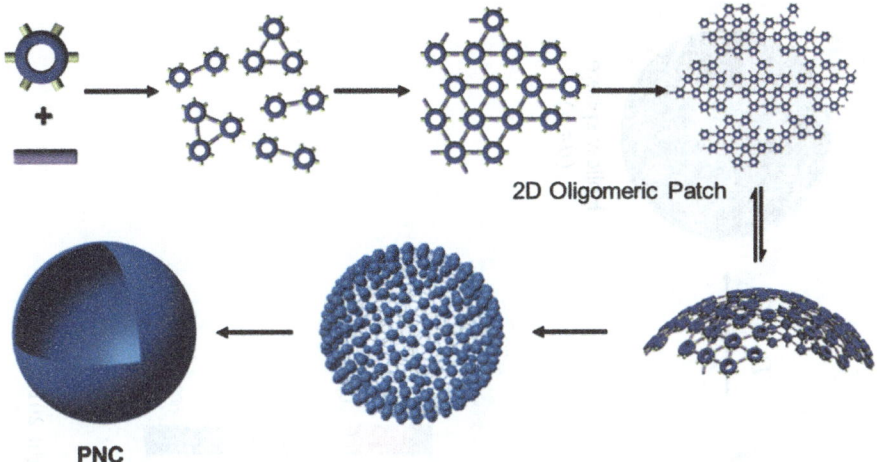

2D Oligomeric Patch

PNC

**Figure 10.3**   Formation mechanism of the **PNC**.[39]
Adapted from ref. 16 with permission from American Chemical Society,
Copyright 2010.

sphere formation is not favored in the early stages. However, when $R$ reaches a critical length, the energy then abruptly decreases with increasing $\theta$; the energy is at its minimum when $\theta = \pi$, which corresponds to a hollow sphere (Figure 10.4). The size of sphere is a trade-off between the bond energy minimum of the sphere and the entropy of the system. A larger sphere is favored energetically, but entropy favors a greater number of smaller spheres. The equilibrium size distribution of the sphere is therefore determined by the minimum free energy of the system. The empirical results qualitatively agree with the theoretical calculations.[15,16]

## 10.2.3   Post-synthetic Modification of CB[6]-based Polymer Nanocapsules

The nanocapsules can be postsynthetically modified both covalently and noncovalently. Although the thiol–ene–linked nanocapsules are extensively cross-linked, there are some remaining loose-end thiols that tend to occur as disulfide loops. These disulfides present an opportunity to post-synthetically modify the **PNC**s. This can be used to modify the dispersibility of the nanocapsules, for example using vinyl ether or methyl iodide. The thiols can also be used to anchor metallic nanoparticles to the surface of the **PNC**s (see page 225 for the catalysis applications of such assemblies).

The CB[6] cavity is available and presents an opportunity for the non-covalent modification of the **PNC**s. Since CB[6] forms high affinity complexes with polyamines, such as spermidine (spmd), the nanocapsule surfaces can be functionalized with spmd-tagged molecules.[15–17,19] Binding experiments with FITC-spmd revealed that around 85% of the CB[6] cavities

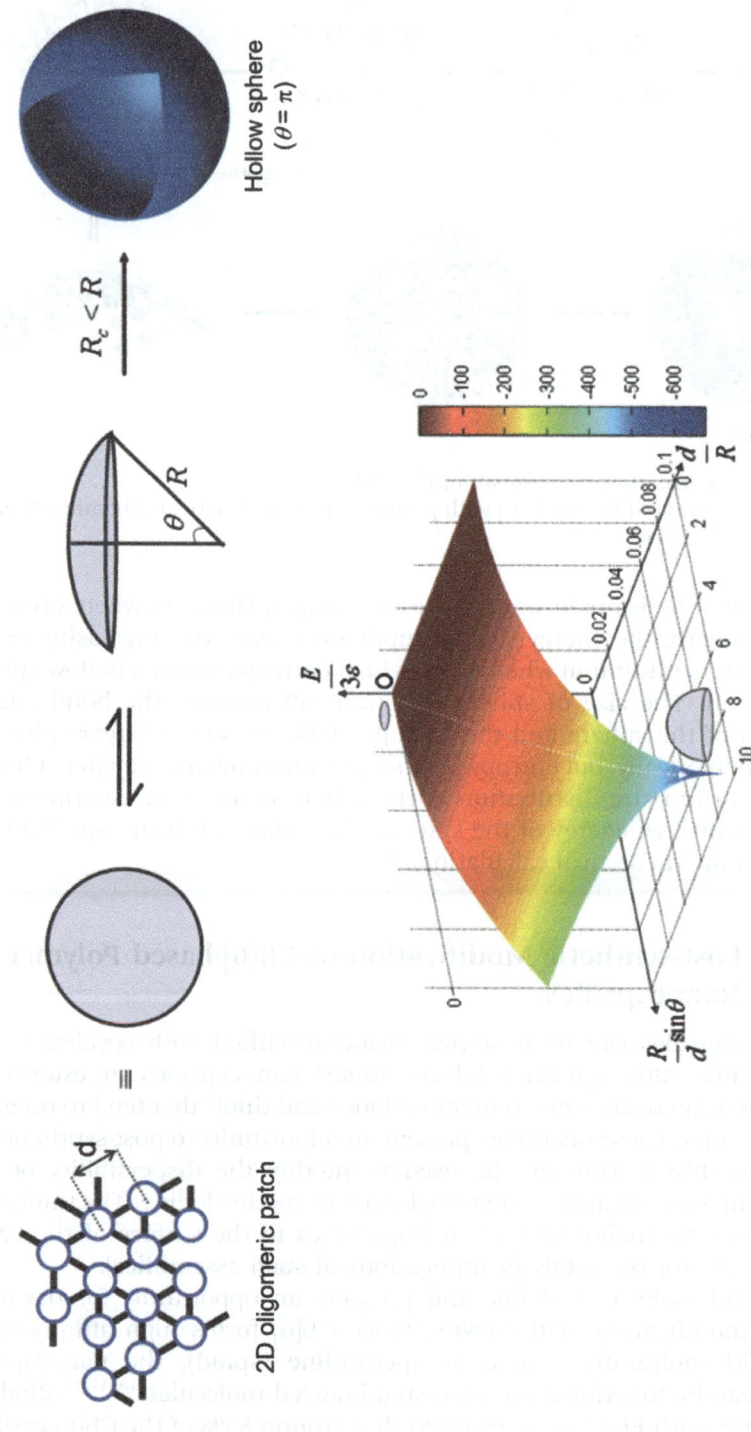

**Figure 10.4**  An energy landscape depicting the stability of the **PNCs**.[39] Adapted from ref. 16 with permission from American Chemical Society, Copyright 2010.

are available for host binding, which implies that the nanocapsules are essentially a single layer thick. Once the nanocapsule is functionalized with spmd-tagged molecules, the interaction is very robust. A negligible amount of free FITC-spmd was found to be released from the surface of the nanocapsule under aqueous conditions. The nanocapsules can be cofunctionalized with other groups such as imaging probes and biorecognition groups. Applications that make use of such functionalized nanocapsules are described in later sections.

The nanocapsules can swell or contract in response to solvent exchange, and this phenomenon can be used to control their permeability.[18] The capsules are first formed in methanol, and the remaining disulfides are capped with vinyl ether to yield nanocapsules with a diameter of $90 \pm 30$ nm. The size increases proportionally with increasing water content; in 90% water in methanol, the diameter of capsules increased to $330 \pm 170$ nm, as measured by DLS. The size change was confirmed by cryo-TEM. When the solvent was exchanged back to pure methanol, the nanocapsule decreased in size to $220 \pm 30$ nm. The expansion and contraction affects the permeability of the nanocapsules. This was demonstrated by introducing a fluorescent dye in the swelled state, where the dye permeates into the capsule. When the solvent is changed and the nanocapsule compresses, the dye becomes entrapped. After the capsule is swelled again, the dye slowly leaks out. Although the kinetics of the process are a little slow, this phenomenon may have uses in controlled release applications.

## 10.3 Applications

Nature's nanocapsules, such as viral capsids and ferritin, are functional. In this section, we show how the **PNCs** are also functional. In particular, we highlight the utility of **PNCs** as a biomedical material.

### 10.3.1 Catalytic Nanoparticle-functionalized Nanocapsules

In addition to the noncovalent modification, these **PNCs** can also be covalently modified through loose-end thiols on the surface, which typically exist as looped disulfides. Although the nanostructures are extensively crosslinked, not every thiol is involved in holding the nanocapsules together. These thiol groups can be used to decorate the nanocapsules' surface with metallic nanoparticles, including Pd, Au and Pt.[20] Such metal-functionalized nanocapsules were prepared in a two-step process (Figure 10.5). First, **PNC1** was methylated with excess methyl iodide, which alkylates some of the disulfides, rendering the nanocapsules positively charged (**PNC2**). **PNC2** is more water dispersible than the as-synthesized **PNC1**. Second, the alkylated capsules are treated with a reducing agent (to cleave the disulfide) and a metal salt (*e.g.*, $K_2PdCl_4$) as the metal source. The metal binds to the thiols on the surface and forms nanoparticles, and the unreacted metal is removed by dialysis. Note that, although metals are known to bind to the portals of

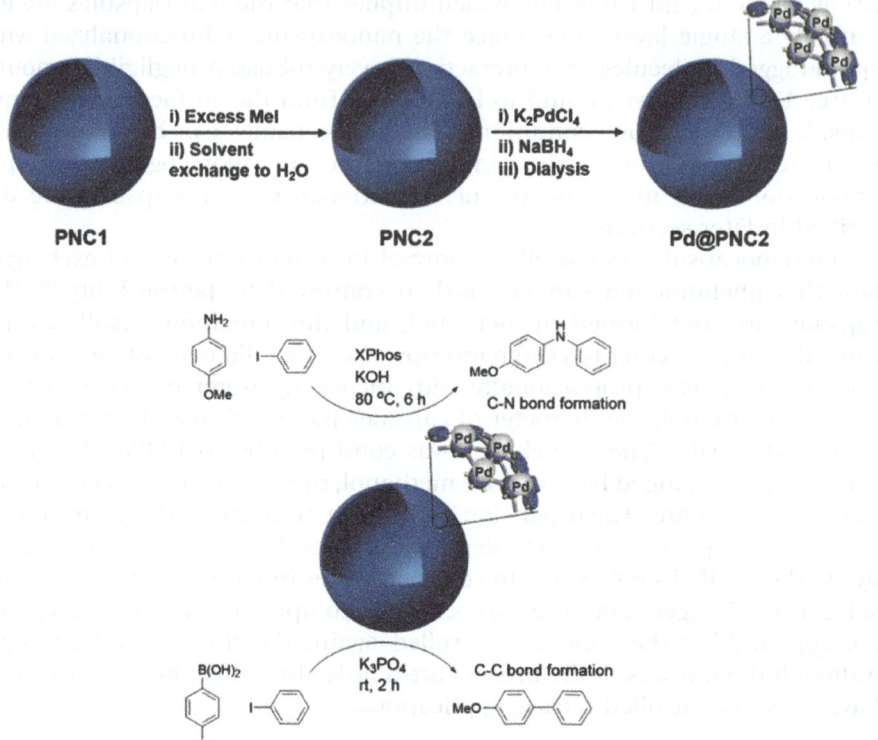

**Figure 10.5** Synthesis and catalytic reaction of Pd@**PNC1**.
Adapted with permission from ref. 20 with permission form John Wiley and Sons, Copyright 2014 Wiley-VCH Verlag GmbH & Co. KGaA, Weinheim.

CB[*n*], IR provided no evidence of such interaction; XPS data proved that the thiol-metal bond was the crucial interaction to form such assemblies. The metal-functionalized capsules are stable in solution for at least 6 months.

Dynamic light scattering studies revealed that the Pd@**PNC2** was $130 \pm 40$ nm in diameter and remained spherical. HR-TEM showed that Pd nanoparticles had formed on the surface and that they had a narrow size distribution ($1.9 \pm 0.2$ nm). The size of the nanoparticle can be effectively controlled by varying the ratio of $K_2PdCl_4$ to CB[6] units in the **PNC2**. To illustrate, when the ratio was set to 2:1, 3:1 and 4:1, the nanoparticles had diameters of $1.7 \pm 0.2$ nm, $1.9 \pm 0.2$ nm and $3.1 \pm 0.3$ nm, respectively. Similar results were observed using Au or Pt as the metal. These results suggest that **PNC2**s provide nucleation sites for the metal nanoparticles.

Since palladium is an incredibly versatile catalyst for C–C and C–N coupling reactions, the utility of the Pd-functionalized nanocapsules as a heterogeneous catalyst was evaluated. Namely, they were used as catalysts for the Suzuki–Miyaura and Buchwald–Hartwig cross-coupling reactions.

In many cases, the substrates were converted to the corresponding products in quantitative yields, as revealed by GC–MS. Reusability was also demonstrated. After the reaction, Pd@**PNC2** was recovered by centrifugation, then resuspended in a new solution of substrate. Even after five cycles, the catalytic activity was undiminished. It appeared that the nanoparticles were stably anchored to the nanocapsules since no leaching of metals was detected by ICP-AES in the reaction solution. Since this catalytic system is heterogeneous, reusable and water dispersible, it may have a promising future as a green catalyst.

## 10.3.2 Stimuli-responsive Polymer Nanocapsules for Targeted Drug Delivery

As previously shown, cargo can be encapsulated inside the **PNC1** by taking advantage of their swelling or compressing in different solvents.[18] Another approach to encapsulate a drug is to simply synthesize the nanocapsules in the presence of the drug, providing that the drug does not interfere with the polymerization chemistry.[17] The drug can be released passively (slowly leaking out of the cargo) or actively by changing the chemistry of the linker between the CB[6] units. Since the cytoplasm of a cell is a reductive environment, due to the presence of strong reducing agents like glutathione, **PNC**s held together by disulfides selectively decompose once internalized into a cell, thereby releasing its cargo. However, the first generation of these nanocapsules was assembled using thiol–ene chemistry, which is incompatible with a disulfide linkage. To work around this, perallyoxyCB[6] was further decorated so that it was per-amine functionalized. Instead of a dithiol linker, a disuccinimide linker containing a disulfide was used. The final nanocapsule, **PNC3**, was formed by amide polymerization. To test whether they could be degraded by reductive conditions, **PNC3** was synthesized in the presence of a dye, and a dye@**PNC3** was then treated with dithiothreitol (DTT). The dye@**PNC3** began to degrade and the dye leaked out, as shown by an increase in fluorescence, in response to DTT, whereas the control nanocapsules **PNC4**, which lacked a disulfide, remained intact.

With reduction-sensitive **PNC**s in hand, the surface of **PNC3** was functionalized with a sugar molecule that would trigger receptor-mediated endocytosis (Figure 10.6). To stably attach the sugar to the surface of **PNC3**, the sugar was labeled with spmd, which has a very high affinity for CB[6].[39] The ability of the sugar-functionalized **PNC3** to deliver a cargo to a cell was tested by loading it with a fluorescent dye. Under a fluorescence microscope, it could be seen that **PNC3** was well internalized into the cell and that the fluorescence increased as the nanocapsules collapsed and the dye was released into the cytoplasm. The analogous capsule **PNC4** did not show a large increase in fluorescence since the dye remained encapsulated. Although the work demonstrated the delivery of fluorescent dye, replacing the cargo with drug molecules can be easily envisioned.

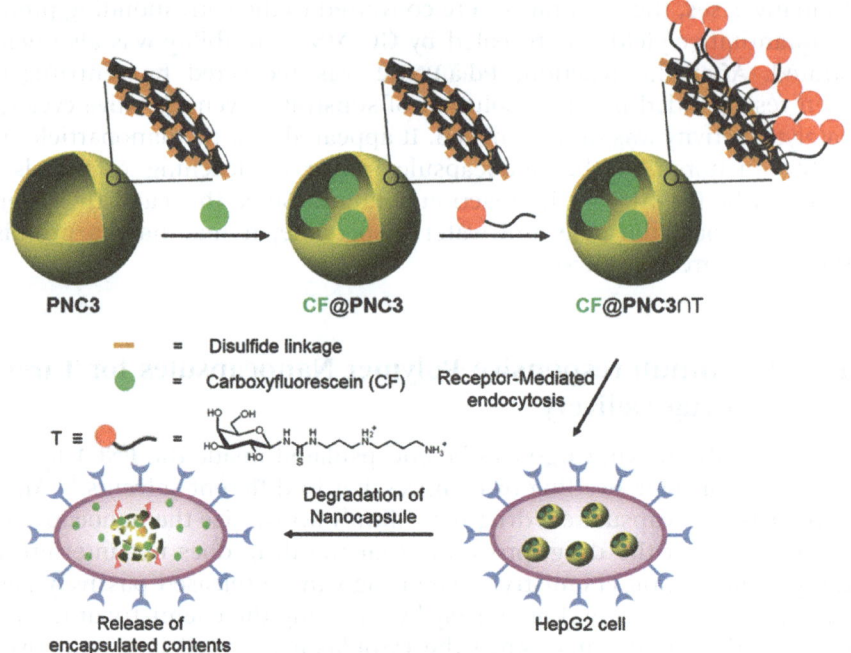

**PNC3**    **CF@PNC3**    **CF@PNC3∩T**

— = Disulfide linkage

● = Carboxyfluorescein (CF)    Receptor-Mediated endocytosis

T ≡ ●~ =

Degradation of Nanocapsule

Release of encapsulated contents    HepG2 cell

**Figure 10.6**    Disulfide-linked **PNC** (**PNC3**) for cellular delivery of cargo. Adapted from ref. 17 with permission from John Wiley and Sons, Copyright © 2010 WILEY-VCH Verlag GmbH & Co. KGaA, Weinheim.

### 10.3.3  Polymer Nanocapsules as a Noncovalent and Modular Bioimaging Platform for Multimodal *In Vivo* Imaging

Not limited to *in vitro* studies, the **PNC**s can also be applied to *in vivo* level studies, for example tumor imaging in cancer-bearing mice.[19] In this work, the many available cavities on the surface of the **PNC2** were also exploited to allow for different imaging modalities to be attached to **PNC**s for multimodal imaging. Namely, a combination of positron emission tomography (PET) and near-infrared imaging (NIR) probes were used because the former provides high sensitivity and the latter ease of use.

Before proceeding to the imaging experiments, we first evaluated the stability of **PNC2** and FITC-spmd@**PNC2** in serum conditions. **PNC2** seemed to swell in size, as judged by DLS (from $110 \pm 30$ to $180 \pm 40$ nm) but were otherwise stable even after 12 h of incubation. The stability of FITC-spmd@**PNC2** in serum conditions was assessed by measuring the amount of free FITC-spmd released from FITC-spmd@**PNC2** during dialysis. More than 90% of FITC-spmd molecules were stably anchored to **PNC2** for 24 h under serum conditions, indicating that the host–guest interaction was stable even in the presence of interfering buffer salts and biomolecules; it suggested that they would be stable *in vivo* as well. This is in contrast to other host molecules such as cyclodextrins, the binding affinities of which are not strong enough for practical use in physiological conditions. Before administering **PNC2** to a live animal, we also evaluated its stability, toxicology and biodistribution profile under *in vivo* conditions. The cytotoxicity of the **PNC2** was evaluated by MTT assays. It turned out that **PNC2** was not significantly cytotoxic since cell viability was 84% at the maximum concentration allowed by the MTT assays. At the mouse level, histological staining showed that various organs were not affected by the nanomaterial so they could be assumed to be safe. Next, the biodistribution of **PNC2** was evaluated using FITC-spmd, Cy5-**PNC2** and FITC-spmd@Cy5-**PNC2**; these molecules allowed the fate of the small molecule and the nanomaterial to be tracked by following the signal from FITC and Cy5, respectively. Consequently, whether the FITC-spmd remained attached to **PNC2** could be evaluated. It was found that the free dye was quickly cleared *via* the kidneys, whereas the dye cleared much more slowly when it was attached to **PNC2**. This result indicated that the dye was stably attached to the surface of **PNC2** since it followed the fate of the nanocapsule. Furthermore, it was found that FITC-spmd@Cy5-**PNC2** accumulated in the liver and kidneys, which is typical behavior for nanomaterials.[40,41]

With the *in vivo* behavior of **PNC2** established, it was employed as a multimodal imaging tool. **PNC2** can be functionalized with spmd-tagged NIR dyes, PET tracers and targeting ligands, on demand, simply by mixing together. The multimodal complex was administered to cancer-bearing mice and compared to individual imaging components (Figure 10.7). On the one hand, the decorated **PNC2** became localized at the tumor site because of the tumor-targeting peptide, and **PNC2** (like many other nanomaterials) took advantage of the enhanced permeability and retention (EPR) effect.[42] On the other hand, the individual imaging species did not localize at the tumor site. In practical terms, the time-sensitive PET imaging was undertaken first, followed by the near-infrared imaging; both imaging modes gave a similar picture of the tumor. This work not only demonstrated the use of **PNC2** for multimodal imaging but also showed that they are promising low-toxicity nanomaterials that may have other biomedical uses, such as acting as a guide for the complete surgical removal of a tumor.

## 10.4 Summary and Outlook

There are several impressive features about these capsules. Among these is the fact that they are self-assembled using irreversible covalent bonds to afford robust nanometer-sized structures. While they are made from very different building blocks, they can be compared to virus capsids and other biological capsules. The CB[6]-based **PNC**s described here, like virus capsids, are self-assembled from many copies of an identical subunit, and they can perform many of functions of a virus capsid, namely cell surface recognition and cargo transport.

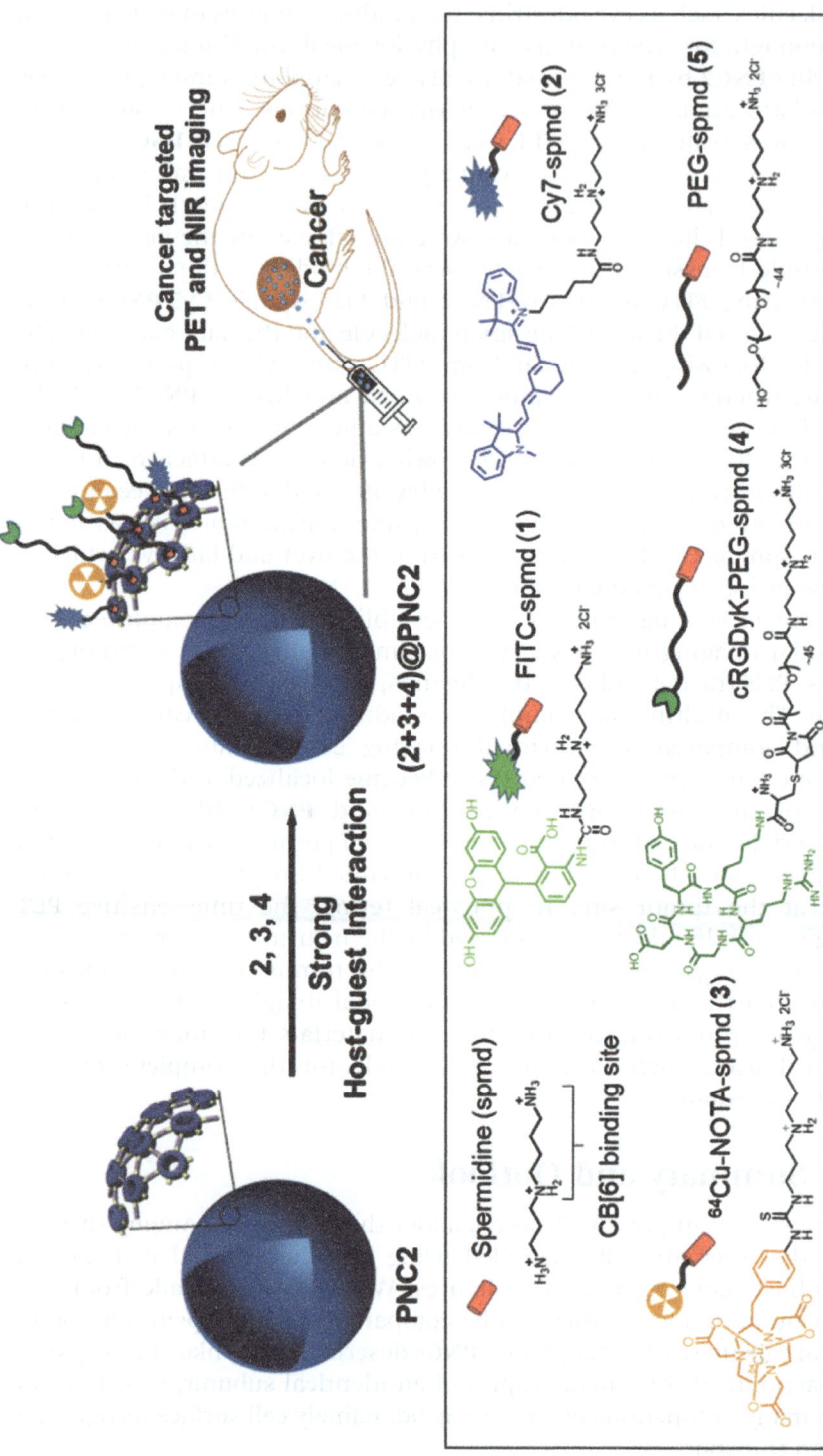

**Figure 10.7** Facile surface modification of **PNC1** using the strong host-guest interactions of CB[6] with spmd-functional tags in a noncovalent and modular manner.
Adapted from ref. 19 with permission from the Royal Society of Chemistry, Copyright 2017.

In the course of this chapter, we demonstrated that CB[6] polymer nanocapsules have many of attributes necessary to be a theranostic platform. Especially, they seem to be an excellent imaging platform since they can be functionalized with multiple imaging modes and targeting ligands. Such multimodal imaging is advantageous because it amplifies the advantages of each imaging mode while mitigating the drawbacks. One of the keys to the success of this strategy is the robustness of the host–guest interaction, which is stable even in the hydrodynamic environment of an animal's bloodstream. Since the nanocapsules are nanometer-sized, they inherently take advantage of the EPR effect to target cancer cells. Furthermore, nanocapsules linked by disulfide bonds can disassemble in the reductive environment of a cell, which has been exploited as a cargo release mechanism. The release profile may be altered by controlling the number of disulfide bridges in the nanocapsules or by employing a different responsive linker chemistry for more dynamic cargo release, not only for drug delivery but also for other controlled release applications.

Another avenue that these nanocapsules could be used for is magnetic resonance imaging (MRI). CB[6] has a moderate affinity for Xe,[43] and a modified CB[6] has been used as a Xe sensor.[44] Dmochowski and co-workers[45] have shown that $^{129}$Xe complexed to unfunctionalized CB[6] experiences a downfield shift in its $^{129}$Xe NMR spectrum and can be detected at as low as 1.8 pM. The $^{129}$Xe@CB[6] complex has also been detected *in vivo*, opening the possibility of CB[6] being used as an MRI contrast reagent.[46] In this chapter, **PNC**s built using many CB[6] units have been described that have been shown to be not only water dispersible but also stable inside an animal's bloodstream. In principle, such nanocapsules should be able to be high-signal MRI contrast agents because they can potentially hold a large number of Xe atoms. So far, studies in this direction have been inconclusive, since incorporation into a nanocapsule seems to reduce the affinity between CB[6] and Xe. Nevertheless, this still seems to be an attractive research goal that will complement the already realized theranostic applications.

**PNC**s decorated with metallic nanoparticles have been used as heterogeneous, water-dispersible and reusable catalysts. So far, the interior of the **PNC**s has not been utilized as a reaction container. In principle, a reaction cascade could be set up where a catalytic reaction occurs on the surface of the nanocapsule; the product of this reaction then selectivity permeates the **PNC**s to undergo a second reaction. Since the size of the gaps between the CB[6] units of the **PNC**s can be tuned by using different linkers and modulating the cross-linking efficiency, size-selective permeation should be possible.

The **PNC**s may be useful in other areas where a well-defined nanostructured material is required. An example of such an application may be light harvesting or other photovoltaic devices.[47] The large surface area of the spherical nanocapsules and the easy on-demand functionalization of the surface (with donor and acceptor molecules, for example) should be attributes for such an endeavor.

# Acknowledgements

This work was supported by the Institute for Basic Science (IBS) [IBS-R007-D1]. K.K. would like to thank all of the collaborators, including Professor W. Sung and his past and present group members for their contributions to polymer nanocapsule research, especially Dongwoo Kim, Kyeng Min Park, Eunju Kim, Kagnkyun Baek, Gyeongwon Yun and Jeehong Kim.

# References

1. E. C. Theil, *Annu. Rev. Biochem.*, 1987, **56**, 289.
2. B. V. V. Prasad, M. E. Hardy, T. Dokland, J. Bella, M. G. Rossmann and M. K. Estes, *Science*, 1999, **286**, 287.
3. E. Donath, G. B. Sukhorukov, F. Caruso, S. A. Davis and H. Möhwald, *Angew. Chem., Int. Ed.*, 1998, **37**, 2201.
4. F. Cavalieri, A. Postma, L. Lee and F. Caruso, *ACS Nano*, 2009, **3**, 234.
5. X. Liu and A. Basu, *J. Am. Chem. Soc.*, 2009, **131**, 5718.
6. A. O. Moughton and R. K. O'Reilly, *J. Am. Chem. Soc.*, 2008, **130**, 8714.
7. H. Huang, E. E. Remsen, T. Kowalewski and K. L. Wooley, *J. Am. Chem. Soc.*, 1999, **121**, 3805.
8. W.-F. Dong, A. Kishimura, Y. Anraku, S. Chuanoi and K. Kataoka, *J. Am. Chem. Soc.*, 2009, **131**, 3804.
9. Y. Hu, X. Jiang, Y. Ding, Q. Chen and C. Z. Yang, *Adv. Mater.*, 2004, **16**, 933.
10. H. M. Jung, K. E. Price and D. T. McQuade, *J. Am. Chem. Soc.*, 2003, **125**, 5351.
11. P. Xu, S.-Y. Li, Q. Li, J. Ren, E. A. Van Kirk, W. J. Murdoch, M. Radosz and Y. Shen, *Biotechnol. Bioeng.*, 2006, **95**, 893.
12. J. Jang and H. Ha, *Langmuir*, 2002, **18**, 5613.
13. F. Lu, Y. Luo, B. Li, Q. Zhao and F. J. Schork, *Macromolecules*, 2010, **43**, 568.
14. D. Sarkar, J. El-Khoury, S. T. Lopina and J. Hu, *Macromolecules*, 2005, **38**, 8603.
15. D. Kim, E. Kim, J. Kim, K. M. Park, K. Baek, M. Jung, Y. H. Ko, W. Sung, H. S. Kim, J. H. Suh, C. G. Park, O. S. Na, D. K. Lee, K. E. Lee, S. S. Han and K. Kim, *Angew. Chem., Int. Ed.*, 2007, **46**, 3471.
16. D. Kim, E. Kim, J. Lee, S. Hong, W. Sung, N. Lim, C. G. Park and K. Kim, *J. Am. Chem. Soc.*, 2010, **132**, 9908.
17. E. Kim, D. Kim, H. Jung, J. Lee, S. Paul, N. Selvapalam, Y. Yang, N. Lim, C. G. Park and K. Kim, *Angew. Chem., Int. Ed.*, 2010, **49**, 4405.
18. E. Kim, J. Lee, D. Kim, K. E. Lee, S. S. Han, N. Lim, J. Kang, C. G. Park and K. Kim, *Chem. Commun.*, 2009, 1472.
19. S. Kim, G. Yun, S. Khan, J. Kim, J. Murray, Y. M. Lee, W. J. Kim, G. Lee, S. Kim, D. Shetty, J. H. Kang, J. Y. Kim, K. M. Park and K. Kim, *Mater. Horiz.*, 2017, **4**, 450.

20. G. Yun, Z. Hassan, J. Lee, J. Kim, N. S. Lee, N. H. Kim, K. Baek, I. Hwang, C. G. Park and K. Kim, *Angew. Chem., Int. Ed.*, 2014, **53**, 6414.
21. R. Hota, K. Baek, G. Yun, Y. Kim, H. Jung, K. M. Park, E. Yoon, T. Joo, J. Kang, C. G. Park, S. M. Bae, W. S. Ahn and K. Kim, *Chem. Sci.*, 2013, **4**, 339.
22. I. Roy, D. Shetty, R. Hota, K. Baek, J. Kim, C. Kim, S. Kappert and K. Kim, *Angew. Chem., Int. Ed.*, 2015, **54**, 15152.
23. K. Kim, J. Murray, N. Selvapalam, Y. H. Ko and I. Hwang, *Cucurbiturils*, World Scientific, Europe, 2018, p. 264.
24. J. Murray, K. Kim, T. Ogoshi, W. Yao and B. C. Gibb, *Chem. Soc. Rev.*, 2017, **46**, 2479.
25. L. Isaacs, *Acc. Chem. Res.*, 2014, **47**, 2052.
26. S. J. Barrow, S. Kasera, M. J. Rowland, J. del Barrio and O. A. Scherman, *Chem. Rev.*, 2015, **115**, 12320.
27. J. Lagona, P. Mukhopadhyay, S. Chakrabarti and L. Isaacs, *Angew. Chem., Int. Ed.*, 2005, **44**, 4844.
28. K. I. Assaf and W. M. Nau, *Chem. Soc. Rev.*, 2015, **44**, 394.
29. E. Masson, X. Ling, R. Joseph, L. Kyeremeh-Mensah and X. Lu, *RSC Adv.*, 2012, **2**, 1213.
30. K. Baek, I. Hwang, I. Roy, D. Shetty and K. Kim, *Acc. Chem. Res.*, 2015, **48**, 2221.
31. K. Baek, G. Yun, Y. Kim, D. Kim, R. Hota, I. Hwang, D. Xu, Y. H. Ko, G. H. Gu, J. H. Suh, C. G. Park, B. J. Sung and K. Kim, *J. Am. Chem. Soc.*, 2013, **135**, 6523.
32. J. Lee, K. Baek, M. Kim, G. Yun, Y. H. Ko, N.-S. Lee, I. Hwang, J. Kim, R. Natarajan, C. G. Park, W. Sung and K. Kim, *Nat. Chem.*, 2014, **6**, 97.
33. M. M. Ayhan, H. Karoui, M. Hardy, A. Rockenbauer, L. Charles, R. Rosas, K. Udachin, P. Tordo, D. Bardelang and O. Ouari, *J. Am. Chem. Soc.*, 2015, **137**, 10238.
34. S. Y. Jon, N. Selvapalam, D. H. Oh, J.-K. Kang, S.-Y. Kim, Y. J. Jeon, J. W. Lee and K. Kim, *J. Am. Chem. Soc.*, 2003, **125**, 10186.
35. N. Zhao, G. O. Lloyd and O. A. Scherman, *Chem. Commun.*, 2012, **48**, 3070.
36. B. Vinciguerra, L. Cao, J. R. Cannon, P. Y. Zavalij, C. Fenselau and L. Isaacs, *J. Am. Chem. Soc.*, 2012, **134**, 13133.
37. J. Kim, K. Baek, D. Shetty, N. Selvapalam, G. Yun, N. H. Kim, Y. H. Ko, K. M. Park, I. Hwang and K. Kim, *Angew. Chem., Int. Ed.*, 2015, **54**, 2693.
38. D. D. Lasic and F. J. Martin, *J. Membr. Sci.*, 1990, **50**, 215.
39. M. V. Rekharsky, Y. H. Ko, N. Selvapalam, K. Kim and Y. Inoue, *Supramol. Chem.*, 2007, **19**, 39.
40. N. K. Tafreshi, S. A. Enkemann, M. M. Bui, M. C. Lloyd, D. Abrahams, A. S. Huynh, J. Kim, S. R. Grobmyer, W. B. Carter, J. Vagner, R. J. Gillies and D. L. Morse, *Cancer Res.*, 2011, **71**, 1050.
41. M. P. Melancon, W. Lu, Z. Yang, R. Zhang, Z. Cheng, A. M. Elliot, J. Stafford, T. Olson, J. Z. Zhang and C. Li, *Mol. Cancer Ther.*, 2008, **7**, 1730.

42. B. R. Smith and S. S. Gambhir, *Chem. Rev.*, 2017, **117**, 901.
43. M. E. Haouaj, Y. Ho Ko, M. Luhmer, K. Kim and K. Bartik, *J. Chem. Soc., Perkin Trans. 2*, 2001, 2104.
44. B. S. Kim, Y. H. Ko, Y. Kim, H. J. Lee, N. Selvapalam, H. C. Lee and K. Kim, *Chem. Commun.*, 2008, 2756.
45. Y. Wang and I. J. Dmochowski, *Chem. Commun.*, 2015, **51**, 8982.
46. F. T. Hane, T. Li, P. Smylie, R. M. Pellizzari, J. A. Plata, B. DeBoef and M. S. Albert, *Sci. Rep.*, 2017, 7, 41027.
47. S. Kundu and A. Patra, *Chem. Rev.*, 2017, **117**, 712.

CHAPTER 11

# Cucurbituril-functionalized Supramolecular Assemblies: Gateways to Diverse Applications

J. MOHANTY,*[a,b] R. KHURANA,[a,b] N. BAROOAH[a] AND A. C. BHASIKUTTAN[a,b]

[a] Radiation & Photochemistry Division, Bhabha Atomic Research Centre, Mumbai-400 085, India; [b] Homi Bhabha National Institute, Training School Complex, Anushakti Nagar, Mumbai-400 094, India
*Email: jyotim@barc.gov.in

## 11.1 Introduction

### 11.1.1 Supramolecular Assemblies

A supramolecular assembly, or so-called supermolecule, is a well-defined complex of two or more chemical entities, held together by noncovalent intermolecular binding interactions in an organized manner and having greater complexity than the individual entity.[1,2] These weak and reversible noncovalent interactions span a wide range of binding energies and comprehend electrostatic interactions such as ion–dipole, dipole–dipole, hydrogen bonding interactions and hydrophobic interactions like van der Waals, π–π interactions, dispersion interactions, *etc.* Engineering functional

Smart Materials No. 36
Cucurbituril-based Functional Materials
Edited by Dönüs Tuncel
© The Royal Society of Chemistry 2020
Published by the Royal Society of Chemistry, www.rsc.org

assemblies endowed with such unprecedented properties requires the exploitation of these intermolecular interactions in achieving the targeted outcomes. Nature itself provides the most spectacular examples of supramolecular assemblies with different functionalities. In the DNA structure, the intricate packing and replication mechanisms, protein–protein interactions and enzyme–substrate complex are all examples of supramolecular assemblies.[2,3] These assemblies establish cooperative/competitive interactions that affect both the mechanism of its formation and the rearrangement, as well as the stability, of the complex. Therefore, thorough mechanistic studies of supramolecular assemblies are important not only for understanding the self-assembly processes but also for designing assemblies for specific applications.

One of the major interests in supramolecular assemblies is the design of structurally well-defined architectures with dynamic and stimulus-responsive properties. Dynamic and adaptive supramolecular assemblies that self-assemble from multiple components find a range of applications in materials and medicines. The noncovalent interactions between a protein and a drug, or between a catalyst and its substrate, or between a macrocyclic host and guest and the self-assembly of nanomaterials, lead to the formation of self-assembled architectures that possess diverse functionalities, such as catalytic, photophysical, electronic or redox properties, and are ideal building blocks for sensors, information storage materials and nanodevices.[4] In the past decades, researchers have made enormous strides toward creating nanoscale assemblies and structures with the aim to achieve applications ranging from targeted drug delivery to the development of functional materials.[1,5–7]

Noncovalently bonded host–guest supramolecular assemblies are of immense importance as they can direct and thread molecular components according to design under appropriate selection criteria. The concurrent association of various guests or hosts by either cooperative binding[8,9] or competitive displacement[10,11] mechanism is another approach to fabricate novel assembled structures with different properties. Typically, host–guest complexes consist of a macrocyclic cavitand molecule that acts as a "host" by forming a complex with suitable "guest" molecules having proper structural features within their cavity. Since such assemblies are held together by weak noncovalent interactions, the preferential involvement of these forces and the ensuing stoichiometric arrangements of the complexes bring out substantial modulation in the molecular properties of the guests and are coveniently controlled by external stimuli.[8,12–16] However, specific control in response to a variety of triggers/stimulators remains challenging as such noncovalent interactions very much act as a part of several substrate-specific/enzymatic interactions in the biological systems.

Among several widely known natural and synthetic macrocyclic hosts (such as cyclodextrins and calixarenes), cucurbit[*n*]uril is a relatively new class of cavitand macrocycles that were originally revealed by Behrend in 1905 but were completely characterized in 1981. They are composed of methylene-bridged glycoluril monomers having highly symmetrical hydrophobic cavities accessible through two identical, partially negatively charged, carbonyl-fringed portals.[17–21] Synthetically, the cucurbituril homologues are prepared by the acid-catalyzed condensation of glycoluril with formaldehyde under optimized conditions.[22–24] As in the case of cyclodextrins and calixarenes, the different homologues of cucurbit[*n*]urils (commonly abbreviated as CB[*n*]; *n* = 5–14, representing the number of glycoluril units in the macrocycle; see Figure 11.1), with varying cavity and portal dimensions and with a constant height (9.1 Å), are known.[17,19–21] Recently, cucurbit[14]uril (remaining in the twisted form and abbreviated as tCB[14]) has been synthesized, having two kinds of cavities (a central cavity and two side cavities), and adopts a folded, figure-of-eight configuration.[25] The CB homologues display varying degrees of water solubility and can be enhanced in the presence of salts, low pH or charged guests.[17,21] All cucurbituril homologues, except the smallest CB[5], can form much stronger host/guest inclusion complexes with a variety of guests, including organic and inorganic compounds, and their binding constants range usually from $10^4$ to $10^{15}$ M$^{-1}$.[21,26] CB6 and CB7 form stable inclusion complexes with guest molecules like diaminoalkanes, benzyl amines, adamantyl amine, methyl viologen cations, fluorescent dyes, surfactants, metal ions, metal nanoparticles *etc.* through complete or partial encapsulation of the guest molecules.[17,19–21,27–29] Redox-active polyoxovanadate, polyoxomolybdate and polyoxotungstate anion clusters interact strongly with the equatorial periphery of CB6, CB7 and CB8.[30,31] Conversely, the cavity of CB8 is large enough to accommodate more than one guest molecule to form multiple/higher-order host–guest complexes.[32–34] In recent years, CB complexes have been shown to exhibit low *in vitro* as well as *in vivo* toxicity, thereby facilitating biologically relevant applications of these macrocycles.[20,35–37] Cucurbit[*n*]uril exhibits excellent complexation properties especially toward cationic guest molecules and have triggered a lot of research interest in numerous applications. Aptly, their usage is attempted in energy storage, photonic devices, drug delivery vehicles, sensors, as well as therapeutics *etc.* This chapter provides a brief account on the advantages of cucurbituril-functionalized molecular assemblies in diverse technological applications.

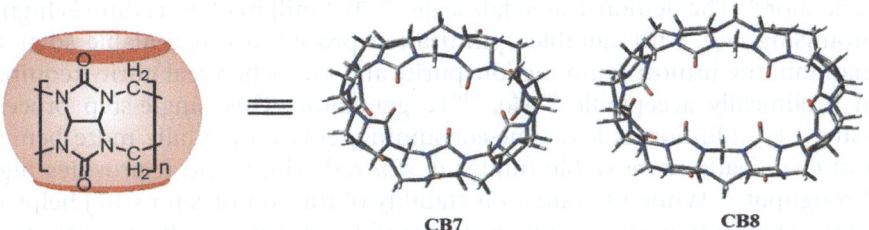

CB7          CB8

**Figure 11.1**   Structural formula of CB*n* (*n* = 5–10) and energy-optimized structures of CB7 and CB8.

## 11.1.2 Cucurbituril-functionalized Molecular Assemblies for Diverse Applications

As discussed *vide supra*, hitherto studies on cucurbituril-based host–guest systems have led to the demonstration of several stimuli-responsive complexes/materials having diverse technological applications. On the one hand, the facile construction of molecular assemblies of cucurbiturils with polyoxometalates, tungstates, perovskites *etc.* and organic chromophores/dyes has led to their applications in radionuclide separation, energy storage, aqueous dye laser, photovoltaic devices, photostable luminescent materials for light-emitting devices, enhanced Li-S batteries, sensor films and biomedical coatings.[31,34,38–46] On the other hand, the recognition-mediated bioassemblies with proteins, surfactants, antibiotics/drugs and their uptake/release of small molecules have resulted in several facile pathways for improved drug binding/release, targeted drug delivery, biosensors, inhibitors, antibacterial, antitumor and other therapeutic activities. In this section, we discuss the construction and attributes of various cucurbituril assemblies with organic, inorganic and biomolecules and elaborate on their respective utilities and benefits in diverse applications.[47–56]

### 11.1.2.1  *Cucurbituril–Polyoxometalate Hybrid Material: A Facile Supramolecular Strategy for Separation and Catalytic Applications*

The fabrication of self-assembled architectures of organic–inorganic hybrid functional materials is an emerging field as researchers find widespread application in areas ranging from electronic devices, optics and photonics to biosensors, drug delivery, advanced catalysis and energy conversion/storage. The construction of a novel noncovalently held cucurbit[7]uril-heptamolybdate hybrid material has been demonstrated, and this procedure has been applied to radioactive Mo ($^{99}$Mo) solution in order to separate the short-lived radiotracer, $^{99m}$Tc,[31] which is in demand for several diagnostic procedures (Figure 11.2). The noteworthy distinction among the anionic charges of $^{99}$Mo$_7$O$_{24}{}^{6-}$ and $^{99m}$TcO$_4{}^-$ and the subtle anion receptor propensity at the peripheral carbons of the biocompatible CB7 macrocycle afforded the essential separation strategy needed to develop this method as a $^{99}$Mo/$^{99m}$Tc generator.[31] The demonstrated lab scale $^{99m}$Tc "milking" procedure is highly promising and, with suitable optimization procedures, is scalable so as to maintain the utmost radionuclidic purity and radiochemical purity required in a clinically acceptable $^{99}$Mo/$^{99m}$Tc generator. This single-step process using CB7 addition is less time-consuming, environmentally more benign and chemically more stable (under *in situ* radiation), and it provides high throughput.[31] While the radiation stability of the complex (or CB7) helps to retain chemical purity as much as possible, it reduces the maintenance intervals for the frequent replacement of the solid CB7- Mo$_7$O$_{24}{}^{6-}$ complex, which is of major practical relevance for the $^{99m}$Tc generator.

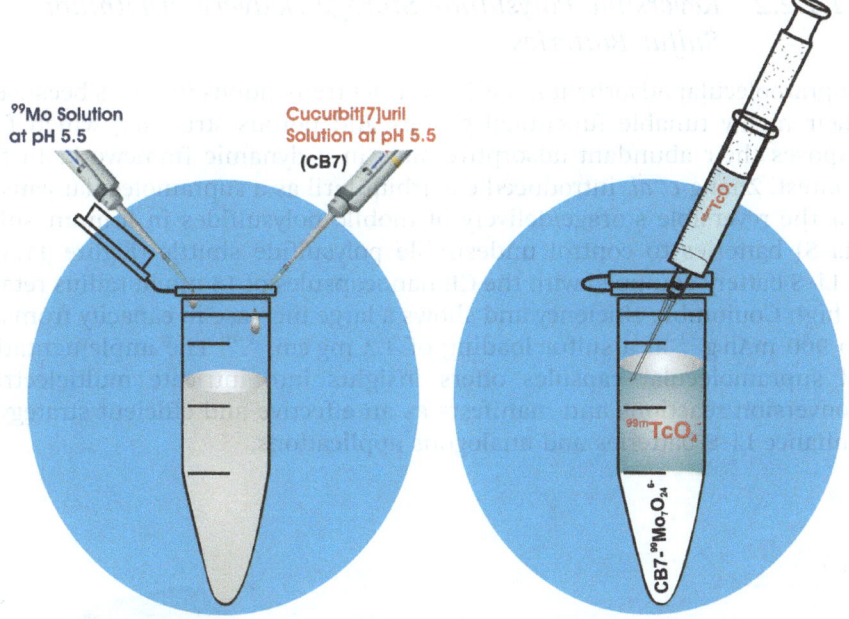

**Figure 11.2** Bench-top demonstration of CB7-mediated $^{99}$Mo/$^{99m}$Tc generator. A laboratory scale arrangement (radioactivity shielding arrangement not shown) to demonstrate the supramolecular strategy for the separation of $^{99m}$Tc from active $^{99}$Mo solution by using CB7-heptamolybdate material as generator bed.
Adapted from ref. 31 with permission from the Royal Society of Chemistry.

Taking advantage of the preceding attributes of the CBs, Fang *et al.* have reported hybrid complexes of cucurbit[6/8]urils (CB6/CB8) and polyoxovanadate as functional porous materials having catalytic, magnetic and recognition properties.[30] This is realized through the interaction of the negatively charged polyoxovanadate with the electron-deficient outer-surface carbons of CB6 or CB8. Thereafter, Han *et al.* reported new hybrid solids of tetravanadate polyanions and CB5/CB6 with a focus on their photocatalytic activity.[31]

On the other hand, developing new, renewable and clean energies has received much attention in recent years. In this regard, the CB7-protected $Cs_{2.5}H_{0.5}PW_{12}O_{40}$ (CsPW-CB7) is prepared as a highly efficient catalyst for direct biodiesel production *via* the transesterification of waste cooking oil. The maximum conversion rate could reach 95.1% under the optimum experimental conditions, which are catalyst of 2 wt%, methanol/oil molar ratio of 11:1, reaction time of 150 min and temperature of 70 °C.[39] In addition, this unique catalyst could be reused more than 5 times with slight deactivation, indicating a desirable recyclability.[39] The results indicated that the CsPW-CB7 catalyst showed good catalytic performance and that its potential application in industrial biodiesel production is excellent.

### 11.1.2.2 Reversible Polysulfide Storage/Delivery in Lithium–Sulfur Batteries

Supramolecular adsorbent materials attract tremendous interests because of their highly tunable functional groups and porous structure, which fully exposes their abundant adsorptive sites in a dynamic framework. In this context, Zhang *et al.* introduced cucurbit[6]uril as a supramolecular capsule for the reversible storage/delivery of mobile polysulfides in lithium–sulfur (Li–S) batteries to control undesirable polysulfide shuttle (Figure 11.3).[40] A Li–S battery equipped with the CB nanocapsules of 15 nm in radius retains a high Coulombic efficiency and shows a large increase in capacity from 300 to 900 mAh g$^{-1}$ at a sulfur loading of 4.2 mg cm$^{-2}$.[40] The implementation of supramolecular capsules offers insights into intricate multielectron-conversion reactions and manifests as an effective and efficient strategy to enhance Li–S batteries and analogous applications.

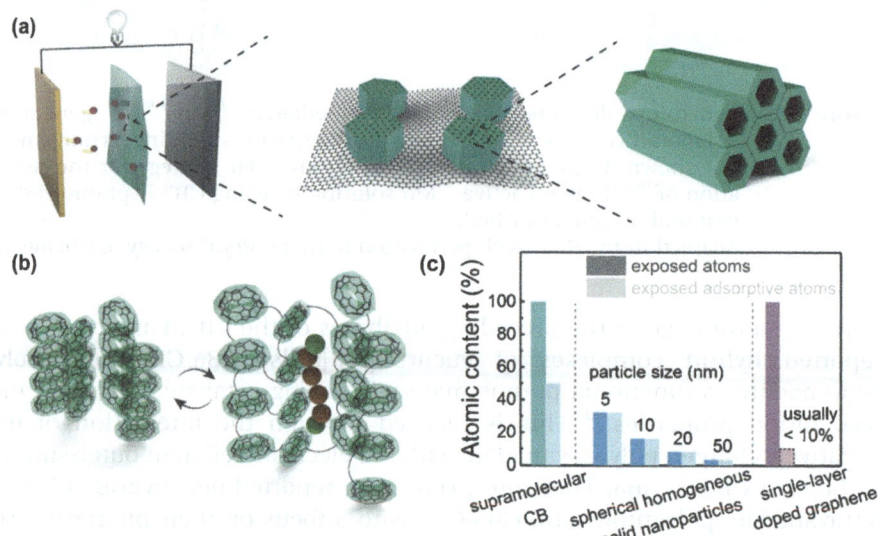

**Figure 11.3**  Conceptual illustration of supramolecular capsules. Schematics for (a) a Li–S battery with a G@CB interlayer to reversibly trap LiPSs (left), CB capsules on graphene substrate (middle), and one-dimensional micropore channels in CB, and for (b) the reversible storage/delivery process of LiPSs in adaptive CB channels. (c) Comparison of exposure ratio of adsorptive atoms between different materials: CB (53% and 50% by weight and atom number, respectively; H atoms are neglected), spherical homogeneous particles without interior exposure (≤33% for 5-nm particles) and doped carbon (exemplified as an ideal single-layer doped graphene) with an atomic doping level usually <10%.
Reproduced from ref. 40 with permission from John Wiley & Sons, Copyright 2017 WILEY-VCH Verlag GmbH & Co. KGaA.

## 11.1.2.3 Water-based Supramolecular Dye Laser

Increased water solubility, enhanced stability and prevention of the aggregation/adsorption of organic dyes in aqueous solution are the crucial parameters for a dye laser system and are nicely demonstrated by the operation of a supramolecularly assisted aqueous dye laser system of rhodamine dyes with CB7 as the macrocyclic host additive.[42,43] The fluorescent dye is quite efficiently protected from water and oxygen, which further reduces intermolecular follow-up (photo)chemistry.[19,43] These combined properties distinguish cucurbiturils from other stabilizing additives and project their widespread applications. The effect of CB7 on the performance of aqueous kiton red (KR) dye solution has been investigated by both broad-band and narrow-band dye laser setups with respect to practically relevant parameters like lasing efficiency, lasing stability and beam quality. A large increase in lasing efficiency is observed with the addition of micromolar concentration of CB7 to the aqueous solution of kiton red dye (Figure 11.4), predominantly

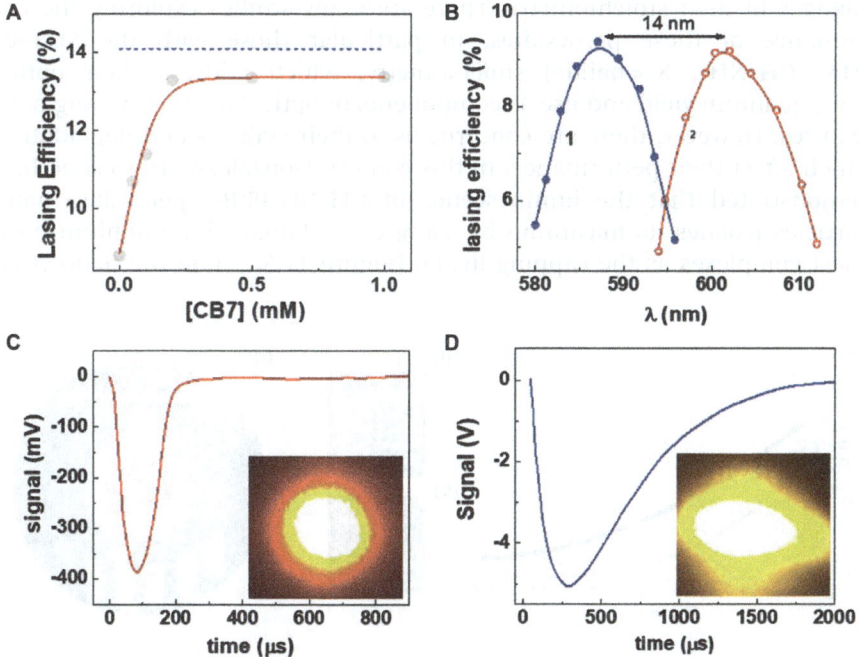

**Figure 11.4** (A) Dependence of the lasing efficiency of KR (200 μM) on CB7 concentration. Shown for comparison (dashed line) is the lasing efficiencies of KR in optically matched ethanol solutions. (B) Tuning curves of KR in EtOH (1) and in the presence of 200 μM CB7 (2) at a pump energy of 6.3 mJ. Thermal deflection signals of KR in water-CB7 (C) and in EtOH (D) systems. Insets of C and D show the laser beam profiles for the respective mediums.

due to the deaggregating action of CB7 on the dye.[19] The resulting supramolecular dye lasers are environmentally more benign, more laboratory-safe and less maintenance-sensitive than the presently employed dye laser systems based on organic solvents. The novel aqueous dye-CB7 systems are not only equally efficient with complementary tuning ranges but also possess superior thermo-optic characteristics and unmatched beam profiles (Figure 11.4), which would jointly enable new and revived photonic applications based on such dye lasers.[19,43]

### 11.1.2.4 Enhanced Luminescence and Photostability of CH₃NH₃PbBr₃ Perovskite Nanoparticles and Organic Chromophores

*11.1.2.4* *Enhanced Luminescence and Photostability of CH$_3$NH$_3$PbBr$_3$ Perovskite Nanoparticles and Organic Chromophores*

Hybrid organic–inorganic lead halide perovskites are of great interest in photovoltaic devices and as luminescent materials for light-emitting devices. Their photoluminescence (PL) spectrum can be modified *via* controlled changes in their stoichiometry. There are many studies exploring the performance of these perovskites, in particular those with the MAPbX$_3$ (MA = CH$_3$NH$_3$, X = halide) stoichiometry, which address their optical gain, quantum yield and use as components of optical devices, among other features. However, there are concerns as to their stability and degradation, which affect their performance. In this context, Gonzalez-Carrero *et al.* have demonstrated that the luminescence of CH$_3$NH$_3$PbBr$_3$ perovskite nanoparticles reaches its maximum by using CB7-adamantyl ammonium host–guest complexes as the capping ligand (Figure 11.5a–c) on the nanocrystal

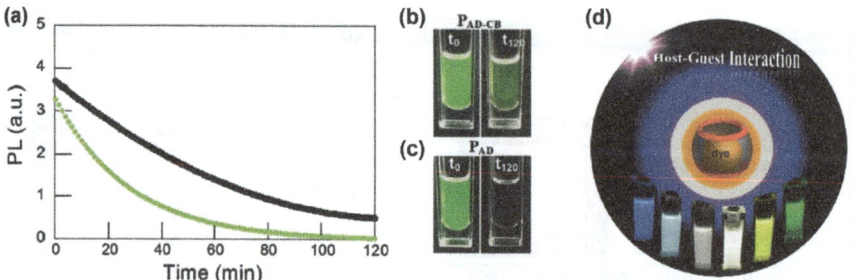

**Figure 11.5** (a) Photoluminescence of P$_{AD–CB}$ (in black) and P$_{AD}$ (in green) dispersed in toluene and in contact with water as a function of the irradiation time; $\lambda_{exc}$ = 350 nm, PL registered at 520 nm under air atmosphere; (b, c) P$_{AD–CB}$ and P$_{AD}$ colloidal dispersions immediately after the addition (left) of water and 120 min afterward (right). Reproduced from ref. 41 with permission from John Wiley & Sons, Copyright 2016 WILEY-VCH Verlag GmbH & Co. KGaA. (d) Schematic representation of cucurbit[8]uril-based tunable luminescent materials in aqueous solution. Adapted with permission from ref. 44 with permission from American Chemical Society, Copyright 2016.

surface.[41] These nanoparticles exhibit astonishing photostability and strong luminescence, having a quantum yield of ~100%.[41]

On the other hand, light-emitting organic materials with tunable properties offer fascinating applications in optoelectronic devices, fluorescent sensors and imaging agents. A facile CB8-based supramolecular approach that greatly decreases the number of required synthetic steps and produces smart luminescent materials in aqueous solution with tunable and dynamical photophysical properties has been developed by Ni *et al.* Because of the peculiar electronic distributions of the chromophore guest (4,4′-[(1E,1′E)-1,4-phenylenebis(ethene-2,1-iyl)]bis(1-carboxyethylpyridinium) bromide) within the rigid hydrophobic cavity of the CB8 host, the color tuning of emissions such as cyan, yellow, green and white light is achieved with better efficiency (Figure 11.5d).[44] Furthermore, the host–guest interaction, which triggers ratiometric fluorescence responses between the blue and yellow emissions, provides evidence that may be used for obtaining pure white light emission by controlling supramolecular assemblies in a cost-effective approach.[44]

### 11.1.2.5 Elusive Excimer Emission from p-Dimethylaminobenzonitrile with Cucurbit[8]uril

The dual emission of *p*-dimethylaminobenzonitrile (DMABN) has fascinated photochemists for decades, but it is through the addition of CB8 in water that its photophysics becomes truly multiemissive because the host templates the dimer to allow the observation of the long sought for excimer emission.[34] The observed excimer emission band (third emission band) from *p*-dimethylaminobenzonitrile, encapsulated inside the CB8 cavity as a head-to-tail dimer template, constitutes a completely new photophysical aspect for this famous chromophore. The properties of the excimer emission are intriguing because it disappears in favor of monomer emission when a smaller host (CB7) is employed, when a slightly sterically modified chromophore is used (*p*-diethylaminobenzonitrile : DEABN) and when the temperature is increased slightly (to 60 °C) (Figure 11.6A).[34] This tunable multiemissive behavior makes DMABN complexes with CB7 and CB8 suitable for applications in differential and ratiometric sensing, including its use as a sensitive optical supramolecular thermometer in the ambient temperature range, the construction of logic gates *etc.*

Supramolecular nanoparticles (SNPs) encompass multiple copies of different building blocks brought together by specific noncovalent interactions. The inherently multivalent nature of these systems allows the control of their size as well as their assembly and disassembly, thus promising potential as biomedical delivery vehicles. Huskens *et al.* have demonstrated dual-responsive SNPs based on the ternary host–guest complexation between CB8, a methyl viologen (MV) polymer, and mono- and multivalent azobenzene (azo)-functionalized molecules (Figure 11.6B,C).[45] UV switching

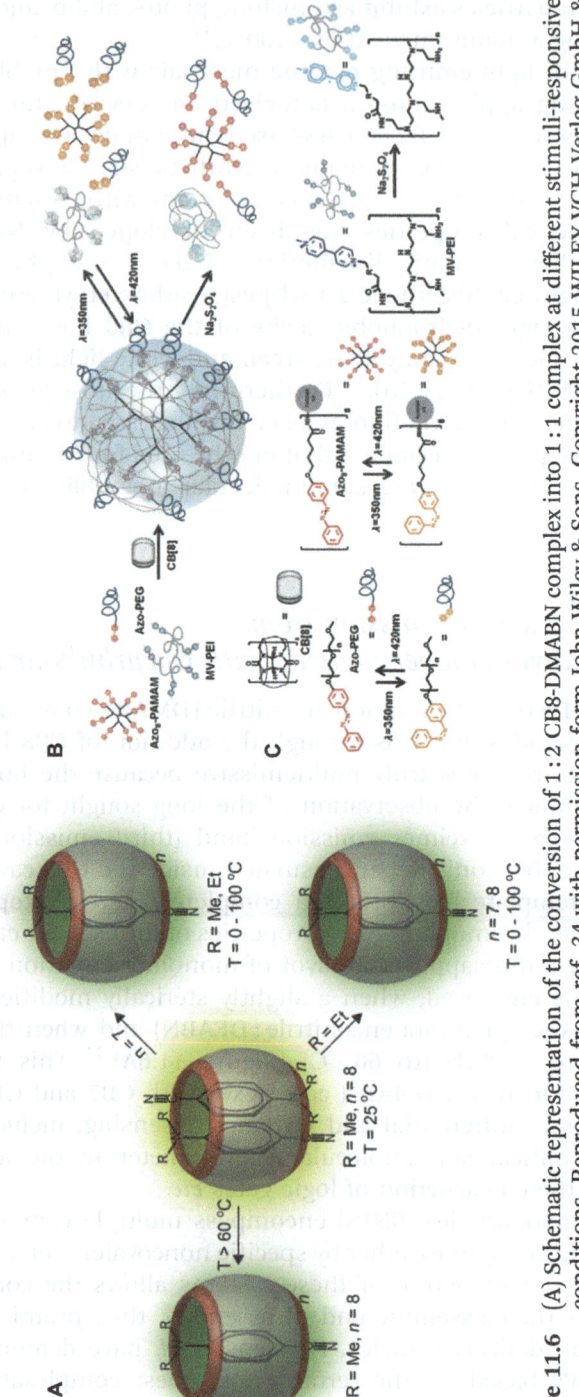

**Figure 11.6** (A) Schematic representation of the conversion of 1 : 2 CB8-DMABN complex into 1 : 1 complex at different stimuli-responsive conditions. Reproduced from ref. 34 with permission form John Wiley & Sons, Copyright 2015 WILEY-VCH Verlag GmbH & Co. KGaA. (B) Schematic presentation of the supramolecular nanoparticle (SNP) self-assembly and triggered disassembly mediated by the formation and disruption of the ternary complex formed between cucurbit[8]uril (CB[8]), methylviologen (MV) and azobenzene (azo) moieties. (C) The supramolecular host (CB8) and responsive guest molecules involved in SNP formation: azobenzene-functionalized poly(ethylene glycol)(azo-PEG), methyl viologen-functionalized poly(ethylene imine) (MV-PEI), and azo$_8$-PAMAM). Reproduced from ref. 45 with permission from John Wiley & Sons, Copyright 2014 WILEY-VCH Verlag GmbH & Co. KGaA.

of the azo groups led to the fast disruption of the ternary complexes but to a relatively slow disintegration of the SNPs. Alternating UV and Vis photo-isomerization of the azo groups led to fully reversible SNP disassembly and reassembly (Figure 11.6B).[45]

### 11.1.2.6 Cucurbit[8]uril-regulated Nanopatterning of Binary Polymer Brushes

Scherman *et al.* have developed a facile and lithography-free approach to creating topologically and chemically defined polymer surfaces under mild and ambient conditions by combining the techniques of monolayer colloidal crystal (MCC)–templated patterning, host–guest complexation and orthogonally controlled supramolecular self-assembly (Figure 11.7).[46] Dual-composition nanopatterned brushes have been prepared by employing CB8-rotaxanes as supramolecular linking agents on gold surfaces so as to controllably "stick" *trans*-azo–functionalized silica colloids onto the surface in a hexagonal arrangement, which serves as a template for the assembly of short polymer brushes (Figure 11.7).[46] The formed MCC is photoresponsive and can be reversibly disassembled upon UV light irradiation to create void spaces for the grafting of a second, longer polymer brush to form a dual pattern on the nanoscale.[46] This facile supramolecular approach provides a platform to prepare nanopatterned composite brushes with sophisticated structures having applications as dual-responsive sensor films and bio-medical coatings.

### 11.1.2.7 Stimuli-responsive Cucurbit[7]uril-mediated BSA and Micellar Assemblies

In a recent report, Mohanty *et al.* has established the construction of a nontoxic nanoassembly of bovine serum albumin (BSA) protein and the CB7 macrocycle, as well as its stimuli-responsive breakage with adamantylamine or change in pH, which restores the protein structure and recognition properties.[47] The assembly showed the efficient loading and controlled re-lease of a standard anticancer drug, doxorubicin (DOX), and the same was validated in live cells (Figure 11.8A).[47] The cell viability studies documented that the DOX-loaded assembly masks the cytotoxicity of DOX and that the toxicity can be revived at the target on demand, triggering its therapeutic activation, and is more effective in cancer cells.[47] In addition, they are highly promising for stabilizing/protecting the native protein structure, a viable approach to prevent/inhibit protein misfolding and aggregation, responsible for several neurodegenerative diseases.

In an earlier study, Bhasikuttan *et al.* demonstrated the dramatic en-hancement of the fluorescence properties of a representative triphenyl-methane (TPM) dyes, *i.e.*, brilliant green (BG) with BSA, by the addition of the macrocyclic host molecule CB7.[8] The resulting cooperative binding of a

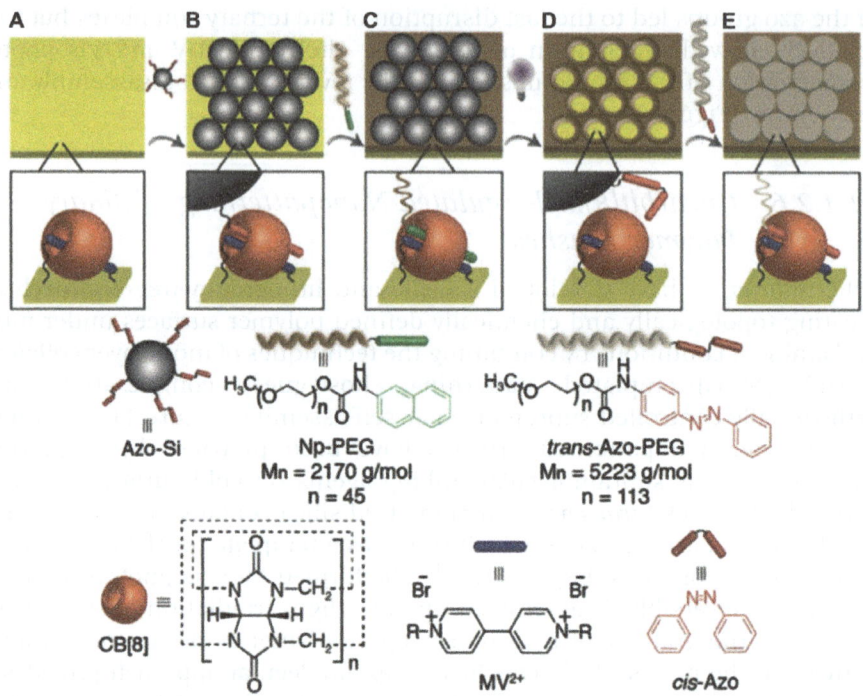

**Figure 11.7** Stepwise preparation of nanopatterned dual polymer brushes *via* the reversible host–guest complexation of CB8. (A) Assembly of azo-Si colloids on the CB8-rotaxane–functionalized surface through the heteroternary complex formation of (*trans*-azo-MV$^{2+}$) : CB8. (B) Preparation of Np-PEG$_{2k}$ brushes on the void spaces around azo-Si colloids on the surface *via* the incorporation of Np functionalities into the CB8-rotaxane cavities. (C) Removal of the azo-Si colloids from the surface by applying UV light (350 nm, 1 min), with the *trans*-Azo photoisomerized to *cis*-azo and being expelled out of the CB8 cavity. (D) Grafting of the azo-PEG2k brushes on the empty spaces vacated by azo-Si colloids. Reproduced from ref. 46, https://doi.org/10.1002/adma.201503844, under the terms of the CC BY 4.0 licence, https://creativecommons.org/licenses/by/4.0/.

dye and potential drug with biological target molecules opens up a new approach to improve medicinal activity or the sensitivity of fluorescent sensor applications by a supramolecular enhancer strategy. Furthermore, the demonstration of the salt-induced p$K_a$ tuning of the supramolecular assemblies has led to the relocation of the guest from the cucurbituril cavity into the biomolecular pocket,[10] indicating potential application in drug delivery.

For regulating the cytotoxicity and improving the efficiency of antitumor drugs, Zhang *et al.* developed a supramolecular chemotherapy by utilizing the clinical antitumor drug, oxaliplatin, which is the specific drug for colorectal cancer treatment.[48] Cytotoxicity of oxaliplatin to the colorectal

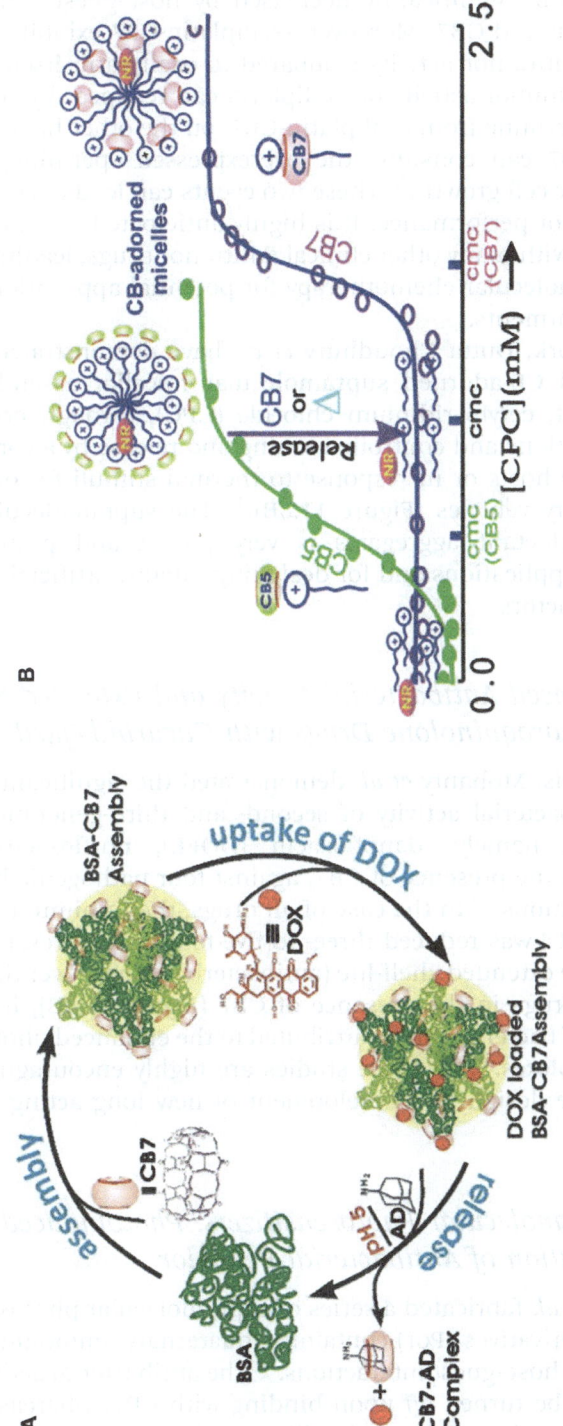

**Figure 11.8** (A) Schematic representation of the formation of CB7-BSA suprabiomolecular assembly, stimuli-responsive uptake and release of DOX. Reproduced from ref. 47 with permission from John Wiley & Sons, Copyright 2017 WILEY-VCH Verlag GmbH & Co. KGaA. (B) Schematic representation of the micellization curve for the CB5- or CB7-adorned CPC surfactant and the release mechanism for the CB5 assembly. Adapted from ref. 49 with permission from the Royal Society of Chemistry.

normal cell could be significantly decreased by host–guest complexation between oxaliplatin and CB7. Moreover, oxaliplatin-CB7 exhibited cooperatively enhanced antitumor activity compared to oxaliplatin itself.[48] On the one hand, the antitumor activity of oxaliplatin can reappear by competitive replacement of spermine from oxaliplatin-CB7; on the other hand, in tumor environments, CB7 can consume the overexpressed spermine, which is essential for tumor cell growth.[48] These two events can lead to cooperatively enhanced antitumor performance. It is highly anticipated that this strategy may be employed with many other clinical antitumor drugs, leading to a new horizon of supramolecular chemotherapy for potential applications in clinical antitumor treatments.

In an earlier work, Dutta Choudhury *et al.* have demonstrated the construction of novel CB-adorned supramolecular micellar assemblies of a cationic surfactant, cetylpyridinium chloride (CPC), through noncovalent host–guest interactions and controlled tuning and release by a combination of the macrocyclic hosts or in response to thermal stimuli for on-demand smart drug delivery vehicles (Figure 11.8B).[49] The supramolecular modulation of the surfactant aggregates is very potent and promising for pharmacological applications and for designing tunable artificial molecular devices or nanoreactors.

### 11.1.2.8 Enhanced Antibacterial Activity and Extended Shelf-life of Fluoroquinolone Drugs with Cucurbit[7]uril

Contributing to this, Mohanty *et al.* demonstrated the significant enhancement in the antibacterial activity of second- and third-generation fluoroquinolone drugs, namely, danofloxacin (DOFL), norfloxacin (NRFL), ofloxacin (OFL), in the presence of CB7, against four pathogenic bacteria at different pH conditions.[50] In the case of all drugs, the minimum inhibitory concentration (MIC) was reduced three- to five-fold in the presence of CB7 (Figure 11.9A). The extended shelf-life (antibacterial activity over time) of the fluoroquinolone drugs in the presence of CB7 (Figure 11.9B), irrespective of the four types of bacteria, can be attributed to the enhanced photostability of their CB7 complexes.[50] All these studies are highly encouraging for the use of CB7 for the design and development of new long acting antibiotic formulations.

### 11.1.2.9 Supramolecular Photosensitizers: Photoinduced Activation of Antibacterial Behavior

Recently, Zhang *et al.* fabricated a series of supramolecular photosensitizers from porphyrin derivatives (Por) containing quaternary ammonium groups with CB7 based on host–guest interactions.[51] The antibacterial activity of Por in the dark could be turned *off* upon binding with CB7, whereas the antibacterial activity under white light illumination could be turned *on*

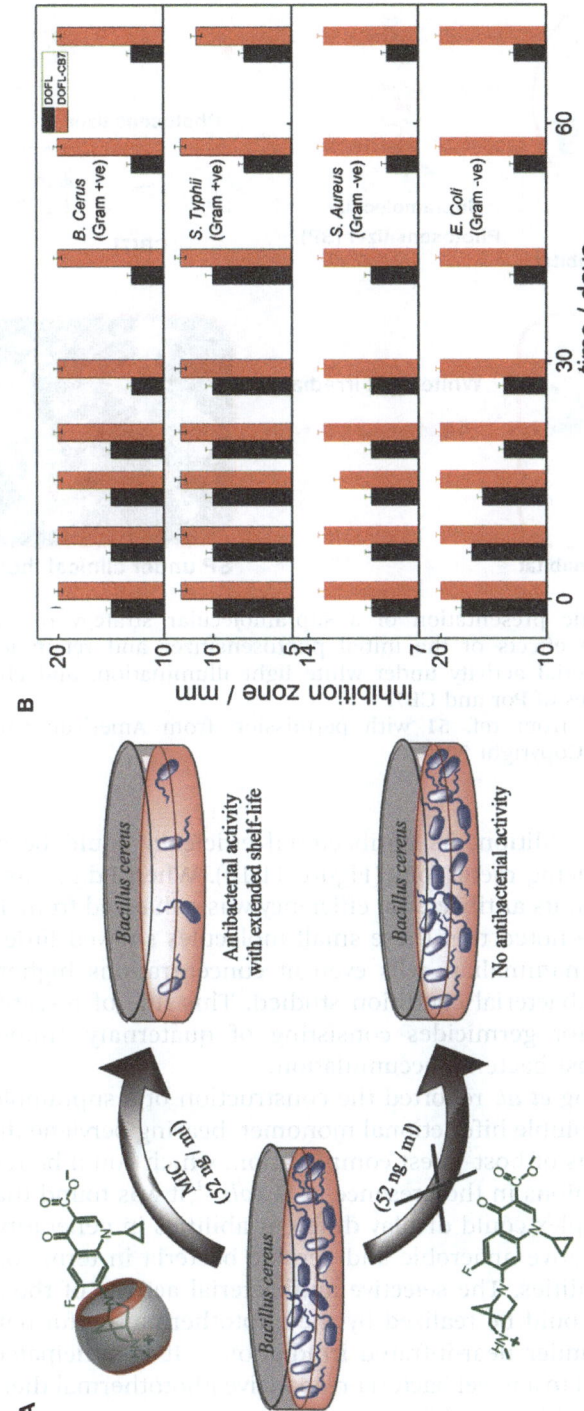

**Figure 11.9** (A) Schematic representation of antibacterial activity of DOFL (52 ng ml$^{-1}$) in the absence and presence of CB7. (B) Bar chart representation of antibacterial activity (in terms of inhibition zone) of DOFL in the absence (black bar) and presence (red bar) of CB7 with time against four bacteria at pH 7.5.

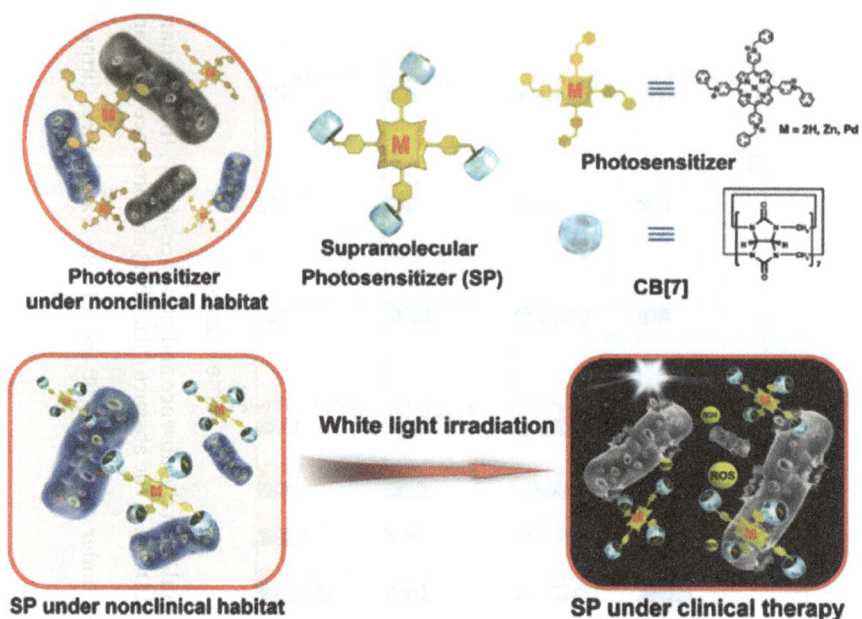

**Figure 11.10**    Schematic presentation of a supramolecular strategy to turn off
                    the side effects of the initial photosensitizer and retain its high
                    antibacterial activity under white light illumination, and chemical
                    structures of Por and CB7.
                    Adapted from ref. 51 with permission from American Chemical
                    Society, Copyright 2017.

(Figure 11.10).[51] In addition, its antibacterial efficiency could be greatly
enhanced by introducing metal ions (Figure 11.10). When Pd(II) was introduced into porphyrin, its antibacterial efficiency was enhanced from 40% to
100%.[51] It should be noted that these small molecules showed little to no
cytotoxicity toward mammalian cells even at concentrations higher than
those under the antibacterial condition studied. This line of research will
provide a strategy for germicides consisting of quaternary ammonium
groups to fight against bacterial accumulation.

Furthermore, Zhang *et al.* reported the construction of a supramolecular
complex of a water-soluble bifunctional monomer–bearing perylene diimide
with CB7 on the basis of host–guest complexation, which could be reduced
to forming radical anions in the presence of *E. coli*.[54] It was found that this
supramolecular complex could display different abilities in generating radical anions by facultative anaerobic and aerobic bacteria in terms of their
various reductive abilities. The selective antibacterial activity of the supramolecular complex could be realized by the photothermal performance of
the radical anions under near-infrared irradiation.[57] It is anticipated that
this method may lead to a novel bacteria-responsive photothermal therapy to
regulate the balance of bacterial flora.

## 11.1.2.10 Cancer Cell Discrimination Through Host–Guest "Doubled" Arrays

Rapid methods for geno- and phenotyping cells are crucial for cancer prognosis and the design of therapeutic strategies for precision medicine. Discrimination between healthy and cancerous cells and then geno-/phenotyping to determine whether the cancer is a slow-growing variant or a highly aggressive form are all-important for optimal treatment. In a recent study, Rotello *et al.* reported a nanosensor that uses cell lysates to rapidly profile the tumorigenicity of cancer cells.[52] This sensing platform uses host–guest interactions between CB7 and the cationic head group of a gold nanoparticle to noncovalently modify the binding of three fluorescent proteins (FPs) of a multichannel sensor *in situ* (Figure 11.11).[52] This approach doubles the number of output channels to six, providing single-well identification of cell lysates with 100% accuracy (Figure 11.11). The unique fingerprint of these cell lysates required minimal sample quantity (200 ng, ~1000 cells), making the methodology compatible with microbiopsy technology and a general means of increasing dimensionality in array-based sensors.[52]

## 11.1.2.11 Supramolecular Assay for Drug Detection in Urine

Colorimetric assays are widely used in life sciences to analyse diverse biochemical functions, from enzyme activity to cytotoxicity, under various conditions including pH, temperature and concentration. Walle *et al.* were able to identify a commercially available dye, neutral red (NR), that is known to bind to CB7 and establish conditions suitable for use in a supramolecular colorimetric assay.[53] The displacement of the dye from the macrocyclic cavity by a therapeutically relevant drug compound, octreotide, was measured, and the correlation between the change in NR absorbance and peptide concentration was established. Through this dependency, Walle *et al.* were able to successfully quantify unknown amounts of the model drug in urine, demonstrating the applicability of this NR:CB7 colorimetric assay for the detection of analytes in biological fluid.[53] It is foreseen that this assay could be applied to the development of a simple point-of-care (PoC) device to monitor biomedically relevant analytes in urine.

## 11.1.2.12 Controllable Capture and Release of Proteins Based on Cationic Perylenebisimides Through Host–Guest Interactions

The self-assembly behavior of the lactose-modified monocationic perylenebisimide derivative PBI-Ion-Lac, the supramolecular host–guest interactions with CB8 and competitive binding with 1-adamantanamine (ADA) have been investigated by Li *et al.*[54] Toward achieving controllable capture and

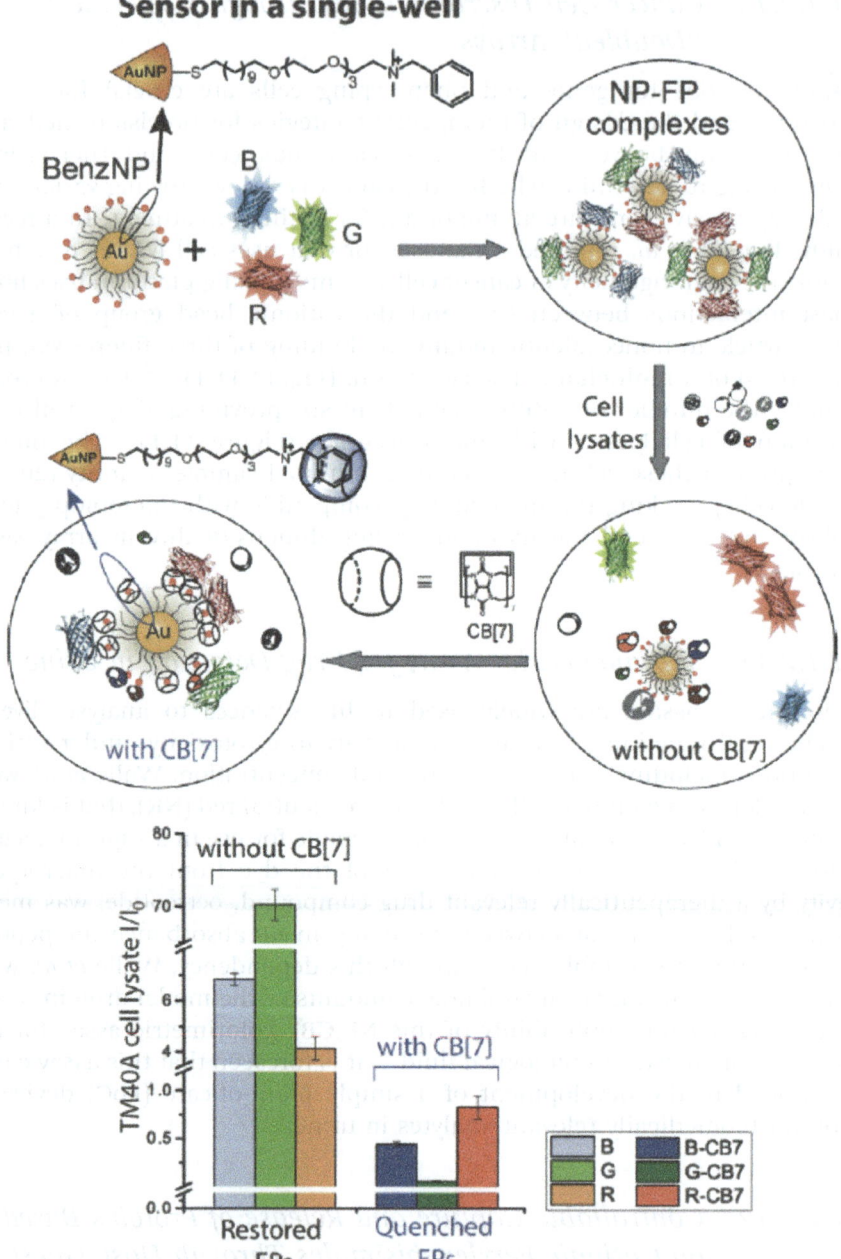

**Figure 11.11**　Six-channel output in a single well. The fluorescence of fluorescent proteins (FPs) is quenched when the benzylammonium-functionalized nanoparticle (BenzNP-FP) complexes are formed. Upon addition of cell lysates, three emission channels are obtained from the released FPs. In the same well, CB7 is added to obtain three additional channels from the three FPs as a result of changed interactions between the analyte and newly formed complex, BenzNP-CB7.

Adapted from ref. 52 with permission from American Chemical Society, Copyright 2017.

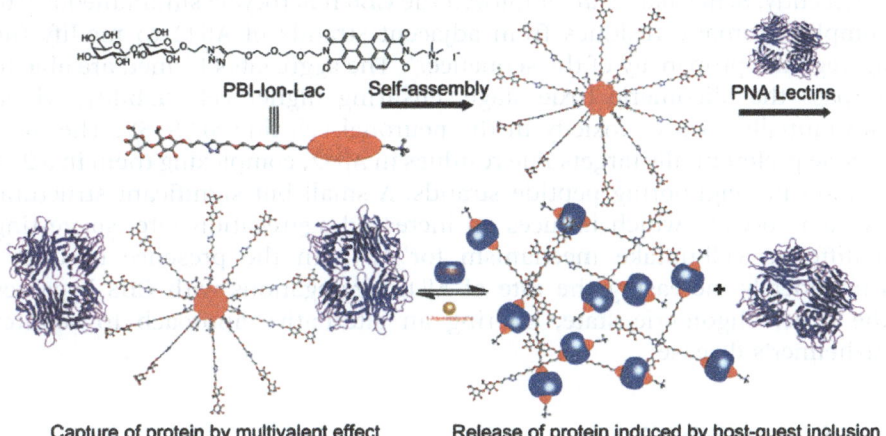

**Figure 11.12** Structural illustrations of the supramolecular host and guest molecules: PBI-Ion-Lac, CB8 and ADA, and the diagrammatic representation of the processes of the capture and release of proteins.
Adapted from ref. 54 with permission from the Royal Society of Chemistry.

release proteins, they have studied the self-assembly of PBI-Ion-Lac–like multivalent glycoclusters showed unique binding interactions with PNA lectins. With further addition of the host molecule CB8, a 1 : 1 supramolecular inclusion complex between PBI-Ion-Lac and CB8 formed, resulting in the self-assembly of PBI-Ion-Lac disaggregates (Figure 11.12).[54] Then, upon adding the competitive guest molecule ADA, the compound PBI-Ion-Lac reformed the supramolecular self-assembly and recovered the binding affinity with PNA lectins (Figure 11.12). Based on the multivalent effects and host–guest properties, a quaternary system for the application of capturing and releasing proteins was achieved.

### 11.1.2.13 Supramolecular Inhibition of Amyloid Fibrillation by Cucurbit[7]uril

Amyloid fibrils are insoluble protein aggregates comprised of highly ordered β-sheet structures, and they are involved in the pathology of amyloidoses, such as Alzheimer's disease. In an earlier study, Kim *et al.* reported a supramolecular strategy for inhibiting amyloid fibrillation by using CB7.[58] CB7 prevents the fibrillation of insulin and β-amyloid by capturing phenylalanine (Phe) residues, which are crucial to the hydrophobic interactions formed during amyloid fibrillation. These results suggest that the Phe-specific binding of CB7 can modulate the intermolecular interaction of amyloid proteins and prevent the transition from monomeric to multimeric states.[58] CB7 thus has potential for the development of a therapeutic strategy for amyloidosis.

Recently, Scherman *et al.* employed the CB8 macrocycle simultaneously to complex aromatic residues from adjacent strands of Aβ42 to modify the aggregation propensity of the sequence.[55] The aggregates formed are able to surpass the oligomeric toxic stage, ensuring higher cell viability, which substantially reduces toxicity in the neuronal cell line SH-SY5Y. The macrocycle preferentially targets Phe residues in Aβ42, complexing them in a 2 : 1 fashion in neighboring peptide strands. A small but significant structural "switch" occurs, which induces an increased aggregation rate, suggesting a different cell-uptake mechanism for Aβ42 in the presence of CB8.[55] Dramatically increasing the rate of Aβ42 aggregations with CB8 bypasses the toxic, oligomeric state, offering an alternative approach to counter Alzheimer's disease.

## 11.2 Conclusion

The design and development of functionalized molecular assemblies constructed through noncovalent host–guest interaction has gained much attention because of their potential applications in energy storage, photonic devices, drug delivery vehicles, catalysts, separation techniques as well as therapeutics. The cucurbituril-encapsulated molecular assemblies of small organic guest molecules exhibit significant modulation in the molecular properties of the guests, which have been exploited to demonstrate various technological and biological applications such as radionuclide separation, highly luminescent and photostable perovskites, improved Li–S batteries, cucurbituril-protected tungstate derivatives for optimized biodiesel production, water-based supramolecular dye lasers, fluorescence-based sensors, molecular capsule, drug delivery systems, antibacterial agents, inhibitors for fibril formation *etc.* The variation of cavity dimensions in the homologues of cucurbituril has revealed many intriguing supramolecular features and their applications as optical thermometer, tunable luminescent materials, stimuli-responsive targeted drug delivery systems, cancer cell/drug detection. The association complexes between the pharmacologically relevant guests and functionalized CB*n* are of particular interest in relation to the improvements of drug stability and solubility and its *on demand* delivery. There is plenty of scope to introduce custom-designed cucurbituril-based functional hybrid materials, which will have different guest binding behavior and would certainly offer new gateways to newer applications.

## Acknowledgements

We acknowledge the coauthors of our published papers cited in the references. We thank Dr. P. D. Naik, associate director, Chemistry Group, BARC, India, for their constant encouragement and support.

# References

1. J.-M. Lehn, *Science*, 1985, **227**, 849.
2. J.-M. Lehn, *Supramolecular Chemistry: Concepts and Perspectives*, VCH, 1995.
3. J. W. Steed and J. L. Atwood, *Supramolecular Chemistry*, Wiley, 2009.
4. E. R. Johnson, S. Keinan, P. Mori-Sánchez, J. Contreras-García, A. J. Cohen and W. Yang, *J. Am. Chem. Soc.*, 2010, **132**, 6498.
5. M. J. Frampton and H. L. Anderson, *Angew. Chem., Int. Ed.*, 2007, **46**, 1028.
6. Y. H. Ko, E. Kim, I. Hwang and K. Kim, *Chem. Commun.*, 2007, 1305.
7. A. V. Davis, R. M. Yeh and K. N. Raymond, *Proc. Natl. Acad. Sci. U. S. A.*, 2002, **99**, 4793.
8. A. C. Bhasikuttan, J. Mohanty, W. M. Nau and H. Pal, *Angew. Chem., Int. Ed.*, 2007, **46**, 4120.
9. S. Dutta Choudhury, J. Mohanty, H. Pal and A. C. Bhasikuttan, *J. Am. Chem. Soc.*, 2010, **132**, 1395.
10. M. Shaikh, J. Mohanty, A. C. Bhasikuttan, V. D. Uzunova, W. M. Nau and H. Pal, *Chem. Commun.*, 2008, 3681.
11. J. Mohanty, N. Thakur, S. Dutta Choudhury, N. Barooah, H. Pal and A. C. Bhasikuttan, *J. Phys. Chem. B*, 2012, **116**, 130.
12. R. Wang, L. Yuan and D. H. Macartney, *Chem. Commun.*, 2005, 5867.
13. J. Mohanty, A. C. Bhasikuttan, W. M. Nau and H. Pal, *J. Phys. Chem. B*, 2006, **110**, 5132.
14. J. Mohanty and W. M. Nau, *Angew. Chem., Int. Ed.*, 2005, **44**, 3750.
15. W. M. Nau and J. Mohanty, *Int. J. Photoenergy*, 2005, 7, 133.
16. S. Dutta Choudhury, J. Mohanty, A. C. Bhasikuttan and H. Pal, *J. Phys. Chem. B*, 2010, **114**, 10717.
17. J. W. Lee, S. Samal, N. Selvapalam, H.-J. Kim and K. Kim, *Acc. Chem. Res.*, 2003, **36**, 621.
18. R. N. Dsouza, U. Pischel and W. M. Nau, *Chem. Rev.*, 2011, **111**, 7941.
19. A. C. Bhasikuttan, H. Pal and J. Mohanty, *Chem. Commun.*, 2011, **47**, 9959.
20. E. Masson, X. Ling, R. Joseph, L. Kyeremeh-Mensah and X. Lu, *RSC Adv.*, 2012, **2**, 1213.
21. J. Lagona, P. Mukhopadhyay, S. Chakrabarti and L. Isaacs, *Angew. Chem., Int. Ed.*, 2005, **44**, 4844.
22. C. Marquez, F. Huang and W. M. Nau, *IEEE Trans. Nanobiosci.*, 2004, **3**, 39.
23. A. Day, A. P. Arnold, R. J. Blanch and B. Snushall, *J. Org. Chem.*, 2001, **66**, 8094.
24. J. Kim, I. S. Jung, S. Y. Kim, E. Lee, J. K. Kang, S. Sakamoto, K. Yamaguchi and K. Kim, *J. Am. Chem. Soc.*, 2000, **122**, 540.
25. X.-J. Cheng, L.-L. Liang, K. Chen, N.-N. Ji, X. Xiao, J.-X. Zhang, Y.-Q. Zhang, S.-F. Xue, Q.-J. Zhu, X.-L. Ni and Z. Tao, *Angew. Chem., Int. Ed.*, 2013, **52**, 7252.

26. S. Liu, C. Ruspic, P. Mukhopadhyay, S. Chakrabarti, P. Y. Zavalij and L. Isaacs, *J. Am. Chem. Soc.*, 2005, **127**, 15959.
27. A. C. Bhasikuttan, S. Dutta Choudhury, H. Pal and J. Mohanty, *Isr. J. Chem.*, 2011, **51**, 634.
28. N. Barooah, J. Mohanty, H. Pal and A. C. Bhasikuttan, *Phys. Chem. Chem. Phys.*, 2011, **13**, 13117.
29. N. Barooah, A. C. Bhasikuttan, V. Sudarsan, S. Dutta Choudhury, H. Pal and J. Mohanty, *Chem. Commun.*, 2011, **47**, 9182.
30. X. Fang, P. Kogerler, L. Isaacs, S. Uchida and N. Mizuno, *J. Am. Chem. Soc.*, 2009, **131**, 432.
31. T. Goel, N. Barooah, M. B. Mallia, A. C. Bhasikuttan and J. Mohanty, *Chem. Commun.*, 2016, **52**, 7306.
32. J. Mohanty, S. Dutta Choudhury, H. P. Upadhyaya, A. C. Bhasikuttan and H. Pal, *Chem. – Eur. J.*, 2009, **15**, 5215.
33. M. Shaikh, S. Dutta Choudhury, J. Mohanty, A. C. Bhasikuttan and H. Pal, *Phys. Chem. Chem. Phys.*, 2010, **12**, 7050.
34. M. Sayed, F. Biedermann, V. D. Uzunova, K. I. Assaf, A. C. Bhasikuttan, H. Pal, W. M. Nau and J. Mohanty, *Chem. – Eur. J.*, 2015, **21**, 691.
35. V. D. Uzunova, C. Cullinane, K. Brix, W. M. Nau and A. I. Day, *Org. Biomol. Chem.*, 2010, **8**, 2037.
36. G. Hettiarachchi, D. Nguyen, J. Wu, D. Lucas, D. Ma, L. Isaacs and V. Briken, *PLoS One*, 2010, **5**, e10514.
37. I. Ghosh and W. M. Nau, *Adv. Drug Delivery Rev.*, 2012, **64**, 764.
38. K. Chen, Y.-S. Kang, Y. Zhao, J.-M. Yang, Y. Lu and W.-Y. Sun, *J. Am. Chem. Soc.*, 2014, **136**, 16744–16747.
39. L. Li, C. Zou, L. Zhou and L. Lin, *Renewable Energy*, 2017, **107**, 14.
40. J. Xie, H.-J. Peng, J.-Q. Huang, W.-T. Xu, X. Chen and Q. Zhang, *Angew. Chem., Int. Ed.*, 2017, **56**, 16223.
41. S. Gonzalez-Carrero, L. Francés-Soriano, M. González-Béjar, S. Agouram, R. E. Galian and J. Pérez-Prieto, *Small*, 2016, **12**, 5245.
42. J. Mohanty, H. Pal, A. K. Ray, S. Kumar and W. M. Nau, *ChemPhysChem*, 2007, **8**, 54.
43. J. Mohanty, K. Jagtap, A. K. Ray, W. M. Nau and H. Pal, *ChemPhysChem*, 2010, **11**, 3333.
44. X.-L. Ni, S. Chen, Y. Yang and Z. Tao, *J. Am. Chem. Soc.*, 2016, **138**, 6177.
45. C. Stoffelen, J. Voskuhl, P. Jonkheijm and J. Huskens, *Angew. Chem., Int. Ed.*, 2014, **53**, 3400.
46. C. Hu, Y. Lan, K. R. West and O. A. Scherman, *Adv. Mater.*, 2015, **27**, 7957.
47. N. Barooah, A. Kunwar, R. Khurana, A. C. Bhasikuttan and J. Mohanty, *Chem. – Asian J.*, 2017, **12**, 122.
48. Y. Chen, Z. Huang, H. Zhao, J.-F. Xu, Z. Sun and X. Zhang, *ACS Appl. Mater. Interfaces*, 2017, **9**, 8602.
49. S. Dutta Choudhury, N. Barooah, V. K. Aswal, H. Pal, A. C. Bhasikuttan and J. Mohanty, *Soft Matter*, 2014, **10**, 3485.

50. H. S. El-Sheshtawy, S. Chatterjee, K. I. Assaf, M. N. Shinde, W. M. Nau and J. Mohanty, *Sci. Rep.*, 2018, **8**, 13925.
51. L. Chen, H. Bai, J.-F. Xu, S. Wang and X. Zhang, *ACS Appl. Mater. Interfaces*, 2017, **9**, 13950.
52. N. D. B. Le, G. Y. Tonga, R. Mout, S.-T. Kim, M. E. Wille, S. Rana, K. A. Dunphy, D. J. Jerry, M. Yazdani, R. Ramanathan, C. M. Rotello and V. M. Rotello, *J. Am. Chem. Soc.*, 2017, **139**, 8008.
53. S. Sonzini, J. A. McCune, P. Ravn, O. A. Scherman and C. F. van der Walle, *Chem. Commun.*, 2017, **53**, 8842.
54. Q. Xu, J.-L. Wang, Y.-L. Luo, J.-J. Li, K.-R. Wang and X.-L. Li, *Chem. Commun.*, 2017, **53**, 2241.
55. S. Sonzini, H. F. Stanyon and O. A. Scherman, *Phys. Chem. Chem. Phys.*, 2017, **19**, 1458.
56. H. H. Lee, T. S. Choi, S. J. C. Lee, J. W. Lee, J. Park, Y. H. Ko, W. J. Kim, K. Kim and H. I. Kim, *Angew. Chem., Int. Ed.*, 2014, **53**, 7461.
57. Y. Yang, P. He, Y. Wang, H. Bai, S. Wang, J.-F. Xu and X. Zhang, *Angew. Chem., Int. Ed.*, 2017, **56**, 16239.
58. H. H. Lee, T. S. Choi, S. J. C. Lee, J. W. Lee, J. Park, Y. H. Ko, W. J. Kim, K. Kim and H. I. Kim, *Angew. Chem., Int. Ed.*, 2014, **53**, 7461.

CHAPTER 12

# Supramolecular Assemblies of Cucurbit[n]urils with Conjugated Polymers and Porphyrins: Effects on Their Photophysical and Photochemical Properties and Their Applications in Photodynamic Therapy

LUCIANO DIBONA-VILLANUEVA, NORY MARIÑO-OCAMPO AND DENIS FUENTEALBA*

Pontificia Universidad Católica de Chile, Facultad de Química y de Farmacia, Laboratorio de Química Biosupramolecular, Vicuña Mackenna 4860, Santiago 7820436, Chile
*Email: dlfuente@uc.cl

## 12.1 Introduction

Supramolecular assemblies formed by noncovalent interactions have great potential for applications in several areas of research, including molecular

Smart Materials No. 36
Cucurbituril-based Functional Materials
Edited by Dönüs Tuncel
© The Royal Society of Chemistry 2020
Published by the Royal Society of Chemistry, www.rsc.org

recognition, self-assembly, metal–organic frameworks, gels, ionic liquids and catalysis, to name a few.[1] Within the many types of supramolecular complexes that can be formed, the inclusion of small molecules within macrocycles[2] has been very relevant in the biomedical fields, for example in the encapsulation of bioactive molecules and drug delivery applications.[3] In general, macrocyclic encapsulation changes some the physicochemical properties of the molecules such as solubility, thermal stability, photostability, reactivity *etc.* The modification of the photophysical and photochemical properties of the encapsulated molecules is important for applications involving light, which can include optical switches, sensors and photocatalysts/photosensitizers. In this chapter, we focus on two types of photoactive macromolecules, conjugated polymers and porphyrin, and their potential impact in a current medical treatment that uses light to induce abnormal or pathogenic cell death, called photodynamic therapy (PDT).[4–10] PDT combines a photosensitizer, oxygen and light to produce a reactive oxygen species (ROS) that oxidize biological molecules such as DNA, lipids and proteins, leading to cell death.[4,6,9,10] ROS are usually generated from the triplet-excited state of the photosensitizer by two general mechanisms: electron transfer to biomolecules (type I) or energy transfer to molecular oxygen to generate singlet oxygen (type II) (Scheme 12.1). ROS include radical species such as superoxide radical anions and hydroxyl radicals, nonradical species such as hydrogen peroxide and excited species such as

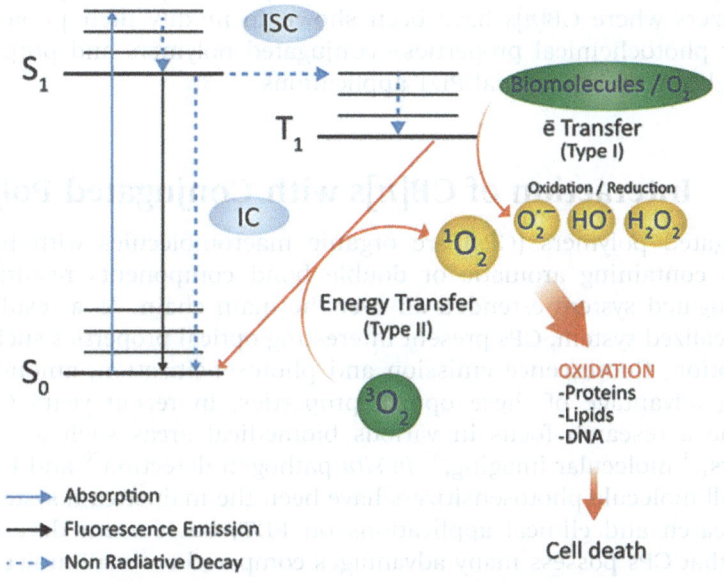

**Scheme 12.1**   Jablonski diagram showing absorption, emission, intersystem crossing and photosensitized reactions (type I and type II).
Reproduced from ref. 22 with permission from John Wiley & Sons, © 2018 WILEY-VCH Verlag GmbH & Co. KGaA, Weinheim.

singlet oxygen.[6] The latter is a long-lived oxidizing species that reacts easily with several amino acids in proteins,[11] unsaturated lipids[12] and DNA bases,[13] and the accumulation of these damages within living cells induces their death. This approach can be used to fight cancer cells, as well as bacteria, fungi, parasites and virus.[4,9,10,14–16] In a clinical setup, it is preferable that the photosensitizer possesses long-wavelength absorption, hopefully in the red or near-infrared (NIR) regions of the electromagnetic spectrum for better penetration of light through the tissues.[4,6,8–10]

There are several approaches to improve the current outcomes of PDT. In this context, supramolecular complexes have shown great potential to prevent aggregation of the photosensitizers and boost the generation of ROS. One family of macrocycles that has garnered a lot of interest in several areas of research over the last two decades is that of the cucurbit[n]urils (CB[n]s).[17–21] Focusing on PDT applications, these macrocycles have important effects in the photophysical and photochemical properties of photosensitizers.[22] In general, the fluorescence emission intensity increases as nonradiative deactivation pathways are restricted in the inclusion complexes.[23–25] This lengthens the lifetime of the singlet-excited state,[24–26] as well as the triplet-excited state.[27–29] As a consequence, singlet oxygen generation is also modified, decreasing in some cases[30] while increasing in others.[28,31,32] Additionally, photoinduced electron transfer from the excited state is also modified depending on the type of CB[n].[28,33] All these properties make CB[n]s very interesting for PDT applications, which have been recently reviewed.[22] Here we will discuss specifically two types of photosensitizers where CB[n]s have been shown to modify their photophysical and/or photochemical properties—conjugated polymers and porphyrins—highlighting their potential PDT applications.

# 12.2   Interaction of CB[n]s with Conjugated Polymers

Conjugated polymers (CPs) are organic macromolecules with backbone chains containing aromatic or double bond components resulting in a π-conjugated system extended all over the main chain. As a result of this π-delocalized system, CPs present interesting optical properties such as NIR absorption, fluorescence emission and photosensitization, among others. Taking advantage of these optical properties, in recent years CPs have become a research focus in various biomedical areas such as chemical sensors,[34] molecular imaging,[35] *in situ* pathogen detection[36] and PDT.[37]

Small-molecule photosensitizers have been the mainstream systems used in research and clinical applications on PDT, but current developments show that CPs possess many advantages compared to small-molecule photosensitizers. For example, CPs show a better absorption cross section, prominent enhanced permeation and retention effect (EPR) and better and easier functionalization that does not necessarily affect the photophysical properties of the chromophore itself.[37]

The photosensitizing mechanisms of CPs can be divided into two types: direct and indirect sensitization. In direct sensitization, the CP conjugated chain is excited by incident light, and the CP acts as the photosensitizer, transferring energy directly to molecular oxygen to generate singlet oxygen through a series of photophysical processes summarized on Scheme 12.2. Indirect sensitization occurs when the light-excited CP transfers energy to a different photosensitizer, which is the molecule that interacts with molecular oxygen to produce an enhanced oxygen sensitization.[37]

As previously mentioned, the optimal absorption wavelength of a photosensitizer must be in the red or NIR regions of the spectrum. In recent years, the research and development of CPs that absorb in the NIR region have grown considerably. Some of the structures that show NIR absorption are polymers based on benzathiadiazole, diketopyrrole and bodipy monomers.[37]

The efficiency of CPs on these applications highly depends on the optical properties of the polymer. In this context, some general problems with CPs can be listed. (1) The conjugated main chain of CPs is usually hydrophobic, causing a poor water solubility. Although this problem can be partially solved by adding hydrophilic or ionic side chains to the polymer structure,[38] the large hydrophobic main chain still limits the water solubility of the CP. (2) Because of the first problem, CPs also present aggregation, interchain

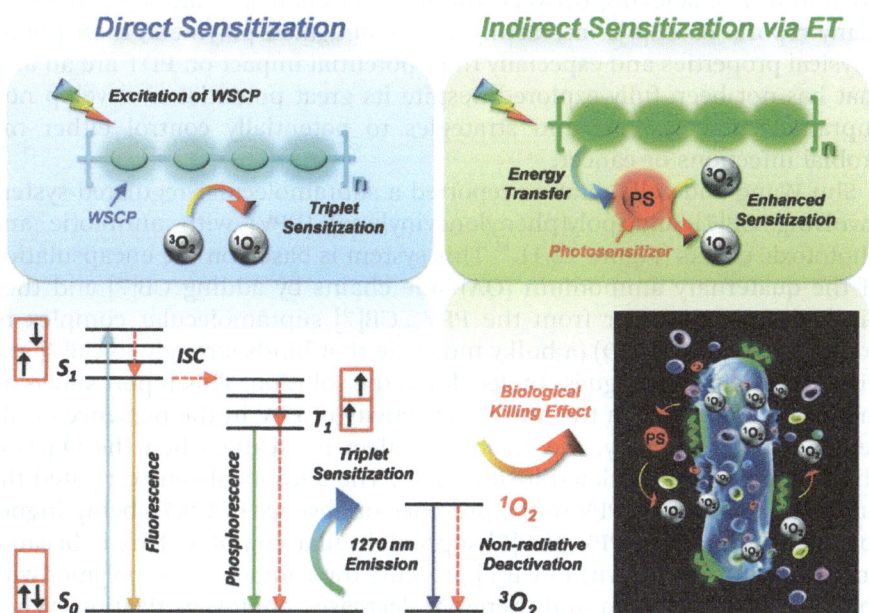

**Scheme 12.2** Two types of mechanisms for singlet oxygen generation by CPs and Jablonski diagram of photosensitization.
Reproduced from ref. 38 with permission from American Chemical Society, Copyright 2012.

interactions and π–π stacking. This could cause a detrimental effect on the photophysical properties of the CP, namely low fluorescence quantum yields,[39] and low singlet oxygen generation. To overcome these problems, various approaches have been studied *e.g.*, hydrophilic pendant groups,[38] use of surfactants[40] and encapsulation of the backbone and side chains of CPs with macrocycles, especially CB[*n*]s.

This last approach has shown very interesting results in terms of the effect of CB[*n*]s inclusion complexes on the photophysical properties of CPs. M. Swager and collaborators reported the formation of a poly-*pseudo*-rotaxane between CB[6], CB[7] and CB[8] onto poly(pyridylvinylene) (PPyV). Notably, the addition of CB[*n*]s to the main chain of PPyV sharply increases the fluorescence intensity of the CP and its solubility and changes the polarity of the microenvironment.[41] It should be noted that the fluorescence intensity and solubility increment depend on the dimensions of the CB[*n*]s portals. According to the authors, CB[7] portal diameter is optimal for binding the CP, showing the largest increase in fluorescence emission.[41]

D. Tuncel and collaborators reported CB[*n*]-threaded fluorene-thiophene–based conjugated polyrotaxanes, observing an enhancement in molar absorptivity, fluorescence quantum yield and fluorescence lifetime for two different conjugated polymers threaded with CB[7] compared with the polymers without the macrocycles. These changes are attributed to the effect of the encapsulation of the polymer backbone by CB[7] and hence the reduction of π–π stacking between the polymer chains.[42] Although there are many reports of CB[*n*]s interacting with conjugated polymers, their photophysical properties and especially their potential impact on PDT are an area that has not been fully explored despite its great potential to develop new supramolecular systems and strategies to potentially control either microbial infections or cancer.

Shu Wang and collaborators reported a supramolecular-regulated system involving CB[7] and poly(phenylenevinylene) (PPV) with antibiotic and phototoxic effects (Figure 12.1).[43] This system is based on the encapsulation of the quaternary ammonium (QA) side chains by adding CB[7] and then displacing the polymer from the PPV@CB[7] supramolecular complex by adding adamantine (AD) (a bulky molecule that binds strongly to CB[7] and displaces the previous guest molecule) to the solution. This report shows an important difference in the antibiotic activity of PPV in the presence or absence of CB[7]; namely, the antibiotic effect is greater when the QA side chains are not encapsulated inside CB[7]. The authors also investigated the production of ROS by PPV in the presence or absence of CB[7], being higher for free PVV than for PPV@CB[7] supramolecular complex. This is because encapsulation of PPV within CB[7] prevents the contact of the polymer with the surrounding oxygen and therefore decreases photosensitization. When AD is added and the CB[7] is displaced from PPV, the generation of ROS is recovered almost completely. These results are validated in *in vitro* antibacterial experiments conducted under white light irradiation, which also show a regulation of the antibacterial behavior of PPV. In the absence of

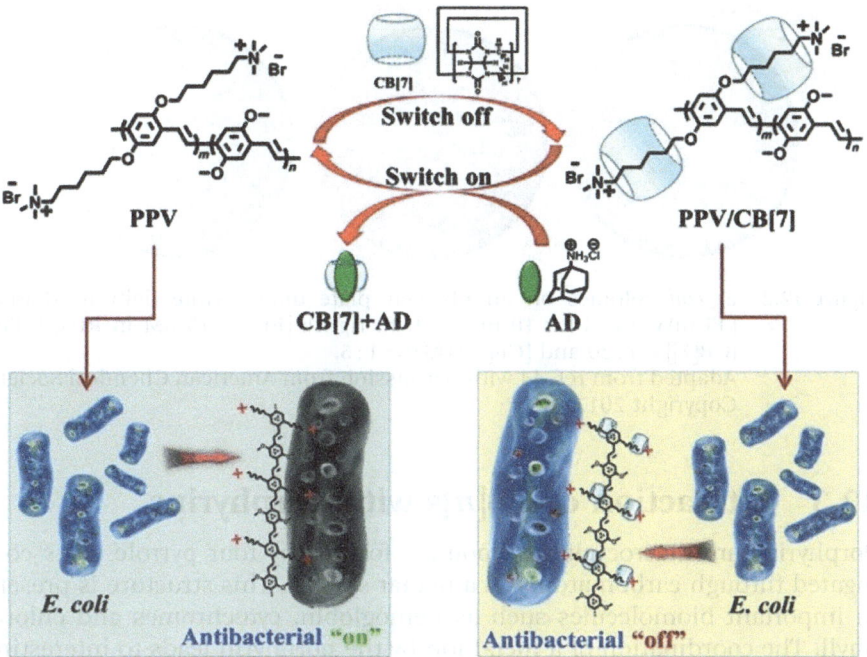

**Figure 12.1** Supramolecular assembly of PPV with CB[7] and disassembly of PPV with CB[7] mediated by AD molecule for reversible control of antibacterial activity of PPV.
Adapted from ref. 43 with permission from John Wiley & Sons.

CB[7], PPV shows a 95% reduction of bacterial growth, whereas the PPV@CB[7] supramolecular system could kill only 40% of the bacteria. Finally, when AD is added, the PPV recovers its antibacterial effect up to 95%.[43]

The same group later introduced a pretreatment step, using triton X as a nonionic surfactant to disaggregate the CP before encapsulation into CB[7].[44] This system is based on poly(fluorene-*co*-phenylene) (PFP), triton X as surfactant and CB[7] and AD to regulate the photoactivity and phototoxicity of the polymer (Figure 12.2). The results show that PFP without pretreatment with triton X is ineffective as an antibiotic system and also presents no regulation of its bactericidal effect when adding CB[7] or, later, AD. This system under white light irradiation and triton X pretreatment shows an effective bactericidal effect and an appreciable regulation effect on both its bactericidal effect and ROS generation.

Current literature on CP@CB[7] systems support the idea that CB[n]s can be used effectively as regulation agents on both antibiotic and phototoxic activities. Innovative formulations could be developed using CPs and CB[n]s, taking advantage of this family of macrocycles in order to enhance solubility and to control the photophysical and photochemical properties of the polymers, with promising applications in antibacterial photodynamic therapy.

**Figure 12.2** *E. coli* colonies on an LB agar plate under white light irradiation (45 mW cm$^{-2}$) for 10 min (left to right): [PFP] = 15 µM in RUs, [PFP]/[CB[7]] = 1 : 20 and [CB[7]]/[AD] = 1 : 5.
Adapted from ref. 44 with permission from American Chemical Society, Copyright 2017.

## 12.3 Interaction of CB[*n*]s with Porphyrins

Porphyrins are macrocyclic compounds formed by four pyrrole units conjugated through carbon atoms in a planar system. This structure is present in important biomolecules such as hemoglobin, cytochromes and chlorophyll. The coordination of a metal ion by the porphyrin leads to interesting electrochemical and photochemical properties. Depending on their structure, porphyrins absorb light in the UV or blue regions of the spectra (Soret band) and far into the red region (Q bands). Their applications range from light-harvesting in photosynthesis (both natural and artificial), solar cells, phototherapies *etc.*[45–48] In general, porphyrins are poorly soluble in aqueous media and aggregate easily, which is why several examples of encapsulation of these molecules within cyclodextrins, calixarenes and CB[*n*]s are reported in the literature.[49] As previously discussed for CPs, porphyrin inclusion complexes with CB[*n*]s show interesting photophysical and photochemical properties that make them potential supramolecular photosensitizers in phototherapies.

The first example of a porphyrin inclusion complex with CB[7] was reported by J. Mohanty, A. C. Bhasikuttan and collaborators: 5,10,15,20-tetrakis(4-*N*-methylpyridyl)porphyrin (TMPyP$^{4+}$) has four pyridinium cation arms, which bind strongly to CB[7], forming a 1 : 4 complex (Figure 12.3). The overall binding constant determined for this stoichiometry was $4.5 \times 10^{19}$ M$^{-4}$, and it was assumed at that time that all CB[7]s bind with the same strength, with a proposed binding affinity of $\sim 8.2 \times 10^4$ M$^{-1}$.[50]

Later on, using fluorescence correlation spectroscopy, the binding process was defined as a stepwise negative cooperativity process, with binding constants for each consecutive CB[7] of $3 \times 10^7$ M$^{-1}$, $1.3 \times 10^5$ M$^{-1}$, $1.6 \times 10^4$ M$^{-1}$ and $2 \times 10^3$ M$^{-1}$. Moreover, the decrease in binding affinity is attributed to faster dissociation rate constants as more macrocycles bind to the porphyrin.[51] The absorption spectra for the complex shows distinct bathochromic shifts of several nanometers for both the Soret and the Q bands.

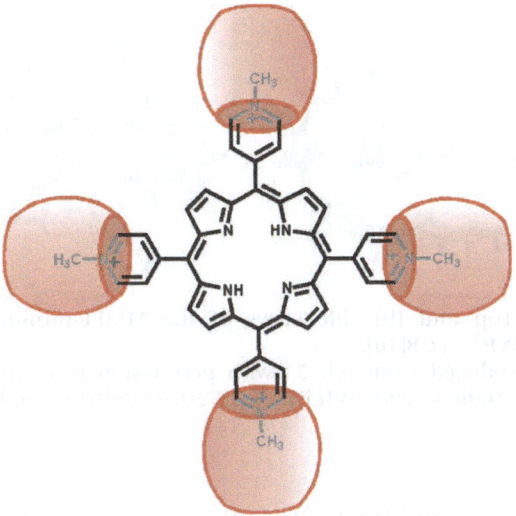

**Figure 12.3** Proposed structure for the 1 : 4 TMPyP$^{4+}$@CB7 complex.
Reproduced from ref. 50 with permission from American Chemical
Society, Copyright 2008.

Fluorescence emission intensity increases, while the emission spectra show
two distinct peaks at 655 nm and 717 nm. Concomitantly, the fluorescence
quantum yield increases, as does the fluorescence lifetime. Such photophysical behavior is likely due to rotational restrictions imposed by the inclusion of the pyridinium arms within CB[7], and it is commonly observed
for other fluorescent molecules.[24,25] It is interesting to note that the triplet-
excited state and singlet oxygen capabilities for this inclusion complex have
not been reported. TMPyP$^{4+}$ has also been described to interact with CB[8]
on the basis of fluorescence measurements; however, no binding constants
have been reported, and the stoichiometry of the complex is not known.[28,32]
For this CB[8] complex, the fluorescence quantum yield does not change
appreciably, but the triplet-excited state lifetime is observed to increase
10-fold. This behavior leads to an increase in singlet oxygen generation,
which is beneficial for phototherapeutic applications.[32]

In spite of its large molecular size, it is possible to encapsulate the entire
porphyrin molecule inside CB[10] (Figure 12.4), which is the largest member
of the CB[n] family, excluding the twisted tQ[13] and tQ[15].[52] Free TMPyP$^{4+}$
or metalated with Zn(II), Fe(III) or Mn(III) has been included inside the cavity
of CB[10] as a 1 : 1 complex, and the macrocycle shows a large ellipsoidal
deformation upon binding to the porphyrin. Furthermore, the complex can
bind a third component, *i.e.*, an aromatic amine, taking advantage of the
preferential inclusion in CB[n]s toward positively charged guests and the
favorable π–π interactions induced by the porphyrin.[53] It is worth
mentioning that the same principles have been recently used to bind an

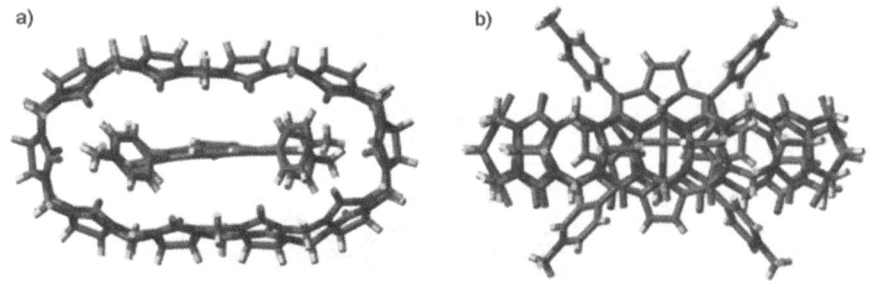

**Figure 12.4**  (a) Top and (b) side views of the MMFF-minimized geometry of TMPyP$^{4+}$@CB[10].
Reproduced from ref. 53 with permission from John Wiley & Sons, Copyright © 2008 WILEY-VCH Verlag GmbH & Co. KGaA, Weinheim.

Mn-porphyrin and an imidazol to CB[10], creating an artificial active site for catalase enzymatic activity.[54] It is also interesting to point out that the photophysical properties of the porphyrin remain unchanged after complexation.[53] This is likely because of the different binding mode of the porphyrin within CB[7], CB[8] and CB[10]. While CB[7] and CB[8] bind directly to the pyridinium arms, restricting their rotation,[32,50] CB[10] binds to the core of the porphyrin, leaving the pyridinium arms free in solution (Figure 12.4).[53] The restriction of the rotation of these arms with the smaller macrocycles likely slows down the nonradiative deactivation pathways for the singlet-excited state, enhancing the fluorescence emission of the molecule.

Most porphyrins reported to interact with CB[n]s are positively charged since this feature favors cation–dipole interactions with the portals of these macrocycles. However, there is also the possibility that anionic porphyrins can self-assemble in the presence of alkaline or alkaline earth metal cations.[55] CB[n]s are known to bind metal cations at the rim, and the strength of the interaction depends on the charge and size of the cation.[56,57] Linking between different CB[n]s is possible due to the formation of exclusion complexes as shown in Figure 12.5.

In other cases, porphyrins modified with cationic side arms are used to build water-soluble rotaxanes or pseudo-rotaxanes based on CB[6] and meso-tetraphenyl porphyrin.[58] Although these types of interactions do not alter the photophysical properties of the porphyrin core, switching behavior induced by pH changes is possible.[58] Supramolecular polymers based on porphyrin–tetramethylcucurbit[6]uril complexes have also been reported.[59,60] Porphyrin-based donor–acceptor dyads have also been synthesized to modify photoinduced electron transfer upon complexation with CB[n]s. An example is the case of porphyrin-viologen dyads where the formation of an inclusion complex with CB[7] notably increases the fluorescence emission of the porphyrin by shifting the reduction potential of the viologen upon

**Figure 12.5**   Cartoon representation of the possible self-assembly manner between $H_2TPPS^{4-}$, $M^{n+}$ and CB[n].
Reproduced from ref. 55 with permission from John Wiley & Sons, Copyright © 2013 WILEY-VCH Verlag GmbH & Co. KGaA, Weinheim.

binding.[61,62] Porphyrin-CB[7] interaction has also been used to restore the luminescence emission of quantum dots. In this case, the interaction of the positively charged porphyrin APTPP with $Cd_{1-x}Zn_xS$ quantum dots quenched the emission by an energy transfer mechanism, which is disrupted after the porphyrin is complexed with CB[7].[63]

As previously discussed, interactions between CB[n]s and porphyrins are very versatile, and the enhancement of their photophysical and photochemical properties upon complexation is amenable to PDT applications.

## 12.4   Porphyrin-based Supramolecular Photosensitizers

Cationic porphyrin derivatives containing naphthalene or benzene units in its four arms have been turned into supramolecular photosensitizers by complexation with CB[7] and tested for their antibacterial photodynamic properties. X. Zhang and collaborators reported the formation of a supramolecular photosensitizer between the naphthalene–methylpyridinium moieties of a synthetic porphyrin (TPOR) with CB[7] that showed enhanced singlet-oxygen generation and potent killing of bacteria upon irradiation.[64] The porphyrins naturally tend to aggregate through π–π stacking, losing their fluorescence emission and decreasing their ability to generate singlet oxygen. However, binding of the CB[7] to the porphyrin arms hinders their aggregation, boosting their fluorescence and singlet-oxygen generation capabilities. Moreover, since these changes in photophysical and photochemical behavior are due to supramolecular interactions, they can be reversed by the addition of a strong competitor such as AD.[64]

More recently, the same group synthesized new quaternary ammonium derivatives of porphyrins that showed selective cytotoxicity and phototoxicity toward bacteria, and both of these properties were switchable upon binding to CB[7] (Figure 12.6).[65] In the dark, the compounds showed significant

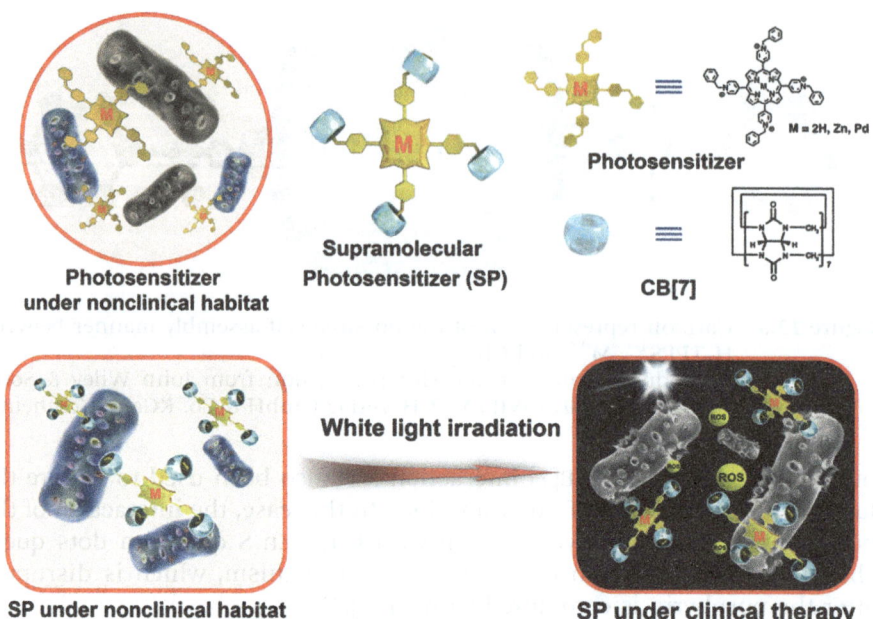

**Figure 12.6**   Schematic presentation of use of a supramolecular strategy to turn off the side effects of the initial photosensitizer and retain its high anti-bacterial activity under white light illumination and chemical structures of Por and CB[7].
Reproduced from ref. 65 with permission from American Chemical Society, Copyright 2017.

cytotoxicity due to their interaction with the negatively charged membrane of the bacteria, while upon binding of CB[7] to the positively charged groups of the molecule, the cytotoxicity was completely inhibited. Irradiation with white light restored their ability to kill bacteria by producing ROS (Figure 12.7). Metalation of the porphyrin by Zn(II) or Pd(II) also enhanced their potency in killing bacteria, and the compounds showed low toxicity toward mammalian cells, making them very promising for clinical applications.[65] It is interesting to note that the Zn-TPOR@CB[7] inclusion complex has also been used as a catalyst for reversible photocontrolled radical polimerization, which further expands the scope of applications for porphyrin-CB[$n$] inclusion complexes.[66]

Recently, the first approach of a covalent conjugation of a trimannosylated-benzoporphyrin directly to monopropargyloxylated CB[7] through click-chemistry was reported (Figure 12.8),[67] which represents a great advance for PDT applications. The porphyrin fragment of the conjugate shows enhanced singlet-oxygen generation compared to the free porphyrin, and the CB[7] fragment was shown to be available for the binding of a model guest; additionally, mannose units can interact with receptors on the surface of bacteria and provide greater solubility for the molecule. In this way, the

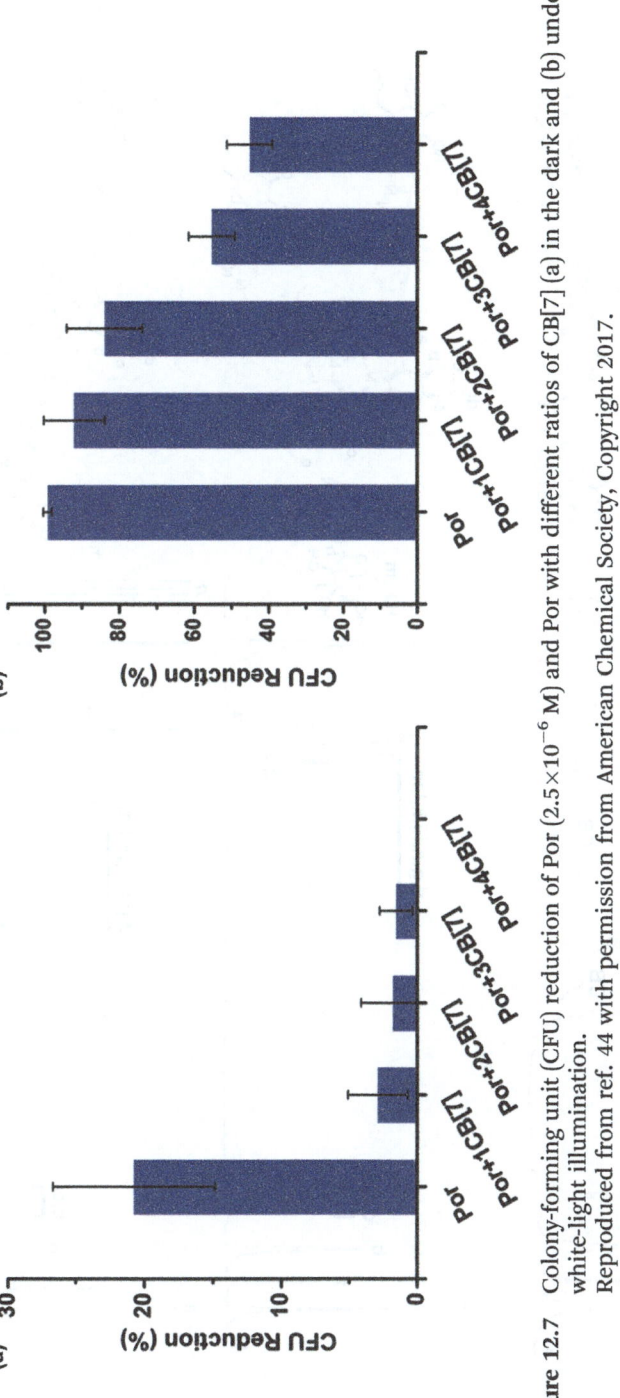

**Figure 12.7** Colony-forming unit (CFU) reduction of Por ($2.5 \times 10^{-6}$ M) and Por with different ratios of CB[7] (a) in the dark and (b) under white-light illumination.
Reproduced from ref. 44 with permission from American Chemical Society, Copyright 2017.

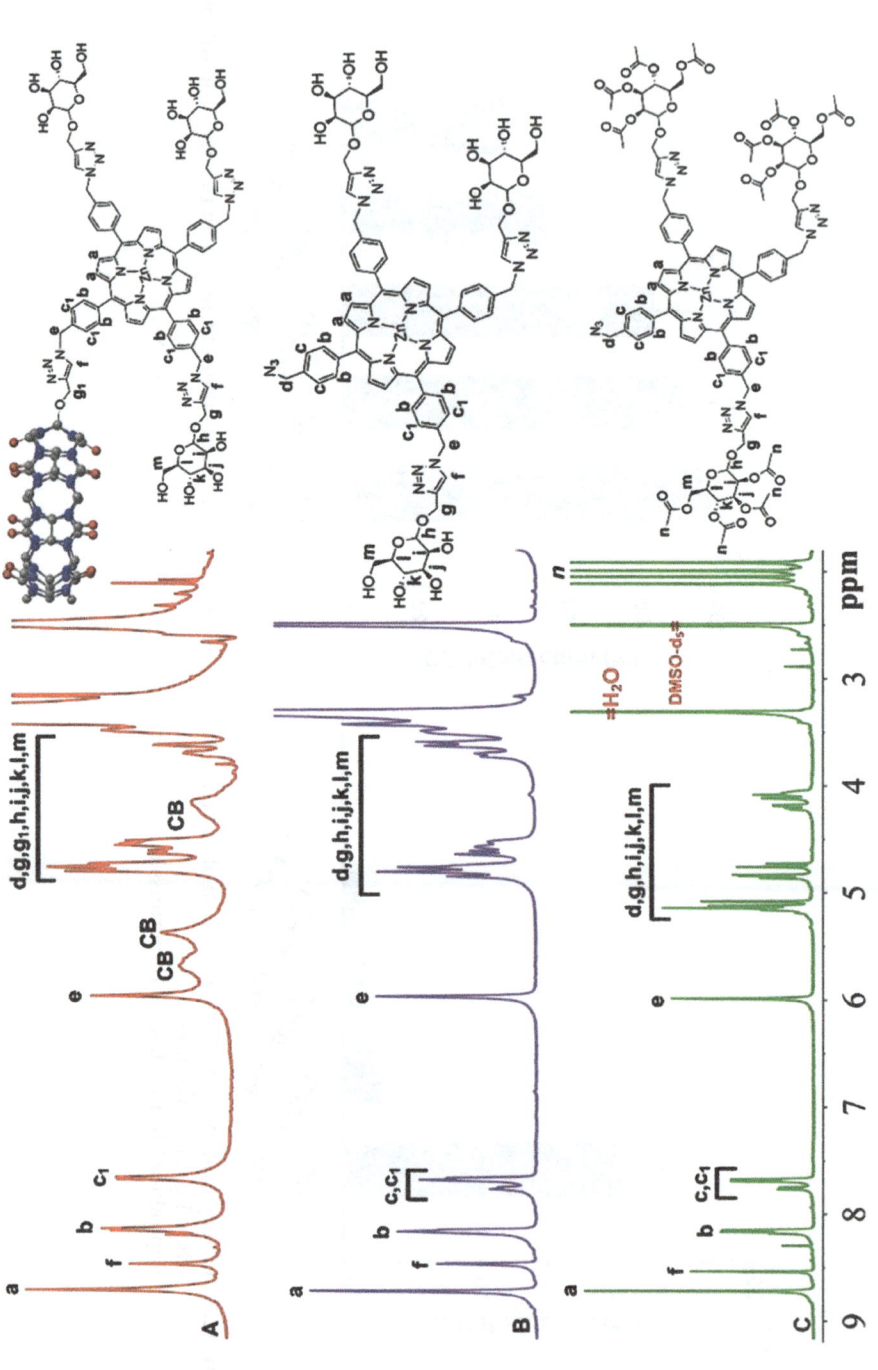

**Figure 12.8** Overlay of the $^1$H NMR (400 MHz, 25 °C) spectra of (A) TPP-3Man-CB7, (B) TPP-Az-3Man and (C) TPP-Az-3AcMan (spectra were recorded in [D$_6$]DMSO). Reproduced from ref. 67 with permission from John Wiley & Sons, © 2018 WILEY-VCH Verlag GmbH & Co. KGaA, Weinheim.

newly synthesized conjugate is expected to have better selectivity, photo-activity and be able to simultaneously transport antibiotics or chemotherapeutic drugs within the CB[7] cavity.[67]

Although cationic porphyrins have been traditionally used with antibacterial purposes, the potential for anticancer applications of these supramolecular photosensitizers is evident, and future work in this area is anticipated.

## 12.5 Porphyrin-containing Hyperbranched Supramolecular Polymers

Supramolecular polymers consist of a series of monomeric units assembled *via* noncovalent and reversible interactions such as hydrogen bonding, π–π interactions and host–guest bonding. This class of supramolecular structures has very interesting properties such as their inherent adaptation, self-healing, recycling and degradability properties. Supramolecular polymers hold great potential for dynamic functional materials applications.[68–70]

In the recent years, a few reports have shown supramolecular polymers using porphyrins as monomeric units. X. Zhang and collaborators reported a naphthyl-substituted porphyrin derivative encapsulated with CB[8] to form a hyperbranched photosensitizer supramolecular polymer. This macromolecular structure is formed by the inclusion of the naphthalene arms within CB[8], as shown on a schematic diagram (Scheme 12.3).[71]

Through ITC experiments, it was determined that TPOR and CB[8] form a strong inclusion complex in a molar ratio of 1 : 2 with a binding constant of $1.6 \times 10^{12}$ M$^{-2}$. This result indicates that the supramolecular polymer is indeed strongly bound together by host–guest interactions. NMR spectroscopy showed upfield shifts for the naphthyl arms, indicating that this is the part

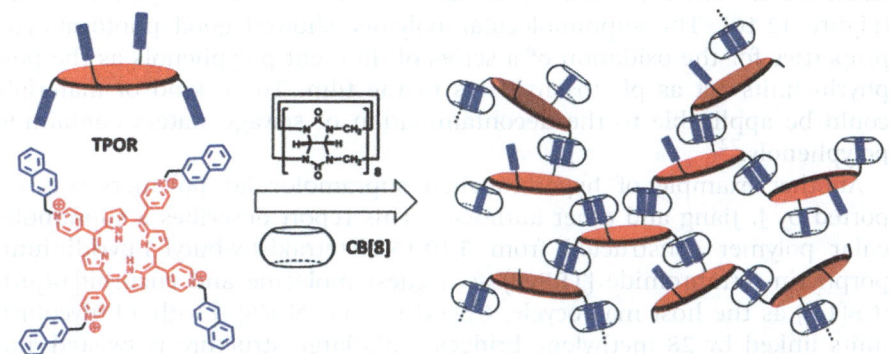

**Scheme 12.3**  Diagram of the supramolecular polymer through self-assembly of TPOR and CB[8].
Reproduced from ref. 71 with permission from the Royal Society of Chemistry.

of the molecule that is encapsulated by CB[8] supporting the structure proposed on Scheme 12.3.[71]

As previously mentioned, porphyrins aggregate in aqueous media, affecting critical photophysical properties of interest for photodynamic therapy applications such as fluorescence intensity and singlet-oxygen generation. Considering that, the authors evaluated the singlet-oxygen generation, absorption and fluorescence emission properties of TPOR and TPOR-2CB[8], and the results are summarized on Figure 12.9. First, the singlet-oxygen generation efficiency was evaluated by electron paramagnetic resonance (EPR) spectroscopy, using 2,2,6,6-tetramethylpiperidine to capture $^1O_2$ to form TEMPO, a paramagnetic probe that shows signals on an EPR analysis (Figure 12.9a). The TPOR-2CB[8] complex generates a stronger EPR signal compared to that of TPOR, in other words, the photosensitizer supramolecular polymer generates more singlet oxygen than the photosensitizer by itself. In terms of kinetics, TPOR-2CB[8] also shows an outstanding behavior generating singlet oxygen much faster compared with TPOR (Figure 12.9b).

The increase in single-oxygen generation can be caused by the disaggregation of the porphyrin units as a result of the inclusion complexes and the subsequent supramolecular polymerization. This phenomenon can be confirmed by the analysis of the absorption and emission spectra of TPOR and TPOR-2CB[8]. Figure 12.9c shows that the addition of CB[8] results in a decrease of absorption around 220 nm due to the encapsulation of the naphthalene moieties in CB[8]. At the same time, the characteristic 425-nm band for TPOR increases, and so does the fluorescence emission. This behavior can be seen as evidence that CB[8] encapsulation causes disaggregation by preventing $\pi$–$\pi$ stacking of the porphyrin cores. Finally, the authors demonstrate the reversible nature of this supramolecular complex by adding AD to the TPOR-2CB[8] solution with great potential for photodynamic therapy.

This complex has also been attached to a naphthalene-modified solid substrate in order to study photocatalysis at the solid–liquid interface (Figure 12.10). The supramolecular polymer showed good photocatalytic properties for the oxidation of a series of different polyphenols as the porphyrin units act as photosensitizers in the film. These kind of materials could be applicable to the decontamination of sewage waters containing polyphenols.[72]

Another example of hyperbranched supramolecular polymers was reported by J. Jiang and other authors.[73] This report describes a supramolecular polymer constructed from 5,10,15,20-tetrakis(N-butyl-4-pyridinium) porphyrin tetrabromide (TBPyP) as a guest molecule and cucurbit[14]uril (CB[14]) as the host macrocycle. CB[14] is a novel CB[n] with 14 glycoluril units linked by 28 methylene bridges. This large structure is twisted and therefore adopts a folded figure-eight-like conformation. As a result of this conformation, the CB[14] can form inclusion complexes with one or two guest molecules due to the two cavities formed.[74]

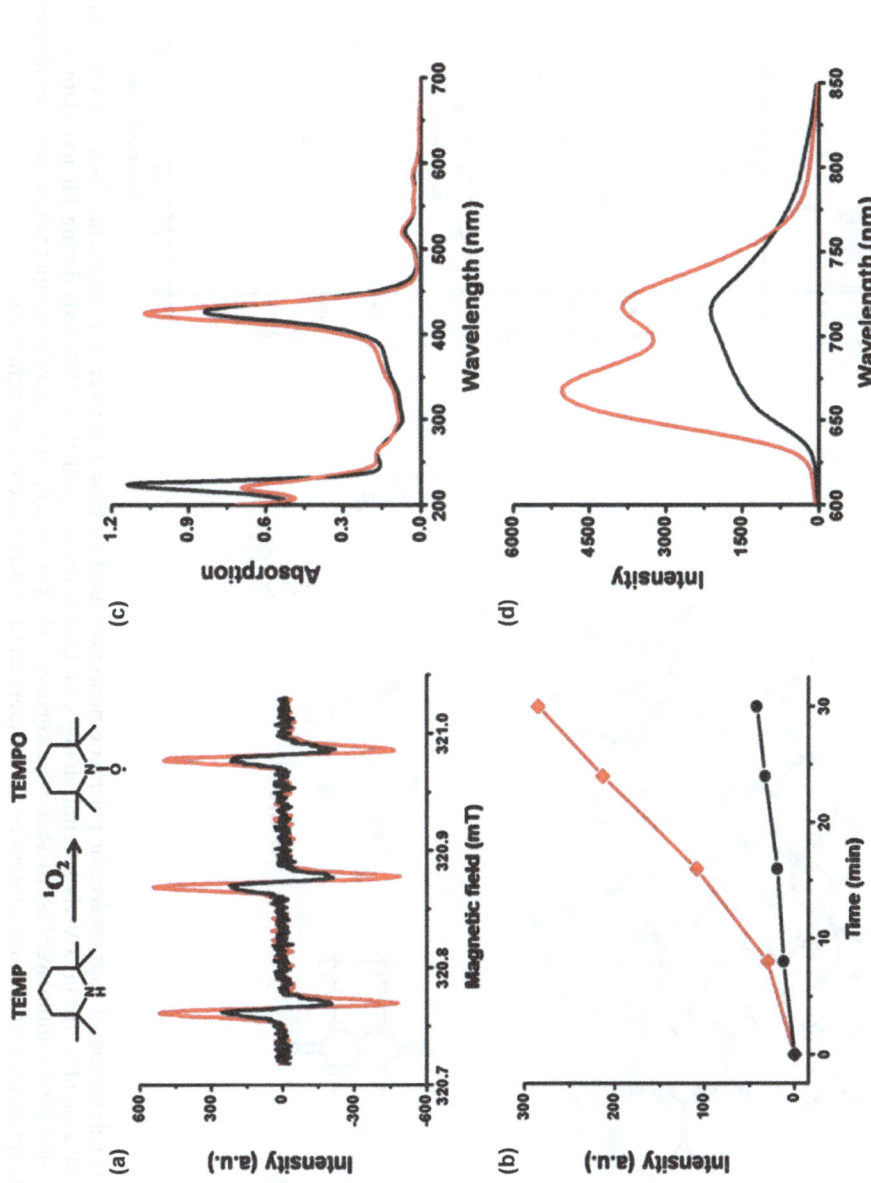

**Figure 12.9**   Photophysical properties of TPOR (black) and TPOR-2CB[8] (red). (a) EPR spectra upon irradiation of 30 min and (b) Change of intensity of EPR peak on different irradiation times. (c) UV–vis Absorption spectra and (d) Fluorescence emission spectra. Adapted from ref. 71 with permission from the Royal Society of Chemistry.

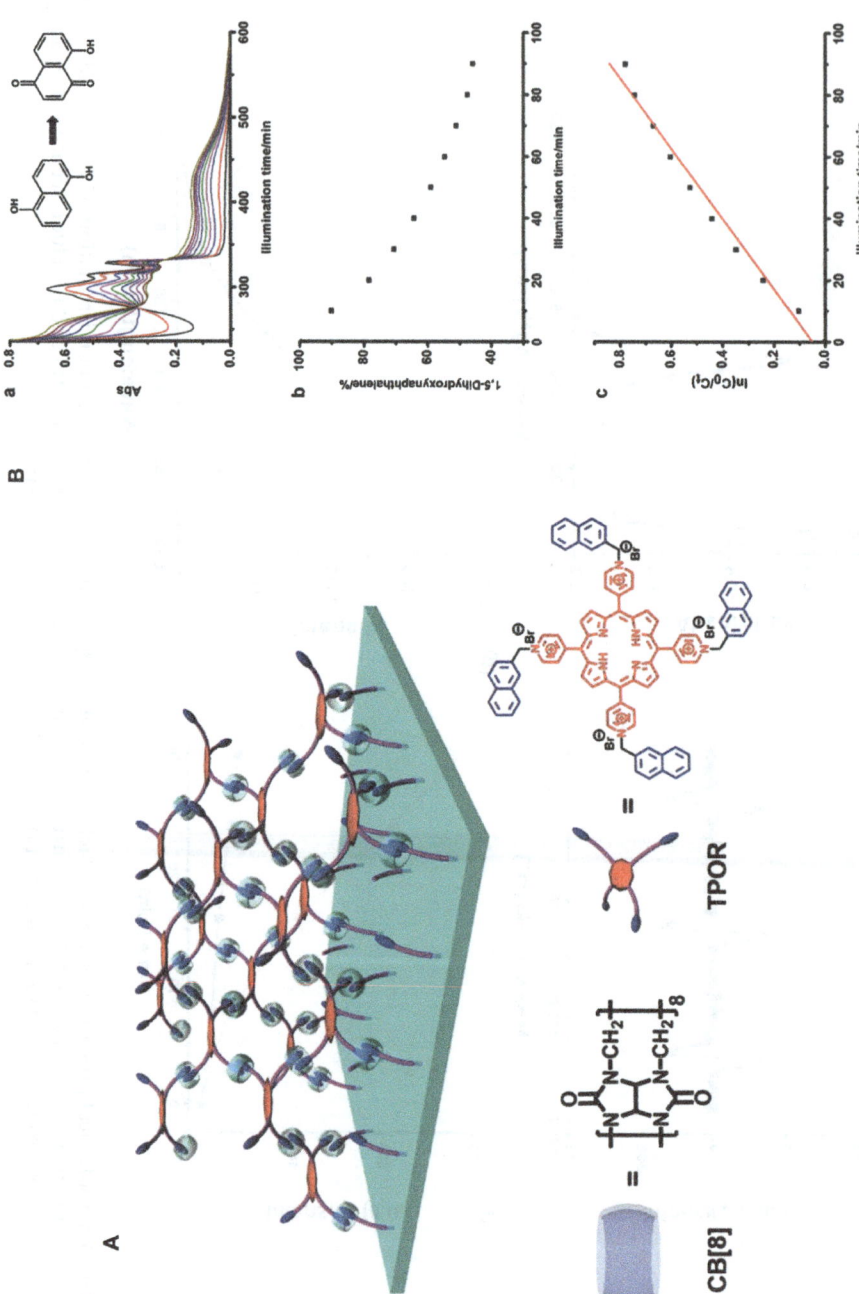

**Figure 12.10** (A) Fabrication of supramolecular polymeric networks based on host–enhanced π–π interaction between CB[8] and TPOR by LbL assembly. (B) (a) UV–vis spectral changes of 1,5-dihydroxynaphthalene (100 μM) during photooxidation catalyzed by a (CB[8]/TPOR)$_5$ multilayer film. (b) Rate of conversion. (c) Plot of ln(C$_0$/C$_t$) of 1,5-dihydroxynaphthalene *versus* the illumination time. Adapted from ref. 72 with permission from American Chemical Society, Copyright 2014.

The inclusion complex and thus the supramolecular polymer are formed by simply mixing CB[14] and TBPyP in aqueous solution. The inclusion complex formation is confirmed by NMR spectroscopy. In this particular supramolecular polymer, an unusual behavior is observed in terms of NMR shifts. The TBPyP alkyl chain experiences an upfield shift, and the proton of the pyridinium moiety of the TBPyP guest labeled as $H_\alpha$ (9.32 ppm) in Figure 12.11a shows an upfield shift after the addition of CB[14], but another pyridinium proton $H_\beta$ (8.94 ppm) shows a downfield shift. These NMR results indicate that both the alkyl chain and the pyridyl moiety of the TBPyP are participating in the inclusion complex but within two different cavity environments of CB[14]. The NMR studies were complemented with molecular modelling using density functional theory (DFT) calculations to provide structural information for the supramolecular polymer. The calculated structure is shown in Figure 12.11b.

The authors also investigated the photophysical properties of TBPyP and 2CB[14]TBPyP to collect more information about the supramolecular polymer formation. As shown in Figure 12.12, the addition of CB[14] to TBPyP in $H_2O$ causes a slight bathochromic shift from 422 nm to 429 nm and a decrease of absorption intensity. Following the addition of CB[14], there is an increase in the emission intensity with a hypochromatic shift in the porphyrin Q band. These phenomena are typical behaviors of a dye forming inclusion complexes with CB[n].[22,24,25] Finally the authors also investigated the reversible transformation nature of this supramolecular polymer, using KCl as the competitive compound to displace TBPyP from CB[14] cavities. The original properties (NMR signals, absorption and emission spectra) were restored after the addition of KCl, indicating the effective formation of the new complex between KCl and CB[14] and the release of TBPyP.[73]

These reports on supramolecular polymers formation based on inclusion complexes between CB[n]s and porphyrins are very promising in the field of PDT. The possibilities for this type of structure, considering different types of CB[n]s and different guest molecules, give the supramolecular hyperbranched supramolecular polymers a wide scope to develop new applications in the upcoming years.

## 12.6 Nonporphyrin Supramolecular Switches for Singlet Oxygen Using CB[n]s

As previously discussed, it is possible to alter the generation of singlet oxygen, a key species in PDT, using CB[n]s. Control over the generation of singlet oxygen is crucial to enhance the effects of PDT, while decreasing side effects such as prolonged photosensitivity to ambient light. In this sense, future applications involving supramolecular switches are likely to play an important role in PDT in the near future. Two examples of this kind of behavior were reported in the literature during the last few years. X. Z. Zhang and F. Sun team explored the effect of CB[8] complexation with a biotin

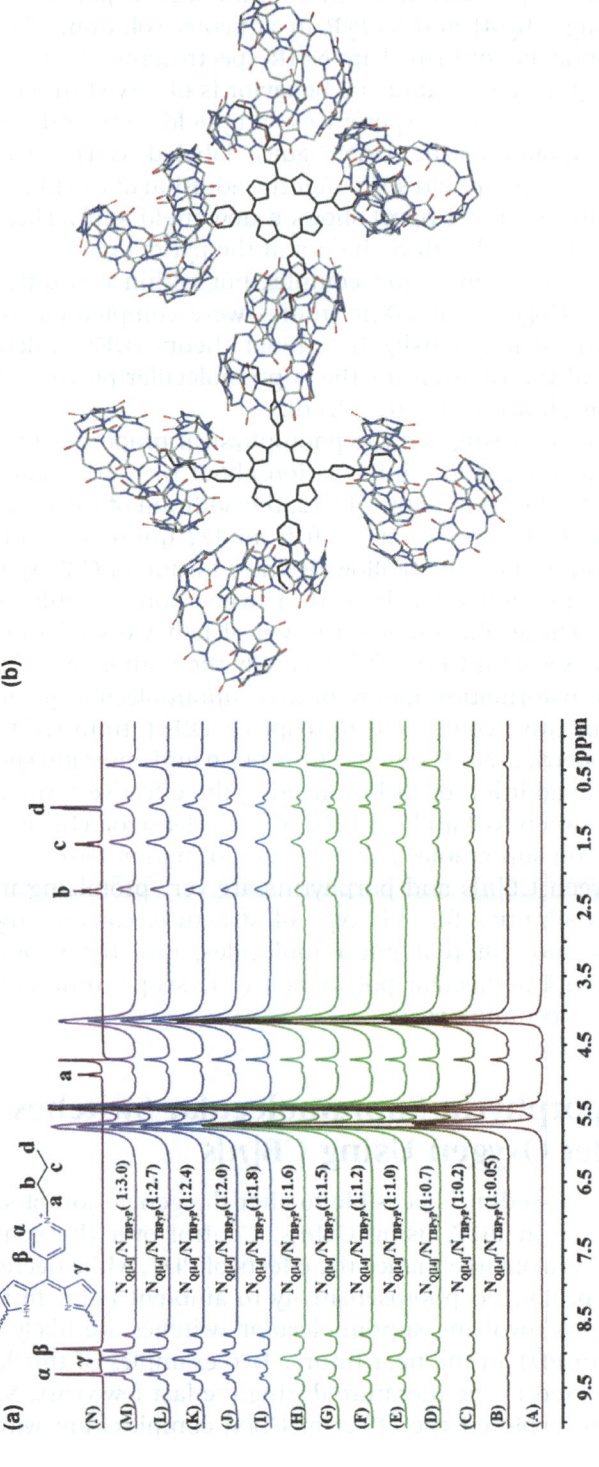

**Figure 12.11** (a) NMR titration spectra for CB[14] (A) in the presence of various equivalents of TBPYP (B-M) and pure TBPyP (N) in $D_2O$. (b) DFT molecular modeling of 2CB[14]@TBPyP. Adapted from ref. 73 with permission from the Royal Society of Chemistry.

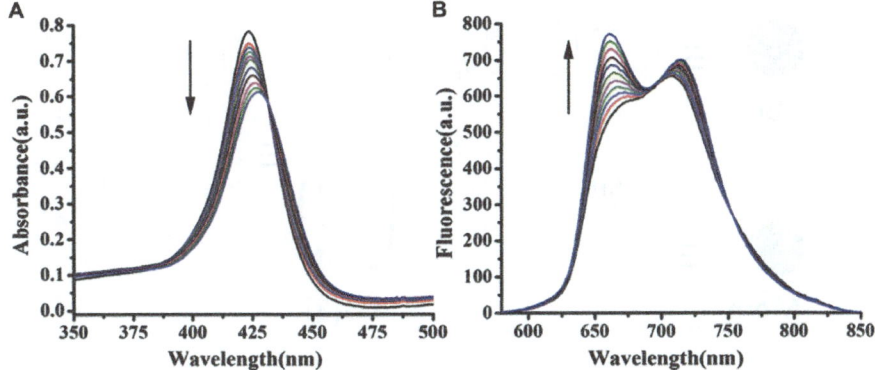

**Figure 12.12**   Electronic absorption (A) and fluorescence emission spectra (B) of TBPyP ($4 \times 10^{-6}$ mol L$^{-1}$) upon addition of increasing amounts (0, 0.2, 0.4, 0.6, 0.8, 0.9, 1, 1.2, 1.4, 1.5, 1.6, 1.8, 2.0 equiv.) of Q[14], with an excitation of 422 nm.
Reproduced from ref. 73 with permission from the Royal Society of Chemistry.

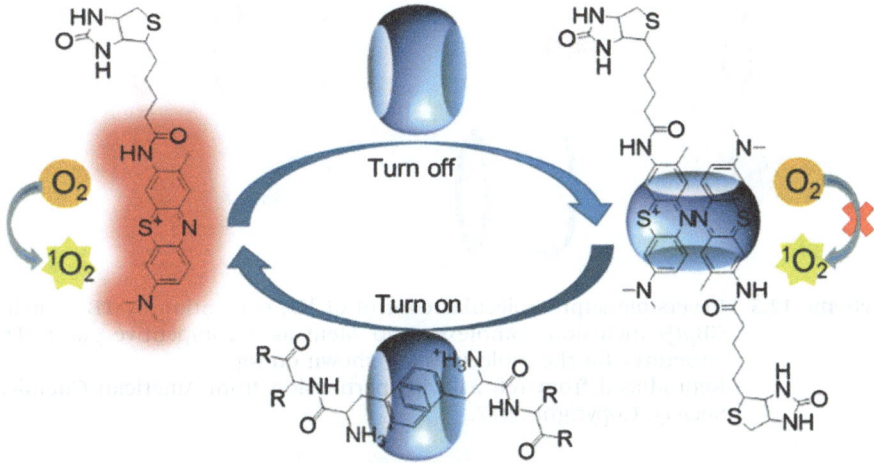

**Scheme 12.4**   Cucurbit[8]uril-regulated activatable supramolecular photosensitizer. Reproduced from ref. 75 with permission from American Chemical Society, 2016.

derivative of Toluidine blue (TBO$^+$). When this compound was encapsulated in CB[8], its fluorescence and singlet-oxygen generation were turned off, while it was possible to restore its PDT activity by displacing the compound with a peptide (Scheme 12.4). In this work, the biotin derivative showed enhanced tumor accumulation of the PS and enhanced PDT activity.[75]

Another example also using TBO$^+$ showed the reversible generation of singlet oxygen by switching the photosensitizer between CB[7] and CB[8], by using self-sorting and competitive binding of a guest (Scheme 12.5).[31]

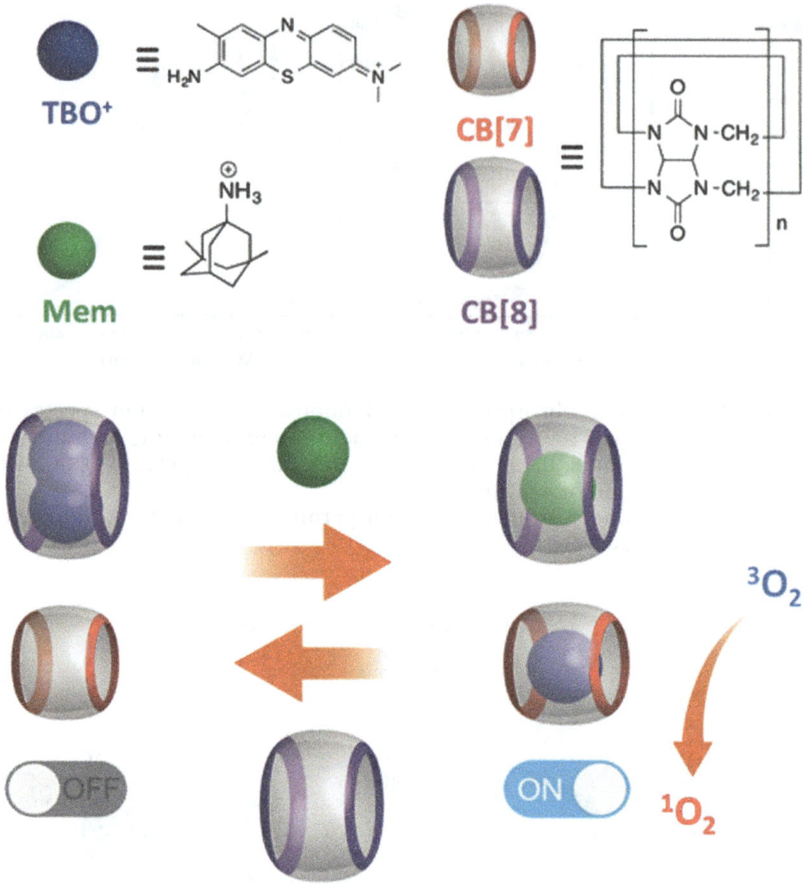

**Scheme 12.5** Reversible supramolecular control of $^1O_2$ generation by TBO$^+$ using
CB[n]s inclusion complexes and Mem as a competitive guest. The
structures for the molecules are shown on top.
Reproduced from ref. 31 with permission from American Chemical
Society, Copyright 2017.

The larger macrocycle turned off the generation of singlet oxygen due to the
promotion of self-quenching of the excited state through the formation of a
2:1 complex, which is the same behavior previously reported.[75] On the other
hand, the smaller macrocycle induced the formation of a 1:1 complex, re-
storing the ability to generate singlet oxygen. It must be mentioned that
CB[7] enhanced the generation of singlet oxygen above the free PS, which is
an advantage of this system compared to porphyrin-based supramolecular
photosensitizers, where self-aggregation is the factor limiting singlet-oxygen
generation. This behavior was the result of stabilizing the triplet-excited
state lifetime of the PS within CB[7], while the conformation of the complex
let the PS partially accessible to oxygen. All the components were biocompatible, which has the potential for PDT applications.

# References

1. J. W. Steed and J. L. Atwood, *Supramolecular Chemistry*, John Wiley & Sons, UK, 2nd edn, 2013.
2. F. Davis and S. Higson, *Macrocycles: Construction, Chemistry and Nanotechnology Applications*, John Wiley & Sons, Hoboken, NJ, 2011.
3. H.-J. Schneider, *Supramolecular Systems in Biomedical Fields*, Royal Society of Chemistry, Cambridge, 2013.
4. S. B. Brown, E. A. Brown and I. Walker, *Lancet Oncol.*, 2004, **5**, 497–508.
5. M. R. Detty, S. L. Gibson and S. J. Wagner, *J. Med. Chem.*, 2004, **47**, 3897–3915.
6. M. R. Hamblin and P. Mróz, *Advances in Photodynamic Therapy. Basic, Translational, and Clinical*, Artech House, Norwood, MA, 2008.
7. S. A. Sibani, P. A. McCarron, A. D. Woolfson and R. F. Donnelly, *Expert Opin. Drug Delivery*, 2008, **5**, 1241–1254.
8. A. E. O'connor, W. M. Gallagher and A. T. Byrne, *Photochem. Photobiol.*, 2009, **85**, 1053–1074.
9. P. Agostinis, K. Berg, K. A. Cengel, T. H. Foster, A. W. Girotti, S. O. Gollnick, S. M. Hahn, M. R. Hamblin, A. Juzeniene, D. Kessel, M. Korbelik, J. Moan, P. Mroz, D. Nowis, J. Piette, B. C. Wilson and J. Golab, *CA Cancer J. Clin.*, 2011, **61**, 250–281.
10. J. M. Dabrowski and L. G. Arnaut, *Photochem. Photobiol. Sci.*, 2015, **14**, 1765–1780.
11. M. J. Davies, *Biochem. Biophys. Res. Commun.*, 2003, **305**, 761–770.
12. E. A. Decker, R. J. Elias, D. J. McClements, *Oxidation of Foods and Beverages and Antioxidant Applications*, Woodhead Publishing, Cambridge, 2011.
13. H. Sies and C. F. M. Menck, *Mutat. Res.*, 1992, **275**, 367–375.
14. T. A. Dahl, W. R. Midden and P. E. Hartman, *Photochem. Photobiol.*, 1987, **46**, 345–352.
15. G. Valduga, G. Bertoloni, E. Reddi and G. Jori, *J. Photochem. Photobiol., B*, 1993, **21**, 81–86.
16. T. Maisch, J. Baier, B. Franz, M. Maier, M. Landthaler, R. M. Szeimies and W. Baumler, *Proc. Natl. Acad. Sci. U. S. A.*, 2007, **104**, 7223–7228.
17. J. Kim, I. S. Jung, S. Y. Kim, E. Lee, J. K. Kang, S. Sakamoto, K. Yamaguchi and K. Kim, *J. Am. Chem. Soc.*, 2000, **122**, 540–541.
18. J. W. Lee, S. Samal, N. Selvapalam, H. J. Kim and K. Kim, *Acc. Chem. Res.*, 2003, **36**, 621–630.
19. K. I. Assaf and W. M. Nau, *Chem. Soc. Rev.*, 2015, **44**, 394–418.
20. S. J. Barrow, S. Kasera, M. J. Rowland, J. del Barrio and O. A. Scherman, *Chem. Rev.*, 2015, **115**, 12320–12406.
21. L. Isaacs, *Chem. Commun.*, 2009, 619–629.
22. J. Robinson-Duggon, F. Perez-Mora, L. Dibona-Villanueva and D. Fuentealba, *Isr. J. Chem.*, 2018, **58**, 199–214.
23. C. Marquez, F. Huang and W. M. Nau, *IEEE Trans. Nanobiosci.*, 2004, **3**, 39–45.
24. W. M. Nau and J. Mohanty, *Int. J. Photoenergy*, 2005, **7**, 133–141.

25. R. N. Dsouza, U. Pischel and W. M. Nau, *Chem. Rev.*, 2011, **111**, 7941–7980.
26. P. Montes-Navajas, A. Corma and H. Garcia, *ChemPhysChem*, 2008, **9**, 713–720.
27. P. Montes-Navajas and H. Garcia, *J. Phys. Chem. C*, 2010, **114**, 2034–2038.
28. J. Caceres, J. Robinson-Duggon, A. Tapia, C. Paiva, M. Gomez, C. Bohne and D. Fuentealba, *Phys. Chem. Chem. Phys.*, 2017, **19**, 2574–2582.
29. E. I. Alarcon, M. Gonzalez-Bejar, P. Montes-Navajas, H. Garcia, E. A. Lissi and J. C. Scaiano, *Photochem. Photobiol. Sci.*, 2012, **11**, 269–273.
30. M. González-Béjar, P. Montes-Navajas, H. García and J. C. Scaiano, *Langmuir*, 2009, **25**, 10490–10494.
31. J. Robinson-Duggon, F. Pérez-Mora, L. Valverde-Vásquez, D. Cortés-Arriagada, J. R. De la Fuente, G. Günther and D. Fuentealba, *J. Phys. Chem. C*, 2017, **121**, 21782–21789.
32. W. H. Lei, G. Y. Jiang, Q. X. Zhou, B. W. Zhang and X. S. Wang, *Phys. Chem. Chem. Phys.*, 2010, **12**, 13255–13260.
33. T. Fuenzalida and D. Fuentealba, *Photochem. Photobiol. Sci.*, 2015, **14**, 686–692.
34. S. W. Thomas, G. D. Joly and T. M. Swager, *Chem. Rev.*, 2007, **107**, 1339–1386.
35. J. Li, J. Liu, C.-W. Wei, B. Liu, M. O'Donnell and X. Gao, *Phys. Chem. Chem. Phys.*, 2013, **15**, 17006–17015.
36. H. Bai, H. Chen, R. Hu, M. Li, F. Lv, L. Liu and S. Wang, *ACS Appl. Mater. Interfaces*, 2016, **8**, 31550–31557.
37. B. Liu, *Conjugated Polymers for Biological and Biomedical Applications*, John Wiley & Sons, Hoboken, NJ, 2018.
38. C. Zhu, L. Liu, Q. Yang, F. Lv and S. Wang, *Chem. Rev.*, 2012, **112**, 4687–4735.
39. K. Liu, Y. Yao, Y. Kang, Y. Liu, Y. Han, Y. Wang, Z. Li and X. Zhang, *Sci. Rep.*, 2013, **3**, 2372.
40. M. Laurenti, J. Rubio-Retama, F. Garcia-Blanco and E. López-Cabarcos, *Langmuir*, 2008, **24**, 13321–13327.
41. N. Willis-Fox, C. Belger, J. F. Fennell, R. C. Evans and T. M. Swager, *Chem. Mater.*, 2016, **28**, 2685–2691.
42. M. Idris, M. Bazzar, B. Guzelturk, H. V. Demir and D. Tuncel, *RSC Adv.*, 2016, **6**, 98109–98116.
43. H. Bai, H. Yuan, C. Nie, B. Wang, F. Lv, L. Liu and S. Wang, *Angew. Chem., Int. Ed.*, 2015, **54**, 13208–13213.
44. H. Bai, H. Zhang, R. Hu, H. Chen, F. Lv, L. Liu and S. Wang, *Langmuir*, 2017, **33**, 1116–1120.
45. D. M. Roundhill, in *Photochemistry and Photophysics of Metal Complexes. Modern Inorganic Chemistry*, Springer, Boston, MA, 1994.
46. U. Siggel, U. Bindig, C. Endisch, T. Komatsu, E. Tsuchida, J. Voigt and J. H. Fuhrhop, *Ber. Bunsenges Phys. Chem.*, 1996, **100**, 2070–2075.
47. M. Biesaga, K. Pyrzynska and M. Trojanowicz, *Talanta*, 2000, **51**, 209–224.

48. M. Makarska-Bialokoz, *Spectrochim. Acta, Part A*, 2018, **200**, 263–274.
49. B. Girek and W. Sliwa, *J. Inclusion Phenom. Macrocyclic Chem.*, 2015, **81**, 35–48.
50. J. Mohanty, A. C. Bhasikuttan, S. D. Choudhury and H. Pal, *J. Phys. Chem. B*, 2008, **112**, 10782–10785.
51. P. Suthari, P. H. Kumar, S. Doddi and P. R. Bangal, *J. Photochem. Photobiol., A*, 2014, **284**, 27–35.
52. Q. Li, S. C. Qiu, J. Zhang, K. Chen, Y. Huang, X. Xiao, Y. J. Zhang, F. Li, Y. Q. Zhang, S. F. Xue, Q. J. Zhu, Z. Tao, L. F. Lindoy and G. Wei, *Org. Lett.*, 2016, **18**, 4020–4023.
53. S. M. Liu, A. D. Shukla, S. Gadde, B. D. Wagner, A. E. Kaifer and L. Isaacs, *Angew. Chem., Int. Ed.*, 2008, **47**, 2657–2660.
54. R. Kubota, T. Takabe, K. Arima, H. Taniguchi, S. Asayama and H. Kawakami, *J. Mater. Chem. B*, 2018, **6**, 7050–7059.
55. W. H. Lei, G. Y. Jiang, Q. X. Zhou, Y. J. Hou, B. W. Zhang, X. X. Cheng and X. S. Wang, *ChemPhysChem*, 2013, **14**, 1003–1008.
56. R. Hoffmann, W. Knoche, C. Fenn and H. J. Buschmann, *J. Chem. Soc., Faraday Trans.*, 1994, **90**, 1507–1511.
57. F. F. Shen, C. Z. Wang, W. X. Zhao, Y. Q. Zhang, S. F. Xue, Q. J. Zhu, Z. Tao, L. F. Lindoy and G. Wei, *CrystEngComm*, 2016, **18**, 4988–4995.
58. D. Tuncel, N. Cindir and U. Koldemir, *J. Inclusion Phenom. Macrocyclic Chem.*, 2006, **55**, 373–380.
59. X. Xiao, J. S. Sun and J. Z. Jiang, *Chem. – Eur. J.*, 2013, **19**, 16891–16896.
60. X. Xiao, W. J. Li and J. Z. Jiang, *Inorg. Chem. Commun.*, 2013, **35**, 156–159.
61. M. Fathalla, N. L. Strutt, J. C. Barnes, C. L. Stern, C. F. Ke and J. F. Stoddart, *Eur. J. Org. Chem.*, 2014, **2014**, 2873–2877.
62. M. J. Lee, M. K. Kim, N. K. Shee, J. Lee, M. Yoon and H. J. Kim, *ChemistrySelect*, 2018, **3**, 256–261.
63. S. Mandal, M. Rahaman, S. Sadhu, S. K. Nayak and A. Patra, *J. Phys. Chem. C*, 2013, **117**, 3069–3077.
64. K. Liu, Y. L. Liu, Y. X. Yao, H. X. Yuan, S. Wang, Z. Q. Wang and X. Zhang, *Angew. Chem., Int. Ed.*, 2013, **52**, 8285–8289.
65. L. H. Chen, H. T. Bai, J. F. Xu, S. Wang and X. Zhang, *ACS Appl. Mater. Interfaces*, 2017, **9**, 13950–13957.
66. L. L. Shen, Q. Z. Lu, A. Q. Zhu, X. Q. Lv and Z. S. An, *ACS Macro Lett.*, 2017, **6**, 625–631.
67. A. Koc, R. Khan and D. Tuncel, *Chem. – Eur. J.*, 2018, **24**, 15550–15555.
68. T. Gulikkrzywicki, C. Fouquey and J. M. Lehn, *Proc. Natl. Acad. Sci. U. S. A.*, 1993, **90**, 163–167.
69. R. P. Sijbesma, F. H. Beijer, L. Brunsveld, B. J. B. Folmer, J. H. K. K. Hirschberg, R. F. M. Lange, J. K. L. Lowe and E. W. Meijer, *Science*, 1997, **278**, 1601–1604.
70. L. Brunsveld, B. J. B. Folmer, E. W. Meijer and R. P. Sijbesma, *Chem. Rev.*, 2001, **101**, 4071–4097.
71. Y. Liu, Z. Huang, K. Liu, H. Kelgtermans, W. Dehaen, Z. Wang and X. Zhang, *Polym. Chem.*, 2014, **5**, 53–56.

72. B. Yuan, H. Yang, Z. Q. Wang and X. Zhang, *Langmuir*, 2014, **30**, 15462–15467.
73. Z. Gao, J. Zhang, N. Sun, Y. Huang, Z. Tao, X. Xiao and J. Jiang, *Org. Chem. Front.*, 2016, **3**, 1144–1148.
74. X. J. Cheng, L. L. Liang, K. Chen, N. N. Ji, X. Xiao, J. X. Zhang, Y. Q. Zhang, S. F. Xue, Q. J. Zhu, X. L. Ni and Z. Tao, *Angew. Chem., Int. Ed.*, 2013, **52**, 7252–7255.
75. X.-Q. Wang, Q. Lei, J.-Y. Zhu, W.-J. Wang, Q. Cheng, F. Gao, Y.-X. Sun and X.-Z. Zhang, *ACS Appl. Mater. Interfaces*, 2016, **8**, 22892–22899.

# Subject Index